The Nature of Culture

Vertebrate Paleobiology and Paleoanthropology Series

Edited by

Eric Delson
Vertebrate Paleontology, American Museum of Natural History
New York, NY 10024, USA
delson@amnh.org

Eric J. Sargis
Anthropology, Yale University
New Haven, CT 06520, USA
eric.sargis@yale.edu

Focal topics for volumes in the series will include systematic paleontology of all vertebrates (from agnathans to humans), phylogeny reconstruction, functional morphology, Paleolithic archaeology, taphonomy, geochronology, historical biogeography, and biostratigraphy. Other fields (e.g., paleoclimatology, paleoecology, ancient DNA, total organismal community structure) may be considered if the volume theme emphasizes paleobiology (or archaeology). Fields such as modeling of physical processes, genetic methodology, nonvertebrates or neontology are out of our scope.

Volumes in the series may either be monographic treatments (including unpublished but fully revised dissertations) or edited collections, especially those focusing on problem-oriented issues, with multidisciplinary coverage where possible.

More information about this series at http://www.springer.com/series/6978

The Nature of Culture

Based on an Interdisciplinary Symposium 'The Nature of Culture', Tübingen, Germany

Edited by

Miriam N. Haidle

Heidelberg Academy of Sciences and Humanities, Research Center "The Role of Culture in Early Expansions of Humans", Senckenberg Research Institute, Frankfurt am Main, Germany

Nicholas J. Conard

Institute of Pre- and Protohistory and Medieval Archaeology, Department of Early Prehistory and Quaternary Ecology, Tübingen, Germany

and

Senckenberg Center for Human Evolution and Palaeoenvironment, Eberhard Karls Universität Tübingen, Tübingen, Germany

Michael Bolus

Heidelberg Academy of Sciences and Humanities, Research Center "The Role of Culture in Early Expansions of Humans", Eberhard Karls Universität Tübingen, Tübingen, Germany

Springer

Editors
Miriam N. Haidle
Heidelberg Academy of Sciences and Humanities
Research Center "The Role of Culture in Early Expansions of Humans"
Senckenberg Research Institute
Frankfurt am Main
Germany

Michael Bolus
Heidelberg Academy of Sciences and Humanities
Research Center "The Role of Culture in Early Expansions of Humans"
Eberhard Karls Universität Tübingen
Tübingen
Germany

Nicholas J. Conard
Department of Early Prehistory and Quaternary Ecology
Institute of Pre- and Protohistory and Medieval Archaeology
Tübingen
Germany

and

Senckenberg Center for Human Evolution and Palaeoenvironment
Eberhard Karls Universität Tübingen
Tübingen
Germany

ISSN 1877-9077 ISSN 1877-9085 (electronic)
Vertebrate Paleobiology and Paleoanthropology Series
ISBN 978-94-017-7424-6 ISBN 978-94-017-7426-0 (eBook)
DOI 10.1007/978-94-017-7426-0

Library of Congress Control Number: 2015952756

Springer Dordrecht Heidelberg New York London

Cover illustration: Left – Description: Chimpanzee using an organic tool. Figure credit: reproduced with kind permission by Ian C. Gilby, Arizona State University. Middle – description: wooden spear from the Lower Paleolithic site of Schöningen, Germany. Figure credit: Copyright University of Tübingen, photo: Nicholas J. Conard. Right – description: venus figurine from the lower Aurignacian deposits of Hohle Fels Cave, Swabian Jura, Germany. Figure credit: Copyright: University of Tübingen, photo: Hildegard Jensen.

Printed on acid-free paper

Springer Science+Business Media B.V. Dordrecht is part of Springer Science+Business Media (www.springer.com)

Preface

This volume is the result of an interdisciplinary symposium organized by the research center "The Role of Culture in Early Expansions of Humans" (ROCEEH) of the Heidelberg Academy of Sciences and Humanities held at Hohentübingen Castle at Eberhard Karls University of Tübingen from June 15–18, 2011. Our goal for the conference was to produce a unified model of cultural evolution integrating ethological accounts of culture in great apes, sea mammals, and birds, as well as to debate the nature of culture as viewed from the perspective of the humanities and social sciences. The resulting model of the expansion of cultural capacities consists of two parts: a theoretical framework tracing the developmental dimensions of cultural performances and a model of the expansion of cultural capacities drawn from ethological and archeological data on information transmission. This volume presents many of the ideas that the participants at the meeting presented and reflects an up-to-date assessment of the state of international research on the evolution of cultural behavior.

We sincerely thank our colleagues who supported this publication with their reviews and many constructive comments. Other than the authors and editors, these include: Nick Ashton (University College London), Anne Delagnes (Université de Bordeaux), Robin Dennell (University of Sheffield), Anna Belfer-Cohen (Hebrew University of Jerusalem), Natalie Uomini (University of Liverpool), Christoph Antweiler (Universität Bonn), Gerald Hartung (Universität Wuppertal), Erella Hovers (Hebrew University of Jerusalem), Isabelle Parsons (University of South Africa), Martin Porr (University of Western Australia), Felix Riede (Aarhus Universitet), Rachel Kendal (Durham University), Luke Premo (Washington State University), Robert Boyd (Arizona State University), Paola Villa (University of Colorado Museum), Gerd-Christian Weniger (Neanderthal Museum Mettmann), Thomas Wynn (University of Colorado), Bennett G. Galef (McMaster University), Jürgen Richter (Universität Köln), and Thiemo Breyer (Universität Köln).

We would like to extend our thanks to the series editors Eric Delson and Eric Sargis, and Fermine Shaly and Jeffrey Taub at Springer, for their support, encouragement, and patience in producing this volume. We are grateful to the Heidelberg Academy of Sciences and Humanities for funding ROCEEH, the University of Tübingen for hosting the symposium, and the Deutsche Forschungsgemeinschaft for providing financial support.

Finally, we hope that the lively discussions, debate, and good cheer that accompanied the meeting in Tübingen will be captured in these papers.

Tübingen
March 2015

Miriam N. Haidle
Nicholas J. Conard
Michael Bolus

Contents

Contributors

Nira Alperson-Afil The Martin (Szusz) Department of Land of Israel Studies and Archaeology, Bar Ilan University, Ramat Gan, Israel

Michael Bolus Research Center "The Role of Culture in Early Expansions of Humans", Heidelberg Academy of Sciences and Humanities, Eberhard Karls Universität Tübingen, Tübingen, Germany

David R. Braun Department of Anthropology, Center for the Advanced Study of Human Paleobiology, George Washington University, Washington, DC, USA; Department of Archaeology, University of Cape Town, Cape Town, South Africa

Nicholas J. Conard Department of Early Prehistory and Quaternary Ecology, Institute of Pre- and Protohistory and Medieval Archaeology, Tübingen, Germany; Senckenberg Center for Human Evolution and Palaeoenvironment, Eberhard Karls Universität Tübingen, Tübingen, Germany

Iain Davidson Department of Archaeology, Flinders University of South Australia, Adelaide, Australia; School of Humanities, University of New England, Armidale, Australia

Volker Gerhardt Institut für Philosophie, Lehrbereich Praktische Philosophie, Rechts- und Sozialphilosophie, Humboldt Universität, Berlin, Germany

Naama Goren-Inbar Institute of Archaeology, Hebrew University of Jerusalem, Jerusalem, Israel

Miriam N. Haidle Research Center "The Role of Culture in Early Expansions of Humans", Heidelberg Academy of Sciences and Humanities, Senckenberg Research Institute, Frankfurt am Main, Germany

Marlize Lombard Department of Anthropology and Development Studies, University of Johannesburg, Auckland Park 2006, South Africa; Wallenberg Research Centre, Stellenbosch Institute for Advanced Study (STIAS), Stellenbosch University, Stellenbosch, South Africa

Shannon P. McPherron Max Planck Institute for Evolutionary Anthropology, Leipzig, Germany

April Nowell Department of Anthropology, University of Victoria, Victoria, BC, Canada

L.S. Premo Max Planck Institute for Evolutionary Anthropology, Leipzig, Germany; Department of Anthropology, Washington State University, Pullman, WA, USA

Claudio Tennie School of Psychology, University of Birmingham, Birmingham, UK; Max Planck Institute for Evolutionary Anthropology, Leipzig, Germany

Thorsten Uthmeier Institut Für Ur- und Frühgeschichte, Friedrich-Alexander-Universität Erlangen-Nürnberg, Erlangen, Germany

Lyn Wadley Evolutionary Studies Institute, University of the Witwatersrand, Johannesburg, South Africa

Andrew Whiten School of Psychology and Neuroscience, Centre for Social Learning and Cognitive Evolution, University of St Andrews, St Andrews, UK

Chapter 1
The Nature of Culture: Research Goals and New Directions

Miriam N. Haidle, Nicholas J. Conard, and Michael Bolus

How do we define and deal with culture? Paleolithic archaeologists view even the crudest human-made stone tools as material expression of cultural behavior. Primatologists claim that chimpanzees (Whiten et al. 1999), orangutans (van Schaik et al. 2003), and possibly also bonobos (Hohmann and Fruth 2003) exhibit some sort of culture. Similar arguments have also been suggested for cetaceans (Rendell and Whitehead 2001) and birds (Bluff et al. 2010). Other researchers, especially from the humanities, often question these claims or even dismiss the proposed evidence of culture in species other than *Homo sapiens* altogether. In June 2011 we hosted an interdisciplinary symposium at Eberhard Karls University in Tübingen organized by the research center "The Role of Culture in Early Expansions of Humans" (ROCEEH) of the Heidelberg Academy of Sciences and Humanities (Fig. 1.1).

At the conference, archaeologists, primatologists, paleoanthropologists and cultural anthropologists discussed and debated these issues. The participants of the conference aimed to move beyond dichotomic statements of culture versus non-culture, which, according to the chosen definition of the central term 'culture', frequently exclude the evidence that makes a productive examination of a development of cultural behavior possible. Instead, the participants followed a synthetic approach that acknowledged different forms of cultural expression in relation to each other. This approach, if it is to be successful, must also be applicable to different forms of evidence: to birds' songs, to chimpanzees' hand clasps, to material finds from the archaeological record as well as to religious practices.

Building on a draft model circulated before the meeting (see Davidson 2016, Fig. 10.4), the members of the Tübingen symposium developed a revised model of the expansion of cultural capacities consisting of two parts (Haidle and Conard 2011; Haidle et al. 2015): (1) a theoretical model for the developmental dimensions of cultural performances, and (2) a scheme for the expansion of cultural capacities drawn from the ethological and archaeological data on the transmission of information. The theoretical model sees the development of behavioral performances in three multifactorial dimensions. While the biological (e.g., anatomy, instincts) and individual (e.g., experience, trained proficiency) dimensions apply for all kinds of behavior, the historical-social dimension is an additional dimension of and a necessary condition for cultural behavior. Cultural behavior is bound to a social context based on non-genetic transmission of information between individuals of a group. In an advanced form, for example, individuals share a negotiated system of values that guides individual behavior. Additionally, the social component of cultural behavior possesses historical depth. Learned practices are, for example, transmitted to other members of a group for generations, but not necessarily in a descendant line, and with sustainable impact on future behavior. Each cultural performance is based on multifactorial developments in the biological, individual and historical-social dimensions. These

M.N. Haidle (✉)
Heidelberg Academy of Sciences and Humanities, Research Center "The Role of Culture in Early Expansions of Humans", Senckenberg Research Institute, Frankfurt am Main, Germany
e-mail: mhaidle@senckenberg.de

N.J. Conard
Department of Early Prehistory and Quaternary Ecology, Institute of Pre- and Protohistory and Medieval Archaeology, Tübingen, Germany
e-mail: nicholas.conard@uni-tuebingen.de

and

Senckenberg Center for Human Evolution and Palaeoenvironment, Eberhard Karls Universität Tübingen, Tübingen, Germany

M. Bolus
Heidelberg Academy of Sciences and Humanities, Research Center "The Role of Culture in Early Expansions of Humans", Eberhard Karls Universität Tübingen, Tübingen, Germany
e-mail: michael.bolus@uni-tuebingen.de

Miriam N. Haidle, Nicholas J. Conard and Michael Bolus (eds.), *The Nature of Culture: Based on an Interdisciplinary Symposium 'The Nature of Culture', Tübingen, Germany*, Vertebrate Paleobiology and Paleoanthropology, DOI 10.1007/978-94-017-7426-0_1

Fig. 1.1 Speakers, the ROCEEH team and some guests of the symposium 'The Nature of Culture', held at Hohentübingen Castle in June 2011 (*left* to *right* from *bottom*): Andrew Kandel, April Nowell, Michael Bolus, Lyn Wadley, Naama Goren-Inbar, Marlize Lombard, Andrew Whiten, Christine Hertler, Miriam Haidle, Claudio Tennie, Anne Delagnes, Angela Bruch, Nicholas Conard, Mark Collard, Stephen Shennan, Thorsten Uthmeier, Shannon McPherron, James O'Connell, Marian Vanhaeren, Iain Davidson, Sibylle Wolf, Annette Kehnel, Michael Märker, Zara Kanaeva, Duilio Garofoli

dimensions are interdependent of each other with the specific environment of conspecifics, agents and objects affected by and affecting the behavior.

From the set of cultural performances of a defined group, its minimal cultural capacity can be deduced. The maximum of the biological, the individual, and the historical-social dimensions of all behavioral performances of that group define the minimum status of these dimensions of the potential cultural capacity. Based on this theoretical model of the developmental dimensions of cultural performances and the deduced potential cultural capacities, we introduce a scheme for the expansion of cultural capacities. The empirical basis of this research lies in ethological and archaeological data on historical-social transmission of different types of information. In the archaeological record, the empirical data is confined to a subset of the original set of cultural performances which have material manifestations and can be preserved over time. These limited remains have been categorized as different types of information using the problem-solution-distance approach which is also applicable to tool behavior by animals, thus bridging ethological and archaeological data (Köhler 1926; Haidle 2012). The participants at the conference identified eight grades of cultural capacities, four of which can be found in non-human animal species. While three of these grades delineate only the presence of some of the fundamental elements of a historical-social dimension in behavior including the use of social information, social learning and traditions, the fourth grade of 'basic cultural capacities' is characterized by the whole set of these elements creating a pattern of behavior with historical-social dimension shared within a group. Four more grades of cultural capacity have so far been documented exclusively in hominin behavior: the modular, composite, complementary and notional cultural capacities (see Haidle 2016). This developmental scheme is not conceived as a progression line. The grades do not replace each other, but extend the formerly existing capacity in the three dimensions and thus the range of cultural performances.

The development of the EECC model of the evolution and expansion of cultural capacities as sketched below (Fig. 1.2) and outlined in detail in a joint article (Haidle et al. 2015) is the product of intensive debate. The course of the discussion at the symposium has been described in Haidle and Conard (2011). The chapters in this volume represent individual contributions to the subject. Some are elaborations of the papers presented at the symposium, some changed the focus according to the discussion, and some pick up the model and work on its details or apply it. The contributions in this volume do not fall along one line of argument, but reflect a multifaceted and sometimes controversial examination of the subject of the expansion of cultural capacities.

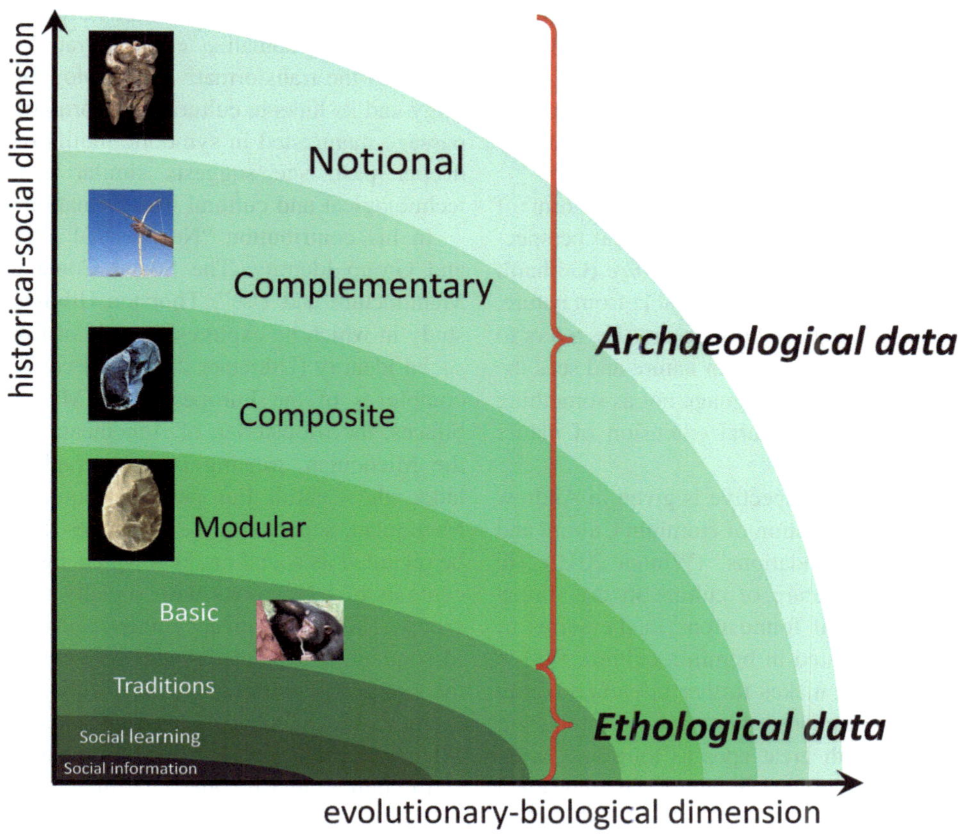

Fig. 1.2 EECC model of the evolution and expansion of cultural capacities in eight grades. The basic four grades ('social information' to 'basic') have been documented in some animal species, while the subsequent four ('modular' to 'notional') have, until now, only been identified in the course of human evolution (modified after Haidle et al. 2015)

The first three chapters in this volume are devoted to general considerations on cultural evolution, cultural performances and capacities, and the link between nature and culture. In "Lessons From Tasmania – Cultural Performance Versus Cultural Capacity" Miriam N. Haidle summarizes the EECC model of the evolution and expansion of cultural capacities and applies it to the example of Tasmanian culture, which has been characterized as 'primitive' and compared with chimpanzee and Middle Paleolithic capabilities (Haidle 2016). A reconsideration of the set of material cultural performances in the Tasmanian ethnographic record and thorough examination using the problem-solution-distance approach shows the presence of modular, composite, complementary and notional cultural capacities. The Tasmanian cultural record is a perfect example of the concept that apparently simple performances cannot be easily equated with archaic or non-modern behavior. Thus, the Tasmanian example complements the evidence of limestone tools from Gesher Benot Ya'aqov presented by Nira Alperson-Afil and Naama Goren-Inbar. These authors address complex behavior in stone tool production, which is situated only within modular cultural capacities (Alperson-Afil and Goren-Inbar 2016). Haidle's chapter picks up the mountaineering principle of cultural evolution developed in Marlize Lombard's contribution (Lombard 2016). It provides the debate with an example that clarifies that the EECC model of the evolution and expansion of cultural capacities does not represent a hierarchical sequence of progressive grades of cultural capacities replacing each other as discussed in Chap. 10 by Iain Davidson (Davidson 2016).

Volker Gerhardt's contribution "Culture as a Form of Nature" elucidates the debate with a philosophical perspective on the very subject of *The Nature of Culture* (Gerhardt 2016). Gerhardt discusses culture as inseparable from nature, more precisely as a part and product of nature. He refers to technology and its active participation in nature and sees the use of signs, symbols and written language not as something completely different, but as a cultural extension of nature based on technology.

A broader primatological perspective is given in Andrew Whiten's chapter on "The Evolution of Hominin Culture and its Ancient Pre-Hominin Foundations" (Whiten 2016). He draws our attention to the nature of culture in the animal world and some very ancient foundations to the series of steps that ultimately culminated in hominin culture. With a focus on great apes, Whiten makes further inferences about the direct evolutionary antecedents of hominin culture, about ancestors humans share with great apes as long ago as 6–14 Ma. Addressing human cultural evolution, he argues that this phase can only be understood in the context of a complex of advances in social and technological cognition, together with other features that include unprecedented encephalization and extended childhood, a topic elaborated upon in this volume by April Nowell (Nowell 2016). Whiten uses a primatological perspective to discuss the deep origins of culture within its wider adaptive niche.

The following four chapters discuss aspects of the archaeological record and their implications for cultural evolution. The chapter by Nira Alperson-Afil and Naama Goren-Inbar: "Scarce but Significant: The Limestone Component of the Acheulean Site of Gesher Benot Ya'aqov, Israel" presents an example of evidence of cultural behavior from the Lower Paleolithic (Alperson-Afil and Goren-Inbar 2016). The limestone assemblage reveals complex life-histories of tools within a single reduction sequence. Percussors, chopping tools, and cores are viewed as interrelated consecutive morphotypes transformed into one another, thus implying behavioral flexibility and contingency. Alperson-Afil and Goren-Inbar present a remarkable instance of complex culture within modular cultural capacity.

Lyn Wadley addresses the difficulties of linking artifacts with cultural capacities and cognition. In the chapter "Technological Transformations Imply Cultural Transformations and Complex Cognition" she draws attention to transformative technology and its implication for other cultural behavior (Wadley 2016). The complex cognitive ability to control material transformations evolved together with the ability to conceptualize cultural transformations. Wadley discusses the transformative technology of Iron Age metallurgy and its links to cultural transformations such as rites of passage manifested in symbolic motifs on artifacts. For the deeper past, she suggests similar connections between technological and cultural transformations.

In his contribution "Neanderthal Utilitarian Equipment and Group Identity: The Social Context of Bifacial Tool Manufacture and Use", Thorsten Uthmeier presents a case study in which he explores the role of bifaces as signals for social identity (Uthmeier 2016). He compares the two main complexes of the European Late Middle Paleolithic with bifaces, the Mousterian of Acheulean Tradition (MtA) and the Micoquian, treating two geographical clusters of the latter, the Central European Micoquian and the Crimean Micoquian, separately. He concludes that bifacial tools can be regarded as social markers which signal social identity in contexts of interactions with socially distant individuals or groups. While Uthmeier understands the MtA and the Micoquian as separated social collectives, he suggests that the two geographical subgroups of the Micoquian represent a single social collective, which consists of at least two extended networks with differing strategies of lithic curation.

In contrast to Uthmeier's optimistic view with regard to the Middle Paleolithic, Michael Bolus in his contribution "Tracing Group Identity in Early Upper Paleolithic Stone and Organic Tools – Some Thoughts and Many Questions" remains pessimistic about the possibility of unambiguously identifying group identity in the early Upper Paleolithic by analyzing stone and organic tools (Bolus 2016). Instead, he highlights the general problems that arise when trying to interpret single elements of material culture. For instance, one of the major problems when dealing with differences in stone and organic tools is to assess if such differences are always a question of different 'styles' and/or 'identities', or rather a question of different tool 'types'. Other than in more recent periods of Pre- and Protohistory, where different types of artifacts are often interpreted as mirroring ethnic identity, ethnic interpretations of this kind are largely absent from Paleolithic research today.

In her contribution "Childhood, Play and the Evolution of Cultural Capacity in Neanderthals and Modern Humans", April Nowell presents an approach dealing with the biological dimension and the expansion of cognitive capacities in the course of human evolution (Nowell 2016). Play is an important factor during the early life history of humans which has a direct impact on the development of social and cognitive learning and hence on the historical-social dimension of cultural capacity. This means that the impact of learning through play on the connectivity of the brain is heightened by slower maturation rates. Thus she argues that extended childhoods of modern humans relative to Neanderthals help to shape the recent phase of cultural evolution. While play likely existed during the childhood of Neanderthals, fantasy play as part of a package of symbol-based cognitive abilities seems to be limited to modern humans, as is suggested by differences in the nature of symbolic material culture of Neanderthals and modern humans.

The chapter "Stone Tools: Evidence of Something in Between Culture and Cumulative Culture?" by Iain Davidson presents different definitions of 'culture' and discusses the role of stone tools in the evolution of culture (Davidson 2016). Davidson criticizes hierarchical models of cultural evolution and assuming that the EECC model of the evolution and expansion of cultural capacities also has a hierarchical structure, he provides an alternative model. As stated above and as exemplified by Miriam N. Haidle (2016), however, the EECC model does not represent a hierarchical sequence of progressive grades of cultural capacities replacing each other. Our model does not imply an inevitable progression, but focuses on expansion of cultural capacities that integrate achievements in earlier states, thus conforming to expectations of the mountaineering principle of cultural evolution discussed in Marlize Lombard's contribution (Lombard 2016).

The last two chapters focus on the aspect of transmission of information as a variable component in cultural behavior. In their contribution "The Island Test for Cumulative Culture in the Paleolithic" Claudio Tennie, David R. Braun, Luke S. Premo, and Shannon P. McPherron question the assumption that the widespread ability to produce Early Stone Age artifacts was grounded on high-fidelity transmission of behavior such as imitation and teaching (Tennie et al. 2016). Instead they suggest regular reinvention of the production and use of simple flake technology within a "zone of latent solutions", defined by a combination of genetic, environmental, and social factors. Tennie et al. introduce a thought experiment, called the Island Test, which may be useful for distinguishing forms of hominin behavior that require high-fidelity transmission from those that do not.

Finally, in her contribution "Mountaineering or Ratcheting? Stone Age Hunting Weapons as Proxy for the Evolution of Human Technological, Behavioral and Cognitive Flexibility" Marlize Lombard raises the question of whether human cultural development can really be seen as being analogous to the effect of a ratchet as in the cumulative cultural approach advocated by Tennie et al. (2009). As an alternative, Lombard introduces the mountaineering model which fits much better to the ups and downs of human cultural development (Lombard 2016). Although path-dependent, the developmental process from a point reached is not necessarily in a progressive line. In the mountaineering scenario the use of sidetracks and loops, but also steps backwards and rapid abseiling to lower levels are possible, as is reinvention. With this contribution, the discussion of the nature of culture comes back to its starting point in this volume and underscores the non-linear nature of cultural evolution.

The chapters presented in this volume cover only a portion of the topics discussed during the symposium. Other speakers who contributed oral presentations and participated in vivid discussions added many more valuable aspects to the understanding of the 'culture' phenomenon. Anne Delagnes, for instance, provided insight into the nature of the earliest hominin cultures which date to the beginning phases of Earlier Stone Age and provide the first archaeological evidence for modular culture. Marian Vanhaeren concentrated on the role of personal ornaments as an example for elements of material culture expressing identity in the Upper Paleolithic and perhaps in the Middle Paleolithic. Steven Shennan highlighted the influence of demographic factors in the evolution of cultural capacities and demonstrated population size and frequency of interactions as influential factors. James O'Connell, who addressed the question of interdependencies with environmental factors during cultural evolution, added a view from behavioral ecology. He stressed the need to contextualize human behavior within its social and economic constraints, rather than using mechanistic resource ranking and cost benefit analysis as an end in and of itself.

In his keynote lecture, Nicholas Conard drew from his excavations in Africa and Eurasia to argue that the emergence of composite, complementary and notional culture does not reflect unique monocentric developments. Instead, he demonstrated how cultural evolution follows a polycentric mosaic pattern via the innovation, spread, modification and disappearance of behaviors. Thus, we should not expect cultural evolution to resemble the flipping of a light switch followed by radiation from a point source. The famous mammoth ivory figurines and musical instruments from Vogelherd on display at Hohentübingen Castle, where the conference took place, represent a unique record of symbolic artifacts, but not the only time and place in human history where art and music evolved (Conard 2007, 2010).

Although all contributions to this volume approach the question of how better to understand 'the nature of culture', and although there are many cross-references between these contributions, every chapter can also stand on its own providing case studies or more theoretical considerations. The editors deliberately did not organize the chapters of the volume following a single line of argument. Instead, it was our intention to allow the contributions to mirror the different and sometimes controversial positions presented during the symposium. We hope that the volume will initiate further discussion and innovative cross-taxa and cross-societal research that will improve our knowledge of the evolution of cultural behavior.

References

Alperson-Afil, N., & Goren-Inbar, N. (2016). Scarce but significant: The limestone component of the Acheulean site of Gesher Benot Ya'aqov, Israel. In M. N. Haidle, N. J. Conard, & M. Bolus (Eds.), *The nature of culture* (pp. 41–56). Dordrecht: Springer.

Bluff, L. A., Kacelnik, A., & Rutz, C. (2010). Vocal culture in new Caledonian crows *Corvus moneduloides*. *Biological Journal of the Linnean Society, 101*(4), 767–776.

Bolus, M. (2016). Tracing group identity in early upper paleolithic stone and organic tools – Some thoughts and many questions. In M. N. Haidle, N. J. Conard, & M. Bolus (Eds.), *The nature of culture* (pp. 79–85). Dordrecht: Springer.

Conard, N. J. (2007). Cultural evolution in Africa and Eurasia during the middle and late Pleistocene. In W. Henke & I. Tattersall (Eds.), *Handbook of paleoanthropology* (pp. 2001–2037). Berlin: Springer.

Conard, N. J. (2010). Cultural modernity: Consensus or conundrum? *Proceedings of the National Academy of Sciences of the USA, 117*, 7621–7622.

Davidson, I. (2016). Stone tools: Evidence of something in between culture and cumulative culture? In M. N. Haidle, N. J. Conard, & M. Bolus (Eds.), *The nature of culture* (pp. 99–120). Dordrecht: Springer.

Gerhardt, V. (2016). Culture as a form of nature. In M. N. Haidle, N. J. Conard, & M. Bolus (Eds.), *The nature of culture* (pp. 19–25). Dordrecht: Springer.

Haidle, M. N. (2012). *How to think tools? A comparison of cognitive aspects in tool behavior of animals and during human evolution.* Cognitive perspectives in tool behaviour Vol. 1. http://tobias-lib.uni-tuebingen.de/frontdoor.php?source_opus=6014. Retrieved June 6, 2014.

Haidle, M. N. (2016). Lessons from Tasmania – Cultural performance versus cultural capacity. In M. N. Haidle, N. J. Conard, & M. Bolus (Eds.), *The nature of culture* (pp. 7–17). Dordrecht: Springer.

Haidle, M. N., & Conard, N. J. (2011). The nature of culture – Summary report on an interdisciplinary symposium held in Tübingen, Germany, 15–18 June 2011. *Mitteilungen der Gesellschaft für Urgeschichte, 20*, 65–78.

Haidle, M. N., Bolus, M., Collard, M., Conard, N. J., Garofoli, D., & Lombard, M., et al. (2015). The nature of culture – An eight-grade model for the evolution and expansion of cultural capacities in hominins and other animals. *Journal of Anthropological Sciences, 93*, 43–70.

Hohmann, G., & Fruth, B. (2003). Culture in Bonobos? Between-species and within-species variation on behavior. *Current Anthropology, 44*, 563–571.

Köhler, W. (1926). *The mentality of apes*. New York: Harcourt Brace.

Lombard, M. (2016). Mountaineering or ratcheting? Stone age hunting weapons as proxy for the evolution of human technological, behavioral and cognitive flexibility. In M. N. Haidle, N. J. Conard, & M. Bolus (Eds.), *The nature of culture* (pp. 135–146). Dordrecht: Springer.

Nowell, A. (2016). Childhood, play and the evolution of cultural capacity in Neanderthals and modern humans. In M. N. Haidle, N. J. Conard, & M. Bolus (Eds.), *The nature of culture* (pp. 87–97). Dordrecht: Springer.

Rendell, L., & Whitehead, H. (2001). Culture in whales and dolphins. *Behavioral and Brain Sciences, 24*, 309–382.

Tennie, C., Call, J., & Tomasello, M. (2009). Ratcheting up the ratchet: On the evolution of cumulative culture. *Philosophical Transactions of the Royal Society B, 364*, 2405–2415.

Tennie, C., Braun, D. R., Premo, L. S., & McPherron, S. P. (2016). The Island test for cumulative culture in the Paleolithic. In M. N. Haidle, N. J. Conard, & M. Bolus (Eds.), *The nature of culture* (pp. 121–133). Dordrecht: Springer.

Uthmeier, T. (2016). Neanderthal utilitarian equipment and group identity: The social context of bifacial tool manufacture and use. In M. N. Haidle, N. J. Conard, & M. Bolus (Eds.), *The nature of culture* (pp. 65–77). Dordrecht: Springer.

van Schaik, C. P., Ancrenaz, M., Borgen, G., Galdikas, B., Knott, C. D., & Singleton, I., et al. (2003). Orangutan cultures and the evolution of material culture. *Science, 299*, 102–105.

Wadley, L. (2016). Technological transformations imply cultural transformations and complex cognition. In M. N. Haidle, N. J. Conard, & M. Bolus (Eds.), *The nature of culture* (pp. 57–63). Dordrecht: Springer.

Whiten, A. (2016). The Evolution of hominin culture and its ancient pre-hominin foundations. In M. N. Haidle, N. J. Conard, & M. Bolus (Eds.), *The nature of culture* (pp. 27–39). Dordrecht: Springer.

Whiten, A., Goodall, J., McGrew, W. C., Nishida, T., Reynolds, V., & Sugiyama, Y., et al. (1999). Cultures in chimpanzees. *Nature, 399*, 682–685.

Chapter 2
Lessons from Tasmania – Cultural Performance Versus Cultural Capacity

Miriam N. Haidle

Abstract Characterized as 'primitive' and compared with chimpanzee and Middle Paleolithic capabilities, the Tasmanian culture has been subject to an intensif debate lasting for more than 40 years. This paper gives a reconsideration of the set of material cultural performances in the Tasmanian ethnographic record and applies the EECC model of the evolution and expansion of cultural capacities to it. Based on the analyses of problem-solution-distances in tool behavior, modular, composite, complementary, and notional cultural capacities are identified. The Tasmanian cultural record is a perfect example that apparently simple performances cannot be easily equated with archaic or non-modern behavior. Instead, the performances have to be seen as products of interdependent developments in biological, historical-social, and individual dimensions embedded in the group specific environment.

Keywords Complementary cultural capacities • Composite cultural capacities • Material culture • Modular cultural capacities • Notional cultural capacities • Problem-solution distance • Simplification • EECC model of the evolution and expansion of cultural capacities

The Primitive Tasmanian Image

Tasmania, today a large island southeast of Australia, has been settled by humans for around 35 kyr (Jones 1995). Around 12,000 years ago, rising sea levels isolated the human population from the Australian mainland. The cultural development since differed from that of Australian aborigines. Some technical achievements and cultural expressions like bone points and the use of volcanic glass as raw material known from prehistoric contexts got lost or were given up until the colonization by Europeans starting in 1773 (Murray and Williamson 2003:316). It is a common feature to characterize the Tasmanian pre-contact material culture as deficient by listing the items which are said to be lacking: bone tools, cold-weather clothing, ground stone tools, hafted tools, bow-and-arrow, boomerangs, spear-throwers, the knowledge to make fire and consequently pottery and metal working as well as canoes, fishery and thus also nets and fishhooks (McGrew 1987:250; Henrich 2004:197). These negative characterizations rather represent the presumptive expectations (cf. Davidson 2007:81) drawn from the Australian record, respectively European assumptions which have not been met, than the real nature or development of the cultural set. However, actually or presumably lacking features like fish-eating (Jones 1978; Allen 1979; Horton 1979; Bowdler 1980; Walters 1981; Taylor 2007) or fire-making (Gott 2002; Taylor 2008) dominated the discussion. The material culture of the aboriginal population of Tasmania by the end of the 18th century is said to have been one of the most 'primitive' in the world (Jones 1977). Sollas (1911) equaled its developmental level to the Mousterian. On the basis of the label of primitiveness and the equation with Paleolithic cultural remains William McGrew (1987) compared a segment of this "impoverished" (ibid.:250) material culture – the toolkit used to access food – with that of wild chimpanzees from Tanzania. His aim was to gain insight in the minimal range of differences between ape and modern human tool behavior. He stated (ibid:247): "The results show surprising similarity in the number of items in the toolkit, raw materials used, proportion of tools made versus those used unchanged, extent of complexity, type of prey, etc. Key contrasts also emerge: only human

M.N. Haidle (✉)
Heidelberg Academy of Sciences and Humanities, Research Center "The Role of Culture in Early Expansions of Humans", Senckenberg Research Institute, Senckenberganlage 25, 60325 Frankfurt am Main, Germany
e-mail: mhaidle@senckenberg.de

Miriam N. Haidle, Nicholas J. Conard and Michael Bolus (eds.), *The Nature of Culture: Based on an Interdisciplinary Symposium 'The Nature of Culture', Tübingen, Germany*, Vertebrate Paleobiology and Paleoanthropology, DOI 10.1007/978-94-017-7426-0_2

tools have more than one type of component and are made using other tools. Overall, however, the gap between the most technically diverse nonhuman toolkit and the simplest human material cultures seems narrow." Sometimes, the characterization of the Tasmanian material culture as primitive goes hand in hand with the implicit or explicit assumption of a more primitive, deteriorating mind. Rhys Jones associated "a simplification in the tool kit" (Jones 1977:202) (actual in the abandoning of bone points and assumed by comparison of Australian and Tasmanian tool kits) and "a diminuation in the range of foods eaten" (ibid.) (based on abandoning of scale fish as part of the diet) with "perhaps a squeezing of intellectuality" (ibid.:203). He summarizes his results with the often cited speculation "Even if Abel Tasman had not sailed the winds of the Roaring Fourties in 1642, where they in fact doomed – doomed to a slow strangulation of the mind?" (ibid.). It is the aim of this paper to illuminate the record of a supposedly increasing simplification in the Tasmanian material culture by applying the EECC model of the evolution and expansion of cultural capacities in the course of human evolution. It will be shown that a simple status or even the ostensible simplification in cultural performance is not contradictory to a concurrent consistence or even expansion of cultural capacities. Thus, the paper addresses central questions of the sapient behavior paradox (Renfrew 1996). The EECC model offers new ways to differentiate between performances and capacities and demonstrates a way to test for potential by analyzing the problem-solution distance in tool behavior as a basic feature instead of looking for outstanding manifestations (cf. Haidle 2014). The approach draws attention to the complexities of the Tasmanian material evidence and opposes unilinear schemes of human history/evolution, that evaluate the simple or sophisticated appearance of a segment of cultural performances in order to classify the performing groups as basic to or high up on an evolutionary ladder (cf. Gamble 1992).

The Three Dimensions of Cultural Performances

The EECC model of the expansion of cultural capacities differentiates between cultural performances as sets of expressions of cultural behavior and cultural capacities as underlying potential faculties which can exceed the actual performances by far (Haidle and Conard 2011; Haidle et al. 2015). Cultural performances are determined by three dimensions: an evolutionary-biological, a historical-social and an ontogenetic-individual one, embedded in and interdependent with the specific environment (Fig. 2.1).

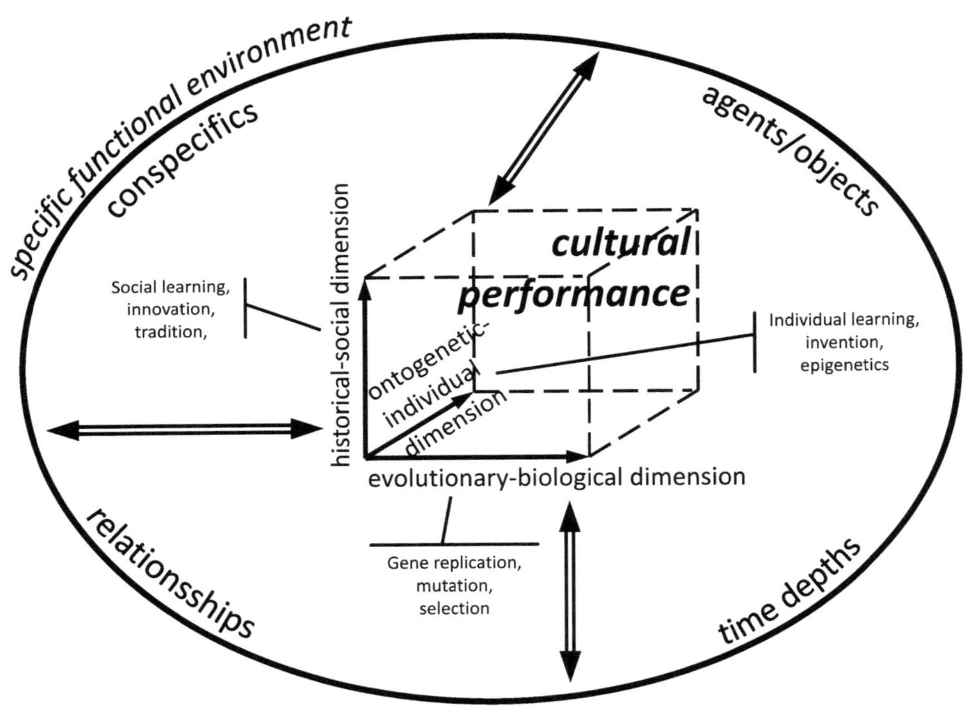

Fig. 2.1 Cultural performances develop in three dimensions: an evolutionary-biological, a historical–social, and an ontogenetic-individual. These are interdependent with each other and the embedding specific environment

On the *evolutionary-biological dimension* the biological potential and constraints for behavior are exposed – in genes, gene expressions, anatomical blueprints and physiological standards of a group of organisms. It is expressed, for example, in the structure of the nervous system and the brain, in sensory perception, in motor and articulation skills, in the form of sociality, and in the abilities to communicate. The evolutionary-biological dimension constitutes the foundation of any behavior. It affects the basic course of life history and the anatomical-physiological potential to perceive and express, create and reflect, learn and remember. For example, the genus *Homo* is characterized by generalized hands facilitating material cultural expressions, by the development of a sound production and perception apparatus enabling language – a specific form of very detailed communication, by an increase of the relative brain size, by changes in brain anatomy, and by shifts in life history giving more time for learning. Regardless which form of training they undergo: dolphins will never be able to talk to us like humans because of their different anatomy of the sound production apparatus; chimpanzees will never succeed in producing sophisticated handaxes because of their different anatomy of the hand. The species-specific characteristic of the evolutionary-biological dimension of behavioral performances enables the finding of species-specific solutions to species-specific problems. Continuity or change in this dimension in general underlies evolutionary mechanisms like gene replication, mutation and selection, but it may be partly modified through long-term interaction with the environment and, as part of it, material culture (Fisher and Ridley 2013).

The Tasmanians can be assumed to share with all other *Homo sapiens* conspecifics the vast majority of biological factors influencing cultural performances. Although developments in the evolutionary-biological dimension like the persistence of lactose tolerance in adulthood (Burger et al. 2007) or the ability to cope with hypobaric hypoxia in high altitudes (Beall 2006) show biological adaptations of *Homo sapiens* after the last glacial maximum that alter the range of cultural performances of some human populations, there is no evidence of a profound change in biological factors associated with cognitive competences. The Tasmanian population would be a noteworthy exception, if a reduction of cultural performances were due to evolutionary-biologically based mental impairment.

The *ontogenetic-individual dimension* of behavior reflects individuals' preferences, aversions, skills and disabilities. The ontogenetic-individual dimension incorporates the potential and constraints of an individual, or of a group of individuals, set by individual talents or poor aptitudes, by the personal social setting and by individual life histories of physical, mental, and emotional experiences. Already in the womb, monozygotic twins with the same genetic complement have different experiences with different epigenetic effects (Petanjek and Kostović 2012). Siblings can be raised in the same family within the same historical-social setting, but nonetheless experience different influences by parents, relatives, friends, teachers etc., by support or deprivation, by diseases, fortuitous timing or traumatic accidents. All these factors affect the additional mechanisms of change, which are operative on this dimension: individual learning, personal inventions, and epigenetics, "factors that influence gene expression without modifying the DNA sequence" (Ledón-Rettig et al. 2013:311). Tasmanian cultural performances are affected by the individual dimension in the same way as those of other humans. Evolutionary-biological and ontogenetic-individual factors are, to a greater or lesser extent, the basis of every behavioral expression, be it self-referential or directed to the environment. Cultural behavior, however, is defined by having a third dimension of factors influencing the behavioral expression: the historical-social.

The *historical-social dimension* represents historical and social potentials and constraints. The set of historically acquired knowledge and skills, customs, views and opinions, makes up a part of the individual's environment that can be acted on, and used as a basis for further innovation. Social access to the knowledge and skills, customs, views and opinions can be affected by population density, active communication systems, child raising habits, teaching systems, systems of religious and political participation, and general group-specific attitudes about learning, innovation and progress (Rogers 1995). The forms and extent of storage, transmission, permutation, and transformation of the knowledge and skills, customs, views and opinions support or hamper the unfolding of cultural performances. The historical-social dimension in general affects the ways in which the evolutionary-biological foundations are used for cultural behavior, but can also affect the properties of some biological bases (Malafouris 2010; Woollett and Maguire 2011). This dimension unfolds via social transmission from stimulus enhancement to the capacity of teaching, via group-wide adoption of innovations and via transgenerational traditions. The historical-social dimension is self-enhancing; cultural behavior influences factors that foster cultural transmission and creativity (Enquist et al. 2008).

The historical-social setting of the Tasmanian population at the time of contact is characterized by a small population size of probably 1500–22,000; most estimates assume less than 10,000 people (Murray and Williamson 2003:314). The ethnographic record of the Tasmanian society is weak (cf. Taylor 2007) and details about behavioral patterns and organizational structure are scarce. Up to nine endogamous regional tribes of 300–500 people have been described split in 48–80 local and exogamous, non-sedentary bands of 30–50 people (Jones 1977; Ryan 1981). They shared a lot of cultural performances and a common toolkit, but showed also some remarkable, but probably environmentally induced differences as in dietary habits (Taylor 2007:21) or the construction and use of shelters (Hiatt 1968:201). Recent linguistic studies have identified 12 Tasmanian languages in five clusters (Bowern 2012).

All of the three major dimensions (evolutionary-biological, historical-social, and ontogenetic-individual) of cultural performances are multi-factorial, and they are not independent. Rather, they influence one another directly or indirectly via reciprocal effects in the context of a *specific environment* (Haidle 2008). The latter is the sum of the cultural (material and non-material) and social aspects of the environment of an organism or a group, plus the section of the natural environment that affects, or is affected by, the organism or the group. The functional relationships of a group of organisms with elements of the specific environment vary according to the resources they perceive in that environment, given the state of their evolutionary-biological, historical-social and ontogenetic-individual dimensions. Although the natural landscape of a lion and *Homo ergaster* may have been the same, their specific environments or realized resource spaces differed markedly. The elements include their conspecifics as well as biotic and abiotic agents (like predators, competitors, climatic conditions), plus objects (like prey or lithic raw material resources) that they affect or are affected by. These factors are linked to the group in focus by specific relations and with a certain time-depth (in perception, conception, and action) in both past and future directions influencing these relationships or behaviors. Cultural performances (and the deduced cultural capacities) are thus neither a mere biological product, nor solely a historical issue; and they are embedded into the specific environment in which the lives of individuals play out. The EECC model is thus an approach that can help to overcome the cultural modernity problem as claimed by Porr (2014): an approach "that recognize[s] humans as socially constituted dynamic beings that are also reflective and cognitively active […]" with genetic factors as one influence within dynamic sets of relationships. "Human beings develop and grow through these relationships, which provide both potentials and constraints. What humans are and can do is not a reflection of internal essences of human nature but is a product of situated growth, reflections, interactions and negotiations" (Porr 2014:264).

The specific environment of the aboriginal Tasmanians has geographically been reduced to Tasmania as an island around 12 ka by the rise of the sea level after the last glacial maximum. Marine scale fish has been part of the diet for a long time but has probably been abandoned around 3500 ka (Jones 1977; but see also Taylor 2007). At the coast the consumption of shellfish continued, completed by the exploitation of seals, waterbirds, and penguins (Allen 1979; Taylor 2007). The settlement of the southwestern part of the island and the use of offshore islands that could be reached by canoe extended the range of the specific environment (Vanderwal 1978; Allen 1979:5). The Tasmanian vegetation is characterized by dense scrub with consequences on technology as long-distance weapons like spear-throwers or boomerangs are not efficient there (Horton 1979:31). The controlled use of fire allowed vegetation changes to improve the landscape for food supply (Hiatt 1968:212, 219) and to ease movements and thus extend the range by opening forest paths (Völger 1973:62). Hiatt (1967:111) lists a wide range of animal and plant species which were used as dietary resources including large and small marsupials, monotremes, seals, marine crustacea and mollusks, freshwater crustacea, coastal, lagoon and land birds respectively their eggs as well as different vegetables from sea and land, and underground storage organs of plants. Wind brakes and small huts in the East, more stable and larger huts in the West gave shelter (Hiatt 1968:201).

Empirically, only single case studies of behavioral performances can be documented, applying an occurrence/presence-only approach. The potential capacity of a group for behavior with historical-social dimension (that is, for cultural behavior) or without historical-social dimension (for non-cultural behavior) can only be estimated from the sum of all behavioral performances of this group and similar groups with which a genetic or historical-social exchange of behavioral patterns would have been possible. The potential minimum range of expression in the three developmental dimensions – the so-called capacity for behavior – must be deduced from the sum of expressions in the different behavioral performances of the groups included. The cultural capacity, e.g., of *Homo heidelbergensis* cannot be directly observed, but must be deduced from the sum of quasi-contemporaneous performances seen in the record of material culture preserved at different archaeological sites associated with *H. heidelbergensis*. The potential cultural capacity of a group, a population or a species, is never completely exhausted by the particular individuals, groups, or populations; rather, different aspects of the capacity are used and expressed.

Tasmanian Cultural Performances in the Late 18th Century

The compilation of Tasmanian cultural performances at the time of the first contact with Europeans has been set up neither systematically nor in detail, and is biased by a focus on material culture. Betty Hiatt (1968) provides an overview that has widely been used as basis in the discussion about Tasmanian cultural simplicity. The discussions about the competence to produce fire (cf. Roth 1899; Hiatt 1968; Völger 1973; Gott 2002; Taylor 2008) or the use of scale fish as part of the diet (cf. Hiatt 1967; Jones 1977, 1978; Allen 1979; Horton 1979; Bowdler 1980; Walters 1981; Bassett 2004; Taylor 2007) highlight the patchiness of our knowledge. Despite the insufficiency of the data, the Tasmanian case may well serve as an example for the assessment of a minimum set of cultural performances in the archaeological record with all its problems of preservation, discovery, documentation, and interpretation.

In his attempt to compare the complexity of behavior of Tasmanians and chimpanzees, William McGrew (McGrew 1987, based on Oswalt 1976) listed 13 items as parts of the tool kit used for food procurement:

- Stones to chop down trees, to notch or bruise the bark of living trees for footholds to make climbing easier (cf. Hiatt 1968:206). Jones (1977:197) mentions stone hammers in combination with a wooden tool used as a chisel to prize bark off trees.
- Wooden chisels to prize shellfish from underwater rock. The small wooden spatulas had been cut and smoothed (Hiatt 1968:209).
- Sticks to dig up roots (cf. Hiatt 1968:209), to dig out prey (cf. Hiatt 1968:206). A three feet long stick, a so-called 'spit' was used by women to catch birds from a hole in the ground (Hiatt 1968:208). It is unclear if this tool can be regarded the same as the combination tool of "digging stick-club-chisel used [by women] for a variety of purposes from digging up vegetable roots, ochre and killing game, to prising bark off trees, the latter being done with the aid of a stone hammer" as referred to by Jones (1977:197). Whether the digging sticks were actually unmodified as Oswalt (1976)/McGrew (1987) classify them is questionable.
- Stones as missiles for the hunt on ducks and swans (cf. Hiatt 1968:208). Jones (1977:197) characterizes them as small spherical pebbles used as missiles (Jones 1977:197).
- Throwing stick to knock prey out of tree.
- Wooden spear to impale prey. The thrusting spears were 10–12 feet long and had a maximum thickness of a man's middle finger. The points were sharpened with flint tools and hardened in fire (Hiatt 1968:205; Jones 1977:196). A picture of two wooden spears is given by C. A. Lesueur (1807/1811, reproduced in Jones 1977:199, Fig. 5).
- Torch to fire grassland to drive prey (cf. Hiatt 1968:212) and to smoke out prey from tree hollows (cf. Hiatt 1968:207).
- Plaited grass rope to climb trees (cf. Hiatt 1968:206). Jones (1977:197) mentions also ropes made from strips of animal fur.
- Tussocks of grass tied together to trip up kangaroos (cf. Hiatt 1968:206).
- Grass baskets to carry shellfish. Hiatt (1968:209) describes the containers as made from leaves of sedgy plants, divided into strips and softened by fire, for collecting and carrying food items, ochre, stone, tinder, string. A picture of a basket is given by C. A. Lesueur (1807/1811, reproduced in Jones 1977:199, Fig. 5).
- Blind made from wood and branches to conceal a hunter.
- Baited blind to conceal bird catchers. Hiatt (1968:209) cites the personal notes of Robinson from 1833: "The natives erect a kind of hut with grass under which they place some fish, fastened by stone, and when the crows come to feed, they do nothing than put out their hands and haul them in. They adopt the same plan for catching ducks except that they bait them with worms."
- Set of spears implanted in the earth on kangaroo trail to stab prey. "The part of the stake stuck in the ground was burnt to prevent it from decaying" (Hiatt 1968:206).

Additional items were depicted in the literature, which were used by Tasmanians for procurement, transportation, and treatment of food, but which McGrew (1987) did not take into account:

- Waddy. A sort of club made from heavy wood and varying in length from two and a quarter feet to four feet (Hiatt 1968:208). Jones (1977:197) mentions these tools as straight to fusiform sticks used as throwing sticks and clubs. A picture of two waddies is given by C. A. Lesueur (1807/1811, reproduced in Jones 1977:199, Fig. 5).
- Stone pestles and mortars to soften vegetable food and to break long bones for marrow (Jones 1977:197).
- Possum-skin pouch bags for carrying valuables such as stone tools, ochre, and ashes of the dead (Jones 1977:197). A picture of a skin pouch bag is given by C. A. Lesueur (1807/1811, reproduced in Jones 1977: 199, Fig. 5).
- Water containers made from seaweed (Hiatt 1968:210; Jones 1977:197). Different kinds of modification of the leaves are described (cf. Hiatt 1968:210): they were cut into circular pieces and formed into a purse with a string run through holes in the margin, or "bent together with a small pin of wood", respectively "skewered together at the sides". An oblong piece could be opened to form a flat bag.
- Sea shells used as drinking vessels. As material for the likely unmodified tools oysters and mutton fish/abalone shells are mentioned (Hiatt 1968:210; Jones 1977:197).
- Sea shells used as cooking device (Hiatt 1968:210).
- Wooden spits to support food against the fire (Hiatt 1968:210).
- Flat stones used to broil food (Hiatt 1968:210).

Stone tools certainly played a major role in the every-day life of Tasmanians, although they are rarely mentioned in detail (cf. Hambly 1931). Simple stone tools from the ethnographic record have been depicted as knives, to take bark from trees, to make points to the wooden spears, to tap sweet sap from trees (Hiatt 1968:212), as scrapers and skinning tools (Hambly 1931:89), and as cicatrizing flakes (Jones 1977:197). The stone technology from the archaeological record has been reported as "being mainly composed of multiplatform and single platform cores, primary flakes, retouched flakes and various sized scrapers (Jones 1977). [...] there are additional items such as small denticulate flakes [...], delicate 'thumbnail' and end scrapers indicating a greater range of stone tool types..." (Cosgrove et al. 2014:180). While there is ethnographic evidence of the use of hammer stones for flaking (Hambly 1931:91), pressure flaked tools from the archaeological record (see e.g. Cosgrove et al. 2014:181, Fig. 14.2) indicate the use of punches probably made from organic material. It cannot be reconstructed from the available sources how the Tasmanians themselves grouped their stone tools in typological and functional classes. It was probably just because of the universal use of not very remarkable stone tools as cutting device that McGrew (1987) forgot to take stone knives on his list of tools for food procurement. Bark vessels (Meston 1935) are an artifact class not specifically associated to food procurement, but increasing the geographical range of it (cf. Vanderwal 1978). Watercraft made from rolls of bark fastened onto each other with straps of bark or tough grass and propelled by poles allowed crossing water between the mainland and islands and large waterways within Tasmania (Hiatt 1968:213–214; Jones 1977:197).

Fire played an important role as tool in the life of aboriginal Tasmanians. Beside the warming factor it was regularly used for cooking/food preparation (cf. Hiatt 1967), to improve the landscape for food supply (Hiatt 1968:212, 219), for hunting/driving prey (Völger 1973:62), for opening forest paths (ibid.), to alter the qualities of wood by hardening spear tips (Hiatt 1968:205) and to prevent stakes in kangaroo traps from decaying (ibid.:206), for softening leaves of sedgy plants to make baskets (ibid.:209), and for cremation of the dead (Hiatt 1969). Fire was carefully maintained; it was common to carry torches and glowing pieces of dry wood and bark rolls (Hiatt 1968:212; Völger 1973:62). The question if aboriginal Tasmanians not only maintained, but also produced fire, is impossible to be definitely answered (cf. Hiatt 1968:211–212; Völger 1973; Gott 2002; Taylor 2008). There are few and not fully reliable reports in the ethnographic record; it is obvious that fire making was not an every-day business. As a possible explanation for a low frequency of fire making Gisela Völger (1973:62) argues that this was normally avoided because of the difficulty to light fire in the wet and windy climate (cf. Gott 2002:653). It may also be that the primary and recurrent function of the frequently described lighted torches was to facilitate the movements of the people and as hunting tool. As by-product they served to maintain fire: using a torch was much more comfortable than lighting fire again and again. Beth Gott (2002) and Rebe Taylor (2008) list the possible ways of lighting a fire for which the early colonial records on the life of aboriginal Tasmanians give some evidence: by percussion, by friction that is rubbing together two pieces of wood, with a fire drill, and with the stick-and-groove method (Roth 1899).

Two forms of huts were reported to give shelter. In the eastern part of Tasmania windbreaks were common. The semi-circular structures were built from stakes interwoven with strips of bark or fastened with leaves of grass and covered with bark (Hiatt 1968:201). From the west coast more substantial, larger and closed houses are reported. Their form resembled a circular dome with a small entrance and a cover made of bark (ibid.). Clothing seems to have been limited to kangaroo skins tied onto the shoulder for warmth (Jones 1977:197), a fact that supported the evaluation of Tasmanian culture as deficient. Vanderwal (1978:123), however, discusses the use of grease and ochre as equal to or even more effective than skin as body protection. The common use of ochre is also testified by reports on baskets and skin bags containing pieces of ochre (Hiatt 1968:209; Jones 1977:197).

The record of Tasmanian cultural life is completed by items which are at first sight not primarily related to subsistence: "In Tasmanian society, there was time for sky-larking, for gossiping, for fighting, for story-telling and dancing, for mourning the dead, for getting ochre, for making shell necklaces and as the rock carvings at Mount Cameron West and other west coast sites show us – for art" (Jones 1977:201). Necklaces from pierced and strung shells as represented by Lesueur (1807/1811, reproduced in Jones 1977:199, Fig. 5), different hair dresses, and patterns of cicatrices (body scars as shown by Copia 1817, reproduced in Davidson 2007:73, Fig. 6.4) were used as personal ornaments. It is unclear whether these items "were [...] markers of status, or group identity, as well as of individual difference" (Davidson 2007:73; cf. Brown 1991). Evidence of visual arts is found in rock paintings and engravings with hand stencils and lines/circles/speckled areas as motifs (de Teliga and Bryden 1958; Brown 1991:101–103) and in painted patterns of circles and lines on bark (de Teliga and Bryden 1958; Brown 1991:103–104). Figurative motifs painted on bark may be influenced by post-colonial contacts (Brown 1991:103). Performing arts are reported in the form of singing and dancing sessions up to several days long with representations of hunts, battles and other mythological or historical themes (Hiatt 1968:215; Horton 1979:32–33). Hiatt (1968:215) sees "a fairly well developed mythology [...]; many of the stories are concerned with where the Tasmanians came from, how fire originated and so on [...]. These could well have been connected with some form of totemic religion and associated ceremonies." The common funeral form was the primary cremation of the dead. The burnt remains were covered by small mounds of grass and/or sticks and sometimes topped by structures of wooden poles, covered with bark and tied together at the top (Noetling 1908:36; Hiatt 1969:104–105). Alternatively, the remains were collected and carried in skin bags (ibid.:105). "The ashes of the dead were also used to smear on the body and face as assign of mourning" (ibid.). Bones were sometimes kept as relics. As another form of treating the dead the deposition in hollow trees is reported (ibid.:106).

The assessment of a simplification in cultural performances within the last 12 kyr is difficult. A comparison with the diverse record from the whole Australian continent (cf. Jones 1977:196) is problematic to define the original status of Tasmanian culture before separation by rising sea levels. As most of the material elements were made from perishable organics their development is hardly to be reconstructed. The archaeological record allows to identify only two reductions that occurred around 4000 years ago: as tools some form of bone points were given up (ibid.; Read 2006) as was the consumption of fish (Jones 1977:196; but see also Taylor 2007). Yet, the stone tool assemblages evolved by an increasing formalization of tool types with reduction of tool size and finer retouch and "a steady and increasing introduction of high quality exotic raw materials such as cherts, siliceous breccias, and spongolites brought in as ready made tools or at least as blanks, from quarries at 50 to 100 km away" (Jones 1977:194). The introduction of the bark canoe, characteristic for Tasmania and most likely an indigenous (re-)invention (Vanderwal 1978:121), allowed the expansion of geographical range in food procurement (Vanderwal 1978:123). The identification of the reasons of cultural changes and their characterization as adaptive or maladaptive is often a matter of perspective (cf. Read 2006). While Jones (1977, 1978; see also Henrich 2004) views the abandonment of fish as part of the diet as well as the cessation of bone point production and the, postulated, associated manufacture of clothing as a maladaptive and passive loss of skills and knowledge, other authors describe them as adaptive achievements (cf. Horton 1979). Allen (1979) points to the low fat content of the available fish and the need of a high-caloric diet in harsher winters, Walters (1981) suggests an optimization of the diet following environmental change. Vanderwal (1978:123) discusses the use of grease and ochre as equally or even more efficient than the use of skins and clothing. In any case, the aboriginal Tasmanians showed a well-developed ability to adapt to new circumstances. The dog was quickly accepted within the Tasmanian cultural setting after its introduction by Europeans (ibid.); post-contact experiences were easily incorporated in the traditional song and dance repertoire.

Assessment of a Minimum of Cultural Capacities from a Set of Cultural Performances

As mentioned above, the cultural capacity of a population (referring to the range of problem-solution-distances that are applied in the set of cultural performances, see below) can be deduced from the set of cultural performances associated with this population and its range of expression in the three developmental dimensions (evolutionary-biological, historical-social, and ontogenetic-individual). The more complete the set of cultural performances is covered, the closer the cultural capacity can be approached. It has to be kept in mind, however, that the reconstructed cultural capacity of a population is always only a minimum one. To get a more detailed picture in the Tasmanian case, the list of performances given by McGrew (1987) with an incomplete focus on subsistence tools has been extended by other performances as given in the literature. This set of cultural performances from the Tasmanian ethnographic record can now be used to assess the cultural capacities at the time of contact in the 18th century AD and to evaluate the postulated deterioration of the mind.

The EECC model of the expansion of cultural capacities in human evolution offers a gradual scheme of development based on the different types of information that is socially transmitted (Haidle and Conard 2011; Haidle et al. 2015). As key character to categorize socially transmitted information and thus cultural performances, the problem-solution distance was chosen, an attribute first mentioned in Wolfgang Köhler's (1926) comparative studies of chimpanzee behavior. An extended behavioral route from perceiving a problem or need to its solution or satisfaction, together with possible loops or sidetracks, is identifiable in all sorts of tool behavior and can be studied in archaeological as well as ethological and ethnographical data; as a basic feature of very different manifestations it can serve as an approach to test for behavioral potential as Renfrew (1996) claimed. In tool use, a goal is not approached directly as when a hungry zebra starts to eat grass, but by moving away from the target object (e.g., a nut) to reach satisfaction (feeding) via an intercalated tool (e.g., a hammer). Haidle (2009, 2010, 2012), Lombard and Haidle (2012) extended the concept of the problem-solution distance and developed a method to systematically assess levels in their complexity by coding them in cognigrams and effective chains. Applying this method, problem-solution distances in tool behaviors in living animals, ancient hominins, and recent humans can be compared directly. An extended problem-solution distance is not exclusive to tool behavior, but inherent also in social behavior. Thus, the capacities to solve a problem identified in tool behavior represent the minimum cultural capacity available to perform all types of cultural behavior. Additional data from non-material cultural behavior can sharpen the picture.

Eight different grades of cultural capacities based on different types of socially transmitted information have been identified so far (Haidle et al. 2015). The grades do not represent an evolutionary ladder on which, by climbing up, the lower stages are left behind. Instead, the grades express a gradual extension of options of action. With increasing potential problem-solution-distances the possible variability and flexibility of actions expands, including the range of options already achieved in earlier stages. The basic four grades ('social information', 'social learning', 'tradition', and 'basic cultural') have been documented in some animal species, while the subsequent four ('modular', 'composite', 'complementary', and 'notional') have, thus far, only been identified in the course of human evolution. Originating in the analysis of tool behavior, they are distinct in the different effective concepts through which attention foci are linked to each other. Up to basic cultural capacities the individual acts on a problem focus directly or with the help of a tool picked up or modified by the subject, but always with the concentration on the central target. With modular cultural capacities as first detected so far in early human manufacture and use of stone tools the problem-solution distance is extended to effective chaining.

Modular cultural capacity is characterized by the effective chaining of independent cultural units which can be combined in different ways to act on and modify one another. A tool can now not only be used to solve a basic problem like the exploitation of food resources, but can also be used to solve secondary problems as the manufacture of tools with the help of another tool (Kitahara-Frisch 1993). As cultural behavior is increasingly decoupled from basic needs, intermediate targets become independent. A hammerstone is not only meant to serve as a knapping device for a special stone artifact to work on a specific food item. Rather it can be used in different situations like stone knapping or operating a chisel or opening a nut. The socially transmitted information extends to behavioral units that are not exclusively bound to specific and acute problems. Instead, the elements of behavioral units (stimulus, concept of solution, goal) are increasingly abstracted from specific purposes and become applicable in different contexts. The execution of modular cultural capacities can occur on various technological levels based on differing knowledge and skills: knapping stone tools directly with simple flake technique, bifacially, following the Levallois scheme, or in an Upper Paleolithic way of blade production does not require more than modular cultural capacity as does the manufacture of a simple wooden spear (Haidle 2009, 2010). The complexity of knowledge and skill and thus the technological standard can be enormously raised without a fundamental change in cultural capacities.

The Tasmanian assemblage of material culture shows evidence of a number of tools manufactured in a modular way by the use of other tools. Spears, waddies, digging sticks, and wooden chisels are made with the help of stone tools produced with hammer stones. And spears, waddies, digging sticks, and wooden chisels are also used in a modular way as they are not made for, and in the course of, a specific event and thrown away when the event finished, but are continuously used in recurrent settings. The modular way of thinking and organizing cultural items refers to processes and actions instead of specific end-products. Chimpanzees also use bark, grass, and twigs, yet as direct raw material for specific tools. The Tasmanians (as other humans with modular capacities) instead could use unmodified or modified modules in the process of solving several secondary problem settings: bark, tough grass, leather straps, and sinews served as binding material, stone tools as cutting devices, pins of wood were used to fix something, and hammer stones to apply punctual force on other items.

With *composite cultural capacity* a new concept of solving problems enters the stage. The problem-solution distance can be extended to a combination of different cultural modules with specific qualities that are fused to form composites with new qualities. While tools made out of many pieces of the same kind, like piles of boxes to be used as a ladder as documented for chimpanzees (cf. Köhler 1926) or the sophisticated baskets of the Tasmanians (cf. Jones 1977:199, Fig. 5), escalate only the properties of the basic element, composite tools show a new combination of different qualities. The socially transmitted information exceeds that of modular cultural capacities through combining separate information on the basic elements of the composite "that may be obtained at different times and in different places" (Ambrose 2010:S139) with information on the newly created functional unit which may be assembled much later (ibid.). Hafted tools and compound adhesives are typical material examples of such composites.

Within the Tasmanian record several items based on composite cultural capacity can be identified. Lighted torches are composites consisting of (at least) portable sticks and fire. The pouch bags and some water containers combine pieces of skin or seaweed pierced at their margins with skin strips or string run through to tighten them. An alternative way of forming composite water containers out of seaweed was performed by fixing it with wooden pins. Shell necklaces represent a combination of pierced shells and string, and the bark canoes were shaped and fastened with straps of bark or grass. The different huts were composed of different materials for building the frame, for filling the interspaces, for binding, and for covering.

Complementary cultural capacity extends the possibilities to solve a problem with a new concept. This allows to develop and to use complementary sets of cultural modules as acting entities with two or more interdependent and exchangeable parts. Bow-and-arrow, needle-and-thread, screw-and-screwdriver, key-and-lock are only some examples of the symbiotic relation of two discrete, but concerted elements working together to fulfill a common task (Lombard and Haidle 2012). The elements of a complementary tool set must be developed in complementary correspondence with each other. The socially transmitted information expands: Different from composite cultural capacity, not only information has to be transmitted about the individual components involved, but also key formal information about the whole concerted system is required. An example of a complementary tool set used by aboriginal Tasmanians was the bark canoe (in itself a composite item) propelled with a wooden pole. The baited blind can also be considered as a complementary set consisting of a sort of grass hut to blind the bird hunter (in itself a composite item) and the bait fastened by a stone. To attract different sorts of prey, different sorts of bait were applied: pieces of fish for crows, worms for ducks.

With *notional cultural capacity* also notional concepts can be developed and used as cultural modules to solve a problem. These notional concepts can be (a) symbolic links (e.g., cross, crescent, and Star of David as symbols of religions), (b) normative definitions (e.g., metric and value systems), or (c) virtual beings (e.g., angels) and characters (e.g., protecting capacities of an amulet). The socially transmitted information exceeds that of all former capacities. It is now based on non-observable concepts, which can be manipulated only in the mind or through imagination, and their effects on real or other notional modules. Notional concepts can stand alone as independent modules, but can also be integrated in composites or act as parts of a complementary set. Material expressions of notional cultural capacity are information carriers. In aboriginal Tasmanian life the hand stencils and geometric patterns engraved and painted on rock and pieces of bark may well have been such information carriers. The well developed mythology was another, though immaterial, expression of the notional cultural capacities.

Conclusions: Lessons from Tasmania

As discussed in the EECC model of the expansion of cultural capacities, the range of cultural performances of different groups has widened out over the course of human evolution coherent with the development of new cultural capacities (Haidle and Conard 2011; Haidle et al. 2015). Nonetheless, a single cultural performance in an advanced grade of cultural capacity may be simpler than another performance in an earlier grade, since different aspects of the full cultural potential can be applied selectively (Hovers and Belfer-Cohen 2006; Lombard and Parsons 2011). The Tasmanian example shows this in an impressive way. The range of pre-colonial Tasmanian performances is rated as very simple based on the scarcity of cultural expression in material objects and the artless appearance of the few recognized manifestations. Yet, although only fragmentarily conveyed the cultural repertoire of aboriginal Tasmanians clearly gives evidence of full cultural capacities in all modular, composite, complementary, and notional aspects. At first glance, the artifact assemblage may appeal as simple and reduced, but the examination should not stuck in assumptions based on more or less subjective perceptions of the accessible record. The assessment of cultural capacities based on the analysis of problem-solution distances of the broad spectrum of cultural performances offers a more systematic approach. However, the examination of only a segment of the repertoire like McGrew's (1987) focus on technology used for food procurement may be misleading, and in this case even applying the problem-solution distance approach would yield a distorted image and underestimation of the cultural capacities behind. Both chimpanzees and humans (e.g., Tasmanians) use tools to access food. But humans (and as such also Tasmanians) developed cultural performances for purposes beyond those documented for chimpanzees so far. Excluding the data of cultural performances beyond food procurement from the analyses means to exclude a complete field of possible differences which were sought for.

The mechanisms of cultural development can be compared with a ratchet (Tomasello 1999) or better with the act of mountaineering (Lombard 2012, 2016). Cultural development is path dependent: the track used so far determines the possibilities of the future route. It is possible to proceed up to a more sophisticated level of performance, but also to move to the side to variations on the same level or down to simpler performances. In cultural development it is even possible to return to an earlier point and to proceed by retracing and then following other routes. Or a hook fixed by another group can be used to continue as shown by several items quickly included in the Tasmanian repertoire which were adopted from the European set of performances brought in by colonists. "Using the mountaineering metaphor, cumulative cultural capacity does not only include those cultural efforts that are built upon the highest level achieved, but also recursions following on seemingly more advanced solutions. Advanced cultural capacities are not necessarily accompanied by a progressive line of ever more sophisticated and complex solutions built on earlier ones, but allow increasing technological, cognitive, and behavioral flexibility from very simple to highly complex solutions depending on changing environmental constraints and cultural decisions. Thus, although the range of cultural performances expands with increasing cultural capacity, cultural evolution is not always linearly progressive. The development of cultural capacity is a systemic process involving the co-evolution of the three dimensions outlined above, and their interactions with the specific functional environment" (Haidle et al. 2015: 51). To understand the evolving system not only single elements which are perceived as interesting should be examined, but as many elements as possible including their interdependencies with and relevance for other elements. As the organism of a fungus cannot be understood by studying mushrooms which appear at the surface for limited time, the examination of specific blade technology or parietal art and their occurrence or absence is not sufficient to define and to explain the behavioral capacities of *Homo sapiens* (Haidle 2014; cf. Renfrew 1996). The Tasmanian cultural record is a perfect example that apparently simple performances cannot be easily equated with archaic or non-modern behavior. "To really comprehend the process of changing humanity, we need to uncouple the 'simple' to 'complex' social evolutionists' paradigm and understand the archaeological record in terms of its cultural variability that reflects the wide range of responses to solving similar social and environmental problems that our species is capable of" (Cosgrove et al. 2014:188). The presumptive abandonment of fish consumption and bone point production in the Tasmanian record can be easily explained without any basic loss of intellectuality which, by the way, cannot be traced in the comprehensive assessment of the cultural capacities on the basis of problem-solution distances in Tasmanian tool behavior.

References

Allen, H. (1979). Left out in the cold: Why the Tasmanians stopped eating fish. *Artefact, 4*, 1–10.

Ambrose, S. (2010). Coevolution of composite-tool technology, constructive memory, and language. *Current Anthropology, 51*, S135–S147.

Bassett, E. (2004). Reconsidering evidence of Tasmanian fishing. *Environmental Archaeology, 9*, 135–142.

Beall, C. M. (2006). Andean, Tibetan, and Ethiopian patterns of adaptation to high-altitude hypoxia. *Integrative and Comparative Biology, 46*, 18–24.

Bowdler, S. (1980). Fish and culture: A Tasmanian polemic. *Mankind, 12*, 334–340.

Bowern, C. (2012). The riddle of Tasmanian languages. *Proceedings of the Royal Society of London B, 279*, 4590–4595.

Brown, S. (1991). Art and Tasmanian prehistory: evidence for changing cultural traditions in a changing environment. In P. Bahn & A. Rosenfield (Ed.), *Rock art and prehistory: papers presented to Symposium G of the AURA Congress, Darwin 1988* (pp. 96–108). Oxford: Oxbow Books.

Burger, J., Kirchner, M., Bramanti, B., Haak, W., & Thomas, M. G. (2007). Absence of the lactase-persistence-associated allele in early Neolithic Europeans. *Proceedings of the National Academy of Sciences of the USA, 104*, 3736–3741.

Cosgrove, R., Pike-Tay, A., & Roebroeks, W. (2014). Tasmanian archaeology and reflections on modern human behavior. In R. Dennell & M. Porr (Eds.), *Southern Asia, Australia and the search for human origins* (pp. 257–264). Cambridge: Cambridge University Press.

Davidson, I. (2007). Tasmanian aborigines and the origins of language. In J. Mulvaney & H. Tyndale-Biscoe (Eds.), *Rediscovering Recherche Bay* (pp. 69–85). Canberra: Academy of the Social Sciences in Australia.

de Teliga, S. & Bryden, W. (1958). A note on Tasmanian aboriginal drawings. *Papers and Proceedings of the Royal Society of Tasmania, 92*, 191.

Enquist, M., Ghirlanda, S., Jarrick, A., & Wachtmeister, C. A. (2008). Why does human culture increase exponentially? *Theoretical Population Biology, 74*, 46–55.

Fisher, S. E., & Ridley, M. (2013). Culture, genes, and the human revolution. *Science, 340*, 929–930.

Gamble, C. S. (1992). Archaeology, history and the uttermost ends of the earth – Tasmania, Tierra del Fuego and the Cape. *Antiquity, 66*, 712–720.

Gott, B. (2002). Fire-making in Tasmania: Absence of evidence is not evidence of absence. *Current Anthropology, 43*, 650–656.

Haidle, M. N. (2008). Verschiedene Welten. Umweltwahrnehmung und Umweltgestaltung im Laufe der menschlichen Evolution. In T. Knopf (Ed.), *Umweltverhalten in Geschichte und Gegenwart. Vergleichende Ansätze* (pp. 30–41). Tübingen: Attempto.

Haidle, M. N. (2009). How to think a simple spear? In S. A. de Beaune, F. L. Coolidge, & T. Wynn (Eds.), *Cognitive archaeology and human evolution* (pp. 57–73). New York: Cambridge University Press.

Haidle, M. N. (2010). Working memory capacity and the evolution of modern cognitive capacities – implications from animal and early human tool use. *Current Anthropology, 51*, S149–S166.

Haidle, M. N. (2012). *How to Think Tools? A Comparison of Cognitive Aspects in Tool Behavior of Animals and During Human Evolution.* Cognitive perspectives in tool behaviour Vol. 1. urn:nbn:de:bsz:21-opus-60146. Retrieved June 6, 2014.

Haidle, M. N. (2014). Examining the evolution of artistic capacities: Searching for mushrooms? In C. Sütterlin, W. Schiefenhövel, C. Lehmann, J. Forster, & G. Apfelauer (Eds.), *Art as behaviour. An ethological approach to visual and verbal art, music and architecture* (pp. 237–251). Oldenburg: Bis-Verlag der Carl von Ossietzky Universität Oldenburg.

Haidle, M. N., & Conard, N. J. (2011). The nature of culture – Summary report on an interdisciplinary symposium held in Tübingen, Germany, 15–18 June 2011. *Mitteilungen der Gesellschaft für Urgeschichte, 20*, 65–78.

Haidle, M.N., Bolus, M., Collard, M., Conard, N.J., Garofoli, D., & Lombard, M., et al. (2015): The nature of culture – An eight-grade model for the evolution and expansion of cultural capacities in hominins and other animals. *Journal of Anthropological Sciences, 93*, 43–70.

Hambly, W. D. (1931). Types of "Tronattas" or stone implements used by the Aborigines of Tasmania. *American Anthropologist, 33*(1), 88–91.

Henrich, J. (2004). Demography and cultural evolution: How adaptive cultural processes can produce maladaptive losses – The Tasmanian case. *American Antiquity, 69*, 197–214.

Hiatt, B. (1967). The food quest and the economy of the Tasmanian aborigines. *Oceania, 38*(2), 99–133.

Hiatt, B. (1968). The food quest and the economy of the Tasmanian aborigines (continued). *Oceania, 38*(3), 190–219.

Hiatt, B. (1969). Cremation in aboriginal Australia. *Mankind, 7*, 104–119.

Horton, D. R. (1979). Tasmanian adaptations. *Mankind, 12*, 28–34.

Hovers, E., & Belfer-Cohen, A. (2006). "Now you see it, now you don't" – Modern human behaviour in the Middle Palaeolithic. In E. Hovers & S. Kuhn (Eds.), *Transitions before the transition. Evolution and stability in the Middle Paleolithic and Middle Stone Age* (pp. 295–304). Heidelberg, New York: Springer.

Jones, R. (1977). The Tasmanian paradox. In R. V. S. Wright (Ed.), *Stone tools as cultural markers: Change, evolution and complexity* (pp. 189–204). Canberra: Australian Institute of Aboriginal Studies.

Jones, R. (1978). Why did the Tasmanians stop eating fish? In R. Gould (Ed.), *Explorations in ethnoarchaeology* (pp. 11–48). Albuquerque: University of New Mexico Press.

Jones, R. (1995). Tasmanian archaeology: Establishing the sequences. *Annual Reviews of Anthropology, 24*, 423–446.

Kitahara-Frisch, J. (1993). The origin of secondary tools. In A. Berthelet & J. Chavaillon (Eds.), *The use of tools by human and non-human primates* (pp. 239–246). Oxford: Clarendon Press.

Köhler, W. (1926). *The mentality of apes*. New York: Harcourt, Brace & Co.

Ledón-Rettig, C. C., Richards, C. L., & Martin, L. B. (2013). Epigenetics for behavioral ecologists. *Behavioral Ecology, 24*, 311–324.

Lombard, M. (2012). Thinking through the Middle Stone Age of sub-Saharan Africa. *Quaternary International, 270*, 140–155.

Lombard, M. (2016). Mountaineering or Ratcheting? Stone age hunting weapons as proxy for the evolution of human technological, behavioral and cognitive flexibility. In M. N. Haidle, N. J. Conard, & M. Bolus (Eds.), *The nature of culture* (pp. 135–146). Dordrecht: Springer.

Lombard, M., & Haidle, M. N. (2012). Thinking a bow-and-arrow: Cognitive implications of Middle Stone Age bow and stone-tipped arrow technology. *Cambridge Archaeological Journal, 22*(2), 237–264.

Lombard, M., & Parsons, I. (2011). What happened to the human mind after the Howiesons Poort? *Antiquity, 85*, 1433–1443.

Malafouris, L. (2010). Metaplasticity and the human becoming: Principles of neuroarchaeology. *Journal of Anthropological Sciences, 88*, 49–72.

McGrew, W. C. (1987). Tools to get food: The subsistants of Tasmanian aborigines and Tanzanian chimpanzees compared. *Journal of Anthropological Research, 43*, 247–258.

Meston, A. L. (1935). Observations on visits of the Tasmanian Aborigines to the Hunter Islands. *Papers and Proceedings of the Royal Society of Tasmania 1935*, 155–163.

Murray, T., & Williamson, C. (2003). Archaeology and history. In R. Manne (Ed.), *Whitewash: On Keith Windschuttle's fabrication of Aboriginal history* (pp. 311–333). Melbourne: Black Inc Agenda.

Noetling, F. (1908). A native burial ground on Charlton estate, near Ross. *Papers and Proceedings of the Royal Society of Tasmania 1908*, 36–43.

Oswalt, W. H. (1976). *An anthropological analysis of food-getting technology*. New York: Wiley.

Petanjek, Z., & Kostović, I. (2012). Epigenetic regulation of fetal brain development and neurocognitive outcome. *Proceedings of the National Academy of Sciences of the USA, 109*, 11062–11063.

Porr, M. (2014). Essential questions: 'Modern humans' and the capacity for modernity. In R. Dennell & M. Porr (Eds.), *Southern Asia, Australia and the search for human origins* (pp. 257–264). Cambridge: Cambridge University Press.

Read, D. (2006). Tasmanian knowledge and skill: Maladaptive imitation or adequate technology? *American Antiquity, 71*, 164–184.

Renfrew, C. (1996). The sapient behaviour paradox: How to test for potential? In P. Mellars & K. Gibson (Eds.), *Modelling the early human mind* (pp. 11–15). Cambridge: McDonald Institute for Archaeological Research.

Rogers, E. M. (1995). *Diffusion of innovations* (4th ed.). New York: Free Press.

Roth, H. L. (1899). Tasmanian Firesticks. *Nature, 59*, 606.

Ryan, L. (1981). *The Aboriginal Tasmanians*. St Lucia: University of Queensland Press.

Sollas, W. J. (1911). *Ancient Hunters and their modern representatives*. London: Macmillan.

Taylor, R. (2007). The polemics of eating fish in Tasmania: The historical evidence revisited. *Aboriginal History, 31*, 1–26.

Taylor, R. (2008). The polemics of making fire in Tasmania: The historical evidence revisited. *Aboriginal History, 32*, 1–26.

Tomasello, M. (1999). *The cultural origins of human cognition*. Cambridge: Harvard University Press.

Vanderwal, R. L. (1978). Adaptive technology in Southwest Tasmania. *Australian Archaeology, 8*, 107–127.

Völger, G. (1973). Making fire by percussion in Tasmania. *Oceania, 44* (1), 58–63.

Walters, I. (1981). Why the Tasmanians stopped eating fish: A theoretical consideration. *Artefact, 6*, 71–77.

Woollett, K., & Maguire, E. A. (2011). Acquiring "the knowledge" of London's layout drives structural brain changes. *Current Biology, 21*, 2109–2114.

Chapter 3
Culture as a Form of Nature

Volker Gerhardt

Abstract Opposite to an impression which is particularly widespread in the humanities and the social sciences, the author starts with the ironic – but nevertheless authoritative – assessment that nature still exists. This being stated, the presumably even more prevalent opinion that culture stands in opposition to nature is criticized. Instead, a case is made that culture is nothing more nor less than a specific way of life that can only develop within nature. Even more: culture can do so exclusively as nature; and culture can operate only with nature. This is exemplified by the origins of technology which are crucial for the evolution of culture. Every technology could primordially only make use of resources provided by nature. Thus it can be proven that in any case culture can only be regarded as part of nature which, as do all parts of nature, differs depending on the exterior and interior preconditions of the respective lives. Therefore, culture is one of the numerous evolutionary products of nature.

Keywords Philosophy • Theory • Consciousness • Language • Signs • Technology • Stone tools • Fire

The Status Quo of Nature

The onset of the self-assertion of history and the social sciences in the 19th century – also to be seen as a reaction to the rising natural sciences – has led to a dominance of the belief in the historical and social nature of all things, so that a reference to nature is viewed as a relapse into mythology.

As beginning, condition, and means, nature is inherent in everybody and everything; every conceivable end of a process that is not explicitly designed to lead to the heavenly realms, is only feasible as a natural state. And yet, the argumentative claim of nature is only accepted in discussions of social sciences if it promises effects critical of society.

An ecologist only needs to abominate the "destruction of nature" and no one thinks to question which "nature" he is referring to. Nature in the 16th century? Nature prior to the artificial irrigation of Egypt? Or Nature prior to the appearance of humans? A sociologist mourns the "estrangement from nature" in the industrial workplace and we forget that his criticism not only assumes a preceding human nature, but that it is also an appeal for its preservation. But how does he want to preserve something that he does not allow himself to speak about? And what is the preservation of nature when all of nature is a product of incessant change? Finally, when a philosopher such as Jürgen Habermas defends the human "environment" against the purpose-rational modern "systems", he is referring to a form of nature, which, according to the premises of his "critical theory", he cannot really define. However, defining it is fundamental if we want to sensibly use the term evolution. Also, the rarely disputed process of evolution declares that nature has its own history.

Generally, in the correlation of reasons of cultural studies and social sciences only those facts are accepted that are dependent on social accessibility. In political contexts, this practice is called "imperial". "Imperial" means that a power also wants to influence the conditions of its acceptance by other powers. This is achieved by ignoring natural conditions in favor of social or political, cultural or spiritual phenomena. It is assumed that "Nature" does not have to be an issue as soon as social relations are considered. Some believe that "Nature" should not even be mentioned as soon as social, political or cultural aspects become part of the

V. Gerhardt (✉)
Institut für Philosophie, Lehrbereich Praktische Philosophie, Rechts- und Sozialphilosophie, Humboldt Universität, Unter den Linden 6, 10099 Berlin, Germany
e-mail: volker.gerhardt@philosophie.hu-berlin.de

Miriam N. Haidle, Nicholas J. Conard and Michael Bolus (eds.), *The Nature of Culture. Based on an Interdisciplinary Symposium 'The Nature of Culture', Tübingen, Germany*, Vertebrate Paleobiology and Paleoanthropology, DOI 10.1007/978-94-017-7426-0_3 debate. Therefore, nature, to give only a personally associated example, is nearly no longer present in *political theories* – unless one thinks ecologically and fights for its preservation.

When we consider the existence of the natural sciences, it is not necessary to further comment on what one is to think of this self-inflicted restriction on so-called self-made causes. What can one say, when it is clear, that even after paying a little attention to the concepts, every meaningful reference to social phenomena at least requires their differentiation from nature? Yet this differentiation only serves the methodological protection of a discipline.

Nature is present in every single process and in every single thing – without nature there would be neither facts nor data. And when we examine the properties and condition of individual occurrences, it is difficult to find anything that does not have a natural component. Thus, nature is also a factor in our actions, our words and our thoughts.

Culture as a Variation of Nature

Upon closer inspection, it becomes clear that every description of social facts must incorporate an infinite number of natural moments in order to guarantee a minimum of credibility. Imagine a sociological analysis of an event that does not require any individuals being present physically and that does not insinuate motives or objects. What is the significance of a statement that is independent of needs and expectations as well as time-constraints? What value do conventions, institutions or religions have that do not need historical dates because they do not rely on geographical places or crowds of people nor the needs and desires of someone somewhere?

Therefore, it is not surprising that nature has not completely disappeared from modern political theories, although it is not the foremost topic. The "natural state" is an exception because it delimits the natural boundaries of the state, because we aim to leave the "natural state" behind. In other cases, such as birth, sickness and death, life and limb – or as the subject of a crisis – the subject itself is enough to get a grip on the basic nature of all things.

Besides, the spheres of history and the humanities, of culture and politics would not lose any of their dignity if we were to admit that they are tied into nature. If we pursue so-called purely sociological phenomena back in time – family in medieval society, cities at the time of Homer, trade prior to the invention of writing or Stone Age technologies – their imbedding in nature becomes obvious. We recognize that the nature of culture is not superficial, but instead determines its inner dynamics. It defines itself not by separating itself from nature but through the *specific demarcation* from it, a circumscription that occurs *with* and *within* nature. This *self-differentiation* of nature makes it possible for nature to stand in opposition to itself and allows room for a continuous increase of its influence.

If nature could be understood as a living entity, then it would grow body parts and organs within society, culture and politics – these are especially useful because they can also oppose it. However, their opposition always remains a part of an ever changing and growing nature.

Nature as an External and Internal Constitution

An example will show that nature does not simply define our physical existence: knapping a stone requires that someone perceives and then appreciates its nature-given material qualities. It is also necessary that one recognizes the nature of the material of the tool with which the stone is shaped – it must be able to cope with its hardness and brittleness. If the tool itself needs to be fashioned, then that, too, needs to be created according to the nature of the raw materials – as the production of iron is dependent on the iron ore. A granite tool may be sufficient to knap a soft stone. Every implementation of a material requires the recognition and acknowledgement of other materials, even if these need to be altered prior to their employment.

A tool-wielding human applies one part of nature to another part of nature and does this according to the natural condition of his own body and the therewith-acquired abilities. The specific efficiency of the body developed throughout evolution; therefore, the resulting cultural developments may also be defined as "nature". Otherwise, we would have to refrain from using the term nature and could, at best, speak of an infinite number of cultural steps. In nature, everything already "is", therefore it is already the result of a mechanical and organic self-cultivation of nature.

Obviously, the process of a person knapping a stone occurs within nature. This may take place inside or outside of a cave, by rain or sunshine, at a leisurely pace or under great duress: The knapper will need air to breathe, enough space for his actions and a solid base for the ideal implementation of his strengths. He needs strength, which he – as an expression of his nature – builds up and regenerates in nature in order to even be *aware* of his strengths.

An active person needs to understand the material he is working with as well as the extent and limits of his own physical constitution – if he wants his projects to be successful. With his hands he can feel, grip, push, pull or strike, but he cannot see, feel or smell with them. Some hand actions only require the use of one finger; others require the strength of the arm or the application of the body's entire mass.

In all that a person attempts to achieve with his hands, he depends on his physical actions, which he (as a person) coordinated as a whole, in order to succeed. In this, he is dependent on his own physical nature, no matter how small his self-awareness is. This confidence allows him to act *as a part of* nature, *in* nature and *with* the resources *from* nature to produce the effects, which can then be valued as cultural achievements.

It is not enough to simply look at a person's physical state as his natural condition. Why is a stone even knapped? Because, or so we assume, a person has recognized, either through personal experience or expectation, that a knapped stone has an advantage. A handaxe can be helpful when working with fibers, furs or wood, when dividing up fruit or a hunted animal, or when protecting oneself from enemies. These activities involve intentions that are a part of human nature: Because man does not grow his own fur he must find and create his clothing and other protection from the elements; man does not have tusks or claws so he needs tools to manipulate his food; at the same time, man is not set apart from the natural competition for resources with other species and other individuals.

The logic of "must" and "need" is based on nature. Within nature, man has enemies against which he "must" defend himself. Therefore, he acquires weapons and because he wants to achieve a certain status among his peers *as a part of* nature *with* nature and *within* nature he also pursues cultural developments – even politics.

The Dense Context of Nature

In the process of a person's continuous development, it is not enough that he just accepts the natural circumstances. He must also identify himself with them; he must also recognize some circumstances as foreign, adverse or hostile. As *a part of, in* and *with* nature it is also important to assert oneself *against* certain parts of nature. The act of shooing away a fly exemplifies this concept: a person *as a part of* nature (in the form of an irritated subject with a specific need), *within* nature (in an environment, which he shares with the fly), *with* nature (with his hand) and *against* nature (the insect) in order to protect something that he requires *to satisfy* his natural needs. Irrespective of the care and precision with which he prepared his meal, it remains a part of nature that he, as a part of nature, wants to enjoy and must protect from a hostile part of nature, in this case the fly. A connection cannot be denser than in this example in which man is connected to other people, animals, plants and other aspects of nature.

Within this context, a person must differentiate nature according to expedient, inappropriate and indifferent factors. This is a natural process in that a person attempts to interact *as a part of, in* and *with* nature. He is a natural being in his sentiments, feelings and desires into his most subliminal motives, and he remains thus although he is so changed through the continuous development of his abilities and talents, that they become *culture*.

The more these abilities and talents are developed and become dependent on newly developed techniques, the more culture becomes prominent. Yet culture does not cause a person to step outside of nature! Instead, the more elaborate his culture becomes, the more he remains connected to nature. Therefore, there is no break or exodus from nature. Culture is an expression of the productive and creative participation of man in nature and its history.

The mechanical engagement of nature is also included in this participation because it has its origins in and satisfies specific needs. Although it seems strange in individual cases and may be harmful in others, it is only possible as a result of cultural evolution – therefore: organically possible. What else shall culture achieve but to shape life in order to facilitate life? Think of the harnessing of fire, which lead to the onset of human culture ca. 2.5 to 3 Ma (Schrenk 2003).

Therefore technology, which plays a crucial role in the evolution of culture, is a significant element of the self-organization of nature into culture. Successful technology is an adaptive intervention in nature, which results in culture. Technology is an active participation in nature. However, the term participation should probably be reserved for a process based on the self-determination of self-conscious and freely acting individuals. At the same time, we, as technological players in nature, can interpret everything we are aware of as a paradigm for what morals and politics expect of independent people. Therefore, it is not incorrect to speak of a participation of technology in nature. That would avert an inappropriate defense of technology as a simple regulation.

Man as an alleged *maître et possesseur de la nature* is wishful thinking or a curse – however one may perceive that to be. But it is a *wish*. Because even though he may reign over parts of nature, the so-called master also remains chained to nature. He follows nature's impulse even in his claim to power. That applies to all cultural activities, particularly politics where it is especially difficult to discipline the unleashed strength of his nature.

Morals and politics may pretend to be autonomous: They will not be able to surpass their participation in nature, whether with efficient anthropo-technological self-discipline, nor with institutional legal techniques, bureaucracy or the military. However, similarly to the individual consciousness, they can take up an "autonomous" position opposite other social processes.

The Problem of Conscious Inner Space

The *consciousness* is a significant action quantity in nature. *Idealism*, *solipsism* and *epistemic skepticism* are the consequences when we only take the consciousness of an individual into consideration. If the consciousness is something that we only allow ourselves and others can only have hypothetically, then the subjective consciousness is a dead-end street that does not allow us to access the world or other people. If we attempt to find the "reason for consciousness" on our own in private reflection then, as the works of Dieter Henrich show quite impressively, we sink more and more into hopelessness (Frank 1991; Henrich 1992). The subjectivity of self-awareness does not lead to a human- *and* world-encompassing consciousness.

The only solution seems to require godly intervention, where the world and individual consciousness are united in one all-encompassing spirit. Plato believed this was the way to make his ideas reality; Descartes bridged the gap between *cogito* and *res extensa*; and Hegel (1986) saw himself forced to mediate the idea and reality in the apotheosis of historical evolution. Hegel did not trust in the solution presented by Kant (1781), that the self-isolation of the consciousness can be overcome through a logical conclusion – a refocus from the inner spaces of consciousness to a necessary external world. He applied an ingenious shift of the intellectual dialectic process into the real-historical process and fell back onto the concept of the spirit, which, in the end, proved to be godly.

Hegel (1986) is not entirely wrong in his critique of Kant (1781). His apparent compelling "refutation of idealism", which uses the analytical equivalent of "inside" and "outside", is based solely on the achievements of the consciousness that proves *itself* without actually having *factual* proof that the developed outside *really* exists. It must be *thought* or *believed*; however, whether it really is there cannot be conclusively proven. Therefore, according to Hegel's conclusion, Kant cannot connect the conscious space to the real world.

The 20th century solution, that the consciousness' *intention* is to exhibit a bridge, determined by the consciousness' function, to the outside world, could not escape the suspicion that something is to be *developed*. That has been shown empirically through the unreduced current epistemic skepticism. It is also obvious due to the inadequacy of the evidence as well as the arguments for the objective achievements of the individual consciousness. In all cases, whether the underlying theories are Husserl's concepts of the concept of *intention*, which gained importance after Anscomb, the consciousness draws conclusions about its own self-exceeding achievements: The structures of understanding and speech allow us to draw conclusions on the reality to which they refer.

Consciousness as a Social Organ

It is not necessary to draw conclusions if we can show that consciousness originally referred to and about others outside the self. This can be shown in the position and effectiveness of the self-conscious (Gerhardt 2012; in part in Gerhardt 2007): It is based on the existence of the world and functions as a medium in the world in everything that it does.

If we take into account the possibility of simple description or realization, whether we think it capable of attention or self-sufficient taciturnity, if it concerns the intention to commune or act or if it is about something to be remembered: In all possible cases, the achievements of the consciousness do not simply stand in theoretical relationship to the world. They are in a very real general interdependence of physical conditions, which include the body and which, under the complex conditions of life, expand into social and cultural entities – without ever ceasing to be a part of nature.

The consciousness participates in this organization of materials and forces. It is an *organ of the organism* and in its achievements, just like the organism itself; it is naturally and irrevocably *individual*. However, like the organism, it does not simply participate in the organization of life, but it also *represents* it – although in a new *technological* and, at the same time, *cultural* way. The consciousness is able to make itself into an organ of a society of sentient beings in its own unique fashion.

The individual consciousness must in all its definitions and classifications be seen as *universal* because it is able to identify and classify a *specific thing* (such as a handaxe) or an *individual event* (recognizing danger) so that its *general characteristic* ("useful", "dangerous", "useless" or "irrelevant") is not just *understood* but can also be *communicated.* Therefore, the conscious becomes a *representative of recognized* as well as the *recognizing* and, through *communicating* this, it presents itself as the registering and coordinating *group sensor*, to which the individual belongs. And wherever this takes place, the consciously acting individual takes on the role of the *authority* and the *representative of social connections*, in which he also recognizes *himself.*

In this achievement, the consciousness is not only the organ of one individual that defines itself in him (and in his world), but it functions as the *organ of the group* in which it is able to develop, in which it can articulate itself, in which it is understood and can be effective. This can go so far, that its social role may cause it to operate against its own organism. Anyone who conforms to the customs of his group, who tries to meet the expectations of his family, the demands of his position or the reason of a task, anyone who sacrifices something for the sport, the arts or the sciences, who is willing to take into account certain restriction for the well-being of the environment or who fights to uphold certain elemental rights for the good of humanity – and really means what he says: Anyone, I say, who commits themselves to a specific cause does so not *just* as an organ of his own organism or as an advocate for himself, but *always also* as an *authoritative representative* of the natural and social entity for which he stands.

Consciousness – Independence Within Context

The question is in which kind of organization can the organizational achievements of our consciousness function as an element and an instrument? The most obvious answer is that it is the organization of the body, which depends on the consciousness to stay alive. The body is dependent on the metabolic exchange with the surrounding nature. It requires the cooperation with its own kind in order to facilitate its own creation, developments and reproduction. It moves through its environment while using it for its own purposes: It uses its senses to search for points of reference which allow it to steer itself through its environment in an ever larger radius, with increasing expense and in coordination with others.

The consciousness is beneficial for man in all of these cases – even though there are numerous examples where it gets in the way of functional expectations: It helps increase the mobility and flexibility in different environments, it adapts to new and even to inhospitable living conditions and increases their usability by converting natural resources. It is (how can it be anything else?) bound to (physical) conditions and (psychological) effects. Yet in this bond it has its own function as an (intelligible) authority of whom one expects clarifying insights and realizations.

Due to his consciousness, man is the only living being that cooks, grills or smokes his food. The selection of enjoyable foods is greatly expanded; many things that were not digestible can now be eaten without negative consequences; perishable foods can be preserved longer. The spectrum of foods processed by the humans' organism becomes continuously richer and broader. The self-cultivating activity broadens the natural basis of human life.

And all that becomes *communicable in an instant*, so that it is not just *currently* effective but can also function in the future! The consciousness is the *social organ,* which most closely approximates a cultural institution, therefore facilitating *short-term learning* and allows man to *learn* through *imaginary experiences* and through *imagined dangers and solutions in connection to a generation* (see: Gerhardt 1999, 2000). The instant becomes the present, which in turn becomes the starting point for the *future* and the *past*. So time itself can become a long path from which different avenues of activity may lead.

Culturally, this will be made productive in a real and very short span of time, the likes of which is not known in evolution. Which other organism has caused so many changes to its environment in its accelerated evolution as man? Which animal has taken up so many different mundane processes and subjected it to its way of life? I only mention the handling of fire.

As man has trusted in the impact of fire, he also begins to trust in his tools and begins to depend, long before he becomes sedentary, on his tamed animals. Cattle breeding and farming clearly demonstrate how much observation of nature, conscious productive work and self-discipline are necessary to make this (what appears to us today as natural, but is in fact a highly) cultivated way of life possible. In case of success, man's own invested consciousness helps him and he learns to conform to all that he has caused to change in his living environment.

So one can state: Among the facts of human culture, the consciousness emerges as a *real effective quantity*. *Technology* and the actions made possible through it create new living conditions that are adapted to human culture, individually and collectively, and therefore are themselves becoming conditions of the evolving human life. They themselves become a part of nature and therefore it is not necessary to draw comprehensive conclusions on a 'real' outside world. Strictly, the term outside world is obsolete; only the universal and effective world exists, in which man integrates himself, even if the consciousness provides him with chances to distance himself from it procedurally. That is the distance that every organism makes for himself through its own organic self-limitation and that can be increased through social experiences and differentiation from other individuals and other "strange" groups.

Yet independent of the functional differentiation between inside and outside, it is important that the *consciousness* is seen as an effective element in the process of the natural evolution of man. To this end, I can only present a few more facts.[1] I will limit myself to two.

---

[1]For further discussions, please refer to the previously mentioned books (Gerhardt 2007, 2012). Further considerations concerning the historical background are presented in: V. Gerhardt, Monadologie des Leibes. Leib, Selbst und Ich in Nietzsches Zarathustra, Philosophische Zeitschrift der Universität Seoul, Korea 2012, 161–232 (Reprint of the introductory paragraph: V. Gerhardt, Die Funken des freien Geistes, Berlin/New York 2011), as well as in: V. Gerhardt, Humanismus als Naturalismus. More about the criticism of Julian Nida-Rümelin's contrast of nature and freedom can be found in: Sturma (2012:201–225); also in: Rohmer and Rabe (2012:159–185).

The Public Interaction of Spirit and Reason

The most important, though not visible from the outside, step in the evolution of mankind must be the *institutionalization* and *instrumentalization* of the consciousness: With the advent of tools and the delegation of labor, it became more and more necessary to *communicate about situations* and *circumstances* that one can refer to with a degree of security and predictability. This necessity for communication makes the act of communication as well as the methods of communication into an *object* or *subject*, which can be argued and bargained with as if it were a *thing*. Something that is communicated, in a general manner so that it is independent of the given situation and active participants, becomes autonomous of the speaker and is given the meaning corresponding to the spoken word. The communiqué takes on *meaning* that is based on the given statement and is no longer dependent on the presence of a specific speaker.

An individual's original achievements, which presents itself in an observation, in an insight or a memory (and that has numerous social conditions through language acquisition, grammar and logic), becomes an *institution* that it must adhere to if it wants to use it for its own interests. It is not the medium, with its distinct characteristics, but the *meaning* that is recognized *as such*. Meaning itself is a *thing* about which one can speak as if it were an object like any other.

What is new is the *independence of the consciousness* and its contents. As a result, it is first possible to speak about "consciousness", the "spirit" or about "reason" as we would speak about organs or substances, authorities and instruments. And when their contents attain "meaning" and "significance", "conceptions", "thoughts" or "ideas" this is due to the *autonomy of the relevant subject*.

The autonomization of that which makes a factual communication possible satisfies the prerequisites for *institutionalization*. Through it, the consciousness itself can be seen as an organization that encompasses everyone who communicates according to its conditions. Under these circumstances, it can be understood as the "spirit" that, like a in a collective consciousness, encompasses everything that can be understood or that constitutes a conclusive idea without being tied to an individual consciousness. Similarly, the consciousness' performance can also be described as "reason", whose functions (understanding, justifying, judging and decision-making) formulate a relation between ends and means and concentrate on the justification and criticism of intentions, goals and suitable actions.

The Meaning of Signs

The material consolidation of the semantic sphere, whose function we have described here, occurs, I presume, in the express *use of signs*. The most significant developmental step in the history of human nature, after the *use of tools* which is closely linked to the manipulation of the hand, is the instrumental application of *signs* (among which I generously include all intentional markings as well as pictures and symbols): That humans no longer depend exclusively on their own *bodily presence* in order to express themselves, but that they use *objects manipulated by tools*, which, in the absence of the acting individual, can have physical *consequences* (such as traps or fish traps) as well as *symbolic meaning* and *consequences*, is the definitive *technological innovation of culture*. In the same way that the body, primarily the arms, legs and hands are used as tools in order to create and implement tools, it is also possible to utilize the mouth, tongue and lips to suddenly create phonetic symbols that stand for themselves in the social space.

Mind you: The vocalization of articulated speech adheres to the rules of the instrumental utilization of an object. According to Ludwig Wittgenstein (1984:§569), "*Die Sprache ist ein Instrument*" ("speech is an instrument"). Man learned to implement it when humans already knew how to handle tools. They made it possible to create a way of life with social insurances and strengthened through wisdom that is based on the perspectives of an open world. The worldwide mobility of early humans gives us the impression that the world during the development of early man was vast and open. I do not think that I have to emphasize this at this point.

But technology played the most important role at close range as well: Picture and clay tablets, steles covered with symbols, painted temple walls, skins with writing on them or written characters scratched onto tortoise shells were the primary haptic carriers of meaning of their time. They required a knowledgeable reader in order to reproduce their meaning but preserved their message independent of the speaker or listener. Their meaning could also be reproduced again and again by copying or retelling the message – in some cases with conceptual exactness – thereby making the meaning independent of the respective materials as well. Therefore, the medium is a prerequisite for the sustainability of the conveyed meaning, at the same time, through its *substitutability*, it contributes to the opinion that meaning is not dependent on any form of material substance.

The successful control of nature, the production of tools, the development in the use of new technologies and the establishment of traditions and practices set the framework for the implementation of the body for the production of meaningful symbols. This includes the millennia old tradition of forming materials into objects that are not necessarily just tools but also precious valuables, cult objects of pieces of jewelry or ornamentation.

It is only necessary to explicitly define their meaning in order to make sure it is committed. In this case, it may have been the technological instruments themselves that, especially in ritual contexts, function as a signature and may have laid the groundwork for the invention of the written language.

The Role of Written Language

The written language is a carrier of meaning that exists detached from the speaking individual and that lends expression to the *spirit*, which is already evident in man's technological and social achievements. It can appear independent of humans. *The self-discipline, made possible through technology, of man, the natural being, is the precondition for the genesis of the human spirit.*

No matter how the historical course of events developed: After the appearance of *written characters*, after the establishment of *public squares* and *gathering halls*, after governing and assembly areas were constructed and, as it seems, as art was instrumental in filling these habitats with meaning produced by the people who created them so that this meaning could also be grasped by higher beings, the spoken language developed and became more precise and distinct. Those who knew how to make use of the spoken word in order to achieve certain effects, the prophets and priests, the singers and wise men, the actors and storytellers, were able to increase their medial autonomy.

Even prior to this, the written language allowed the individual and entire societies to remember *specifically*. It was possible to document who owned which property, the quantity of tribute paid by a specific person and the exact conditions according to which he received grazing rights or was made custodian of his wife's property.

The written language made it possible to preserve for the future the agreed upon meaning of a process. Therefore, meaning seemed to stand still while new events and agreements moved through its space incessantly and continuously created new facts.

It seems as though one is able to watch the material consolidation of the semantic sphere if we imagine that the written language made it possible to communicate events exactly across large distances and multiple generations. An original bill of sale, a royal decree, carved into stone or cut into a simple piece of wood, a biography of kings and their heroic achievements or a protocol of the movement of the heavenly bodies: These occurrences are made permanent through the single documentation for future readers and audiences.

An individual scribe recorded the events. Yet they reach far into the future where they were accessible to *anyone* who found and read the accounts. And not just to those who could read. For the creation of a self-sufficient sphere of meaning, it was enough that the laws, contracts, and observations exist and were accessible. The propagation of the read content occurred by word of mouth. And because it was always possible to go back and reread it and was spread by oral reports or stories, the written word also became a bridge into the sphere or *memory*, of *comprehension*, of *thought* and the *spirit*. It is more than simple "writing" or "language" because it is something that we can comprehend "communally".

In closing, I quote Wittgenstein (1984:§36) to *contradict* him: "*Wo unsere Sprache uns einen Körper vemuten läßt, und kein Körper ist, dort, möchten wir sagen, sei ein* Geist". ("Where our language lets us assume a body, but a body does not exist, there, we may say, is the *spirit*"). And so it is! Since an "assumption" itself is not *speech*, at the very least it is *thought*, we can say that in this way of *thinking* we are in possession of the power of *speech* as well as the *spirit*.

References

Frank, M. (1991). *Selbstbewußtsein und Selbsterkenntnis. Essays zur analytischen Philosophie der Subjektivität.* Stuttgart: Reclam.

Gerhardt, V. (1999). *Selbstbestimmung. Das Prinzip der Individualität.* Stuttgart: Reclam.

Gerhardt, V. (2000). *Individualität. Das Element der Welt.* München: C. H. Beck.

Gerhardt, V. (2007). *Partizipation. Das Prinzip der Politik.* München: C. H. Beck.

Gerhardt, V. (2012). *Öffentlichkeit. Die politische Form des Bewusstseins.* München: C. H. Beck.

Hegel, G. W. F. (1986). *Wissenschaft der Logik I*; *Wissenschaft der Logik II*; *Enzyklopädie der philosophischen Wissenschaften I.* In E. Moldenhauer & K. M. Michel (Eds.), *Werke in 20 Bänden, volumes 5, 6 and 8.* Frankfurt am Main: Suhrkamp.

Henrich, D. (1992). *Der Grund im Bewußtsein. Untersuchungen zu Hölderlins Denken (1794–1795).* Stuttgart: Klett-Cotta.

Kant, I. (1781). Critic der reinen Vernunft. Riga: Johann Friedrich Hartknoch.

Rohmer, S., & Rabe, A. M. (Eds.). (2012). *Homo naturalis.* Zur Stellung des Menschen innerhalb der Natur. Freiburg i. Br.: Verlag Karl Alber.

Schrenk, F. (2003). *Die Frühzeit des Menschen. Der Weg zum Homo sapiens.* Fourth, revised edition. München: C. H. Beck.

Sturma, D. (Ed.). (2012). *Vernunft und Freiheit. Zur Praktischen Philosophie von Julian Nida-Rümelin.* Berlin/Boston: De Gruyter.

Wittgenstein, L. (1984). *Philosophische Untersuchungen.* Werkausgabe in 8 Bänden – Band 1: Tractatus logico-philosophicus. Tagebücher 1914–1916. Philosophische Untersuchungen (pp. 225–618). Frankfurt am Main: Suhrkamp.

Chapter 4
The Evolution of Hominin Culture and Its Ancient Pre-hominin Foundations

Andrew Whiten

Abstract This chapter examines the nature of culture in the broad evolutionary context of animal behavior, thus delineating the ancient foundations of the series of steps that eventuated in hominin culture. Focusing then on primates, further conclusions are drawn about the direct evolutionary antecedents of hominin culture in the most recent ancestors that humans share with great apes. Hominin cultural evolution is finally examined in the context of a complex of advances in social and technological cognition and other features that include unprecedented encephalization and extended childhood. The 'nature of culture' is dissected through two complementary conceptual schemes: a broad pyramidal evolutionary model extended in other chapters in this volume, and a three-element comparative analysis considering in turn social learning processes, cultural contents and the spatio-temporal distribution of traditions.

Keywords Apes • Early humans • Social learning • Traditions • Imitation • Emulation • Oldowan • Acheulean

Introduction

Studies focused on the evolution of culture have developed into an enormously rich growth area in recent years, driven by an exciting variety of discoveries across numerous sister disciplines including archaeology, primatology, animal behavior and the cognitive sciences, and by cross-fertilization and integration across these communities and others (Laland and Galef 2009; Shennan 2009; Whiten et al. 2011, 2012; Mesoudi 2011; Whiten 2013a). The present volume represents a further important step in these multidisciplinary endeavors.

In this chapter, I offer three related contributions to this cross-disciplinary cause. First, I outline a broad perspective on the nature of culture in the animal world and some of the major distinctions within it. These imply some very ancient foundations to the series of steps that eventuated in hominin culture. Second, I focus in on the animals most closely related to ourselves, the great apes, to make further inferences about the direct evolutionary antecedents of hominin culture. This brings us closer to present times, but the inferences are still about ancestors we share with these species as long ago as about 6–14 Ma, that are relatively ancient compared to those analyzed in the greater part of this volume. In the third part of the chapter, I focus on these more recent times and address human cultural evolution. I argue that various lines of evidence suggest this phase can only be understood in the context of a complex of advances in social and technological cognition, together with other features that include extended childhood and unprecedented encephalization. Such an analysis sets culture in its wider adaptive niche and draws on discoveries in a range of different disciplines.

The Evolution of Culture in the Animal Kingdom

Whiten and van Schaik (2007) offered a quite simple evolutionary model to express key hierarchical steps in the evolution of hominin culture. It takes the form of a pyramid (Fig. 4.1). At the base level is social information transfer, defined by animals acquiring information from others, as opposed to through their own individual efforts. Such effects, which are of course fundamental to any conception of culture (although not sufficient for it), appear increasingly to be widespread in the animal kingdom and have been identified

A. Whiten (✉)
Centre for Social Learning and Cognitive Evolution, School of Psychology & Neuroscience, University of St Andrews, St Andrews, KY16 9JP, UK
e-mail: aw2@st-andrews.ac.uk

Miriam N. Haidle, Nicholas J. Conard and Michael Bolus (eds.), *The Nature of Culture: Based on an Interdisciplinary Symposium 'The Nature of Culture', Tübingen, Germany*, Vertebrate Paleobiology and Paleoanthropology, DOI 10.1007/978-94-017-7426-0_4

not only in many vertebrate species but invertebrates too; as with all levels in the pyramid, I discuss this further, below. At the next level up come traditions, in which socially acquired information passes from individual to individual to become characteristic of a population of animals, such as a group, sub-group, family or community, over an extended period of time. Through repeated transmission events of this kind, innovations arising in different groups can spread locally such that the groups eventually display different traditions. The reason this level is represented as a smaller entity than the pyramid base is that by no means all socially acquired information leads to the emergence of traditions: much of it is more ephemeral. For example, once a monkey discovers a new fruiting tree, that information may be acquired socially by others, but a week later when the fruit is all gone, there will be no tradition of visiting the tree.

For some researchers in animal behavior, 'traditions' are equated with 'culture' in their writings: once the criteria for the existence of traditions are met, the case is classed as culture and the repeated instances of social information transfer are referred to as 'cultural transmission'. Other authors instead distinguish culture as a higher-level concept that requires additional criteria to identify it, although the criteria advocated vary (see, for example, Hill 2009; Perry 2009). Galef (1992) suggested that particularly sophisticated social learning processes known to be important in the human case – imitation or teaching – ought to be involved before the culture label is applied. Whiten and van Schaik (2007) instead drew the tradition/culture distinction in terms of the richness of the repertoire of traditions the animal in question sustains: in this sense, 'a single tradition does not a culture make'. This is why the third, cultural level in the pyramid is a yet smaller entity than the traditions one. There are extensive examples of traditions now identified in the animal kingdom (Fragaszy and Perry 2003; Whiten et al. 2012) as illustrated further below, but these are often single traditions in the species of interest. What Whiten and van Schaik (2007) were drawing attention to is that in our closest relatives – great apes such as the chimpanzee and orangutan – there is evidence of numerous traditions spanning a considerable range of types of behavior, such as tool use, vocalizations, grooming techniques and courtship gambits. Such multiple traditions are what we typically refer to when asked to distinguish different human cultures around the world: we start to describe a list of different traditions, such as in food habits, technology, communication customs and so on. We do the same when we contrast cultural differences across different eras, such as the cultures of Europe 30 ka versus today.

Finally, atop the pyramid is cumulative culture: culture that has the quality of building on what was achieved in past generations, refining and advancing it. Such cumulation is very much the narrative of human cultural evolution across

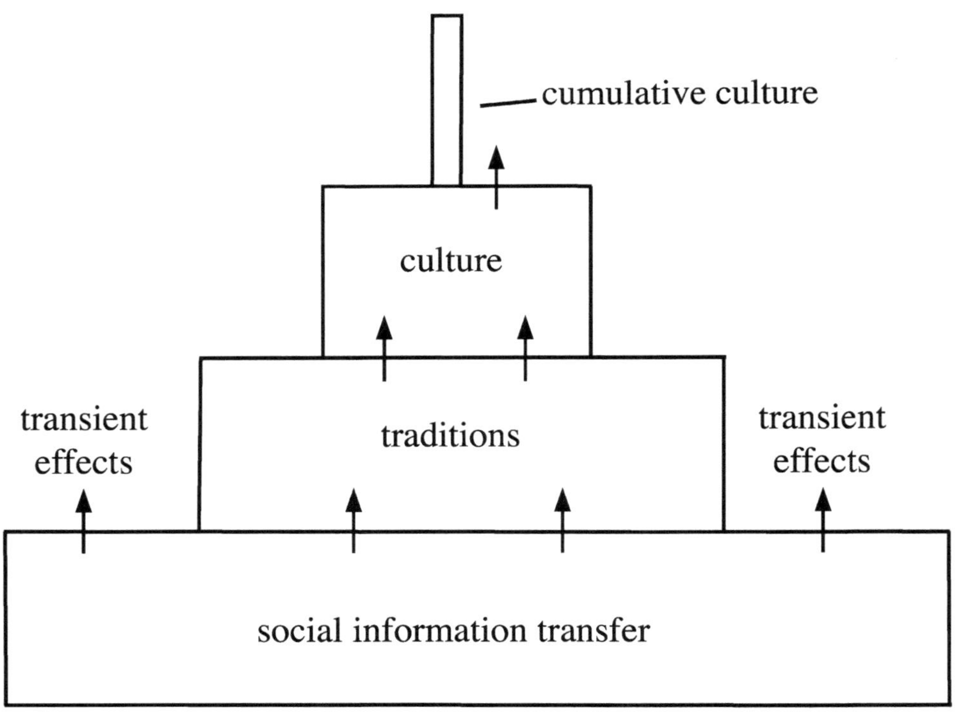

Fig. 4.1 Culture pyramid. The base of the pyramid represents social information transfer, widespread in the animal kingdom. The second level represents the longer-term subset of social learning consequences that are traditions. The third level distinguishes the smaller-yet set of cultural phenomena defined by multiple traditions, and the fourth level denotes the complexities resulting from cumulative cultural evolution (after Whiten and van Schaik 2007)

the most recent millennia (Whiten et al. 2011) and is the subject matter of much of this volume. However, cumulative culture appears rare – some argue nonexistent – in other species (Galef 1992; Tomasello 1999). Accordingly, the cumulative culture level of the pyramid is yet smaller than the 'basic culture' one characteristic of some other species.

Note that the relative sizes of the levels illustrated in the pyramid are notional. At present, we have no way of quantifying the size of each and attempting to do should probably be regarded as a futile exercise. Rather, the pyramidal structure, with its higher levels becoming progressively smaller, expresses the important principle that each builds on the lower one, of which it represents only a subset. The distinctions between levels should also be thought of as less discrete than as portrayed in the figure. For example, if we ask what distinguishes the traditions level from that below it, we can refer to such definitions as that of Fragaszy and Perry (2003): "A distinctive behaviour pattern shared by two or more individuals in a social unit, which persists over time and that new practitioners acquire in part through socially aided learning". Note here that although the minimum criterion of two individuals makes sense even from the perspective of everyday human usage ("my sister started it and I copied her and now we both share this funny tradition of peeling our apples in a single strip"), traditions become more substantial entities the more broadly spread they are, such as the case of Acheulean tools among geographically very dispersed hominin populations. Similarly, 'persists over time' is generally accepted by researchers in this field – but itself is quite vague. Some traditions pass across many generations, others are shorter lived. We might consider people to have developed a 'tradition of lunching together on Fridays' once they have repeated it for just a month or two. Some traditions may be ephemeral and grade into mere fads or fashions. All this indicates that traditions may be best thought of as graded entities. The Whiten and van Schaik notion of cultures as defined by multiple traditions faces a similar issue: a community of animals with two distinguishable traditions is suddenly 'cultural'. But as noted above, what Whiten and van Schaik were highlighting is that in some species there is evidence of a considerable richness of multiple traditions that begs to be recognized. It is thus rather the 'central tendencies' of the different levels in the pyramid that are of most interest, rather than that there are truly sharp divisions between them.

Evidence for Animal Social Learning, Traditions and Culture

There is not the space or scope here for a comprehensive review, but I offer some examples and reviews to allow the reader to appreciate the scope of the evidence concerning the different levels of the pyramid.

Social Information Transfer

In principle, this base layer to the pyramid should perhaps be divided into two, conceptually distinguishable possibilities. The upper one would be constituted by social learning, which is often taken to be the most basic process that will underlie the larger-scale phenomena of traditions and culture that lie above it in the pyramid. Probably many readers assumed that 'social information transfer' as described by Whiten and van Schaik equated with such social learning. However, a simpler possibility that would form a new base level of the pyramid would entail social information transfer without the necessity of social learning. An example might be an animal being attracted to a location at which others are feeding and using this information to target its behavior at the same locus, without retaining this information into the future (learning). For a species that typically exploits ephemeral resources, the locations of which are very difficult to predict in the long term, devoting neural resources to learning in this situation might not be selected for (or might even be selected against).

In practice, I am not aware that clear evidence bearing on this potential 'social information transfer without learning' level exists. Most researchers are interested in exploring how widely in the animal kingdom the phenomenon of social learning is distributed. And as time goes by, such social learning has been identified in rigorous field and laboratory experiments and other studies in a diversity of animal taxa that include primates (Price and Whiten 2012), other mammals (Thornton and Clutton-Brock 2011), birds (Slagsvold and Wiebe 2011), fish (Laland et al. 2011), and insects (Leadbeater and Chittka 2007). Perhaps the most common functional context demonstrated for such social learning concerns foraging, including what to eat (e.g., rodents; Galef and Stein 1985), where to find it (e.g., chimpanzees; Haun et al. 2012), what routes to follow to get it (e.g., guppies; Laland and Williams 1997), what actions to use to access it (e.g., mongooses; Müller and Cant 2010; chimpanzees; Horner et al. 2006), how to use a tool to this end (e.g., chimpanzees; Whiten et al. 2005), or fashion a tool to do so (e.g., chimpanzees; Price et al. 2009). Other functions include the learning of song (e.g., songbirds; Catchpole and Slater 1995; Byers et al. 2010; whales; Garland et al. 2011), learning which predators to fear (e.g., blackbirds; Curio et al. 1978; and see Griffin and Haythorpe 2011), and choosing the best mate (e.g., guppies; Dugatkin and Godin 1992); see Zentall (2012) for a recent review. Even this short and selective list of examples indicates how widely social learning is distributed not only taxonomically, but also in the contexts in which it provides functional benefits across the animal kingdom, compared to learning individually. Of course, not all social learning employs the same learning processes: these vary in their cognitive nature (Hopper and Whiten 2012). However, some basic forms of social learning appear increasingly widespread in the animal kingdom, suggesting a very ancient origin for this crucial foundation level in the culture pyramid.

Traditions

Social learning is necessary for the existence of traditions but insufficient in itself to generate these phenomena. However, one might expect that any species capable of social learning ought to have the potential for traditions, insofar as this requires only that a socially learned act or information is sustained long enough to be learned by another conspecific and thence spread more widely (although the principle noted earlier still applies: socially learned, ephemeral information will not be expected to form the basis for any traditions, so the base layer in the culture pyramid will remain the broadest).

Like social learning itself, traditions are being identified in an ever widening circle of species that includes primates (Whiten 2012), cetaceans (Garland et al. 2011), birds (Byers et al. 2010; Slagsvold and Wiebe 2011), fish (Laland et al. 2011) and insects (Battesti et al. 2012).

As in the case of raw social learning, the functional contexts of traditions are diverse, with the commonest identified lying in domains concerning foraging. It can be hard to confirm the reality of suspected cases of foraging traditions in the wild because of the difficulty of excluding environmental and/or genetic explanations for differences between groups living in different locations (Laland and Janik 2006), but 'diffusion experiments' in more controlled captive conditions have done this with clarity (Whiten and Mesoudi 2008). For example, Whiten et al. (2005) seeded two different means of using a stick tool to extract edible items from a naturalistic foraging problem in two separate groups of chimpanzees, training one individual from each group in her own distinct technique, then reuniting her with her group to act as a potential model. The two techniques spread in their respective groups to become traditions that were still in place two months later. Similarly, Dindo et al. (2009) found that two different manual techniques for extracting edible items from an 'artificial fruit' spread with impressive fidelity in the groups in which they were initially seeded.

Such experiments remain in their infancy in the wild, but Müller and Cant (2010) were able to introduce artificial food items to pairs of mongooses that included a juvenile not yet able to feed on the items and the older individual to whom they naturally apprenticed themselves. The study did not experimentally seed alternative actions in different populations, but the adults used either of two techniques to break the presented item to obtain the edible part inside, biting it or manually smashing it, and when the apprentices matured and were tested themselves, they tended to adopt the technique they had witnessed their adult companion perform earlier.

Multiple-Tradition Cultures

It is possible and even likely that a large number and variety of animal species will technically enter the category of multiple-tradition cultures, based upon the demonstrated existence of more than a single tradition. However, the number of these studies remains small at the present time and the most elaborate examples come from primates, particularly the great apes.

It is likely no accident that our closest relatives, chimpanzees, have to date revealed the greatest number and diversity of traditions among non-human animals. Whiten et al. (1999, 2001) drew together accumulated records of behavior from nine long-term study sites across East, Central and West Africa, incorporating over 150 years of observations in total. This study identified a total of 39 putative traditions, defined as behavior patterns common at certain sites yet absent from others without apparent genetic or environmental explanations, and these encompassed a variety of kinds of behavior including tool use, food processing, grooming styles and courtship. There are some significant genetic differences between sites exhibiting different behavioral profiles (Langergraber et al. 2010), but this would be expected from a cultural species in which migrating individuals spread both their genes and their cultural repertoire together, as would likely have been apparent across hominin populations until all but the most recent times when inter-population transfer has leapt very large distances. Genetic causes of behavioral differences amongst chimpanzees remain to be shown, whereas there is strong and substantial experimental evidence that such differences can be caused through cultural transmission. One source of such evidence is the kind of diffusion experiment outlined above, of which an extensive suite has now been completed for varieties of tool use and other food processing techniques (Whiten 2011; see Fig. 4.2). These experiments are complemented by studies tracing the spontaneous emergence and spread of group differences that have endured for years, such as in hand-clasp grooming (Bonnie and de Waal 2006).

A case of particular cross-disciplinary interest for archaeologists concerns the use of stones and wood as natural hammering materials, used to open hard shelled fruit. This is spread across several hundred kilometers of West Africa, but absent in other areas including all the East Africa

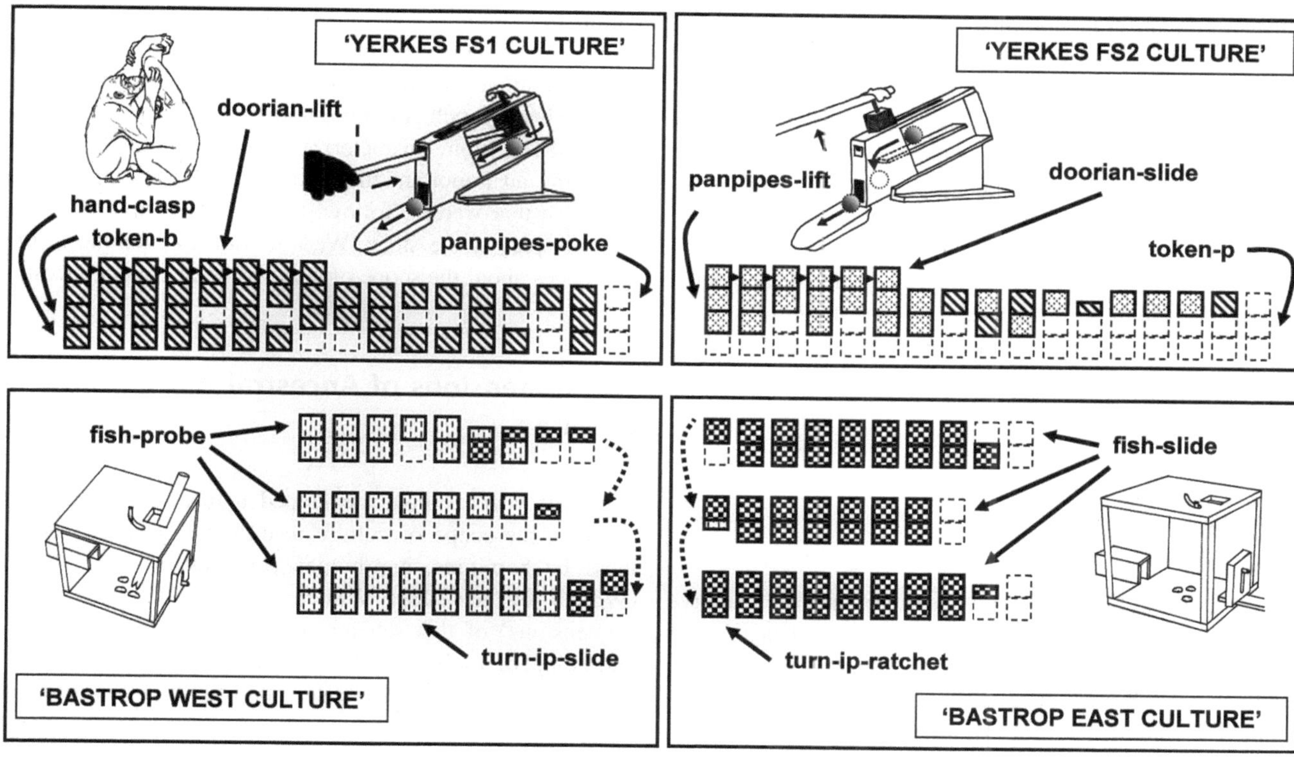

Fig. 4.2 Spread of experimentally seeded, multiple traditions generating four chimpanzee 'cultures' (after Whiten 2011). At each pair of locations, alternative techniques were experimentally seeded in a single individual. They then spread in the local community. Each column represents a single chimpanzee, with different hatching patterns corresponding to the alternative techniques seeded in the leftmost individual in each case. At Yerkes, Row 1 = lift versus slide methods to open door in 'doorian fruit', run as a diffusion chain (Horner et al. 2006); Row 2 = poke versus lift panpipes techniques spread in an open (unconstrained) diffusion (Whiten et al. 2005); Row 3 = bucket versus pipe posting option for tokens in an open diffusion (Bonnie et al. 2007); Row 4 = hand-clasp grooming, which arose and spread spontaneously in only Yerkes FS1 community. At Bastrop, Row 1 = fish-probe versus fish-slide techniques; Row 2 = turn-ip-slide versus turn-ip-ratchet techniques, used to extract food from two different devices; each technique spread to a second group (*middle*) and then a third (*bottom*) (Whiten et al. 2007)

populations. Possible environmental explanations have been excluded by investigations showing that in areas where the behavior is absent – in West Africa even just on the other side of a large river – the relevant species of nuts and suitable hammer materials are available (Boesch et al. 1994). The prospect that the difference is a genetically-determined, innate one is also denied by an experiment showing that East African, orphaned chimpanzees living on an island sanctuary would readily learn to crack nuts (Marshall-Pescini and Whiten 2008a) and by differences in this technology amongst neighboring groups in the wild (Luncz et al. 2012; Luncz and Boesch 2014).

Using the same approach as the chimpanzee studies, van Schaik and colleagues (van Schaik et al. 2003; Krützen et al. 2011) have identified multiple tradition cultures in orangutans, leading Whiten and van Schaik (2007) to suggest that such a state of affairs is a basal great ape characteristic, casting back the origins of this phenomenon to around 14–16 Ma. To date, this capacity seems less expressed in gorillas and bonobos, but these taxa have been less studied, and perhaps studied by primatologists lacking the interests of those who have studied chimpanzees and orangutans. The principal candidates for multiple tradition cultures outside the primates are cetaceans such as killer whales (Whitehead and Rendell 2015), although here the task of excluding environmental and genetic explanations for putative cultural population differences is yet more challenging, given the aquatic environment and the constraints in observation (Allen et al. 2013).

Related comparisons have more recently come from primates other than the great apes. One that may be intriguing to archaeologists interested in lithic traditions is stone-handling among Japanese macaques (Leca et al. 2007). The monkeys are provisioned at certain sites in Japan and spend some of the leisure time that results handling batches of small stones in over 30 different ways such as clacking, rolling or scraping them, and these vary across different locations. Superficially, this could be said to constitute a complex multiple-tradition culture, with each handling method that varies independently of others counted as a tradition. However, the narrowness of the context, all limited to the strange phenomenon of stone handling, contrasts with the variety of traditions that constitute the cultures of apes. Variation that more closely resembles that of chimpanzees to date has more recently come from spider monkeys (Santorelli et al. 2011), which in some ways can be seen as a case of New World convergence on the niche of chimpanzees. However, whereas many of the traditions identified in chimpanzees concern tool use, most of the spider monkey ones are social in nature. Similarities and differences in such 'contents' of cultures are important to consider in relation to the evolution of hominin culture, and are discussed in more detail below in relation to tool use, particularly its percussive components.

Cumulative Culture

It is at the level of cumulative culture that the putative examples from non-human species become meager, and indeed are dismissed by some authors, such as Tennie et al. (2009; see also Tennie et al. 2016). According to the latter, cumulative culture is simply unique to humans. Some of the more elaborate tool use repertoires of chimpanzees have been argued by others to exemplify cumulation (e.g., Sanz et al. 2009) but even so it is clear that there is a vast gulf between such minimal putative cases, and the cultural achievements of even those human hunter-gatherers whose material cultures are of the lightest grade and easily carried in nomadic camp-shifts. The early rise of such cumulation has increasingly been traced through archaeological studies of the Oldowan and Acheulean lithic cultures (Whiten et al. 2011).

In recent years, experiments have begun that pursue explanations for this gulf by contrasting cultural transmission in chimpanzee and human children. These studies employ tasks in which there are both simple solutions and more complex ones, the latter potentially building on the former. Marshall-Pescini and Whiten (2008b) and Whiten et al. (2009) found that while children would typically progress from observational learning of simple solutions to more complex ones, chimpanzees would tend to remain 'stuck' on the simple ones, despite the ability of control chimpanzees to perform the complex one if not already focused on the initial one. However in another example of this kind of experiment comparing chimpanzees, children and capuchin monkeys, Dean et al. (2012) found that this 'satisficing' scenario would not explain their results; instead, they provided evidence they took to suggest that the crucial underlying differences lie in children's spontaneous readiness to teach and to share cooperatively, together with a higher-fidelity copying capacity.

The analyses summarized above show that comparative studies have told us much about the foundations for cumulative hominin culture represented by the lower three layers of the cultural evolution pyramid of Fig. 4.1. As reviewed above, studies of great apes have yielded evidence of surprisingly rich, multiple-tradition cultures, suggesting this phenomenon characterized our common ancestors, providing an important foundation for the steps in cultural evolution that were to follow. However, multiple traditions are only part of the story. We now turn to examine further inferences about the scope of culture in our ancient ancestry.

The Dimensions of Ancestral Ape Culture

In recent years, I have developed a three-part scheme to compare the scope of culture in different species, whether these be hominins, non-human primates or other species (Whiten et al. 2003; Whiten 2005). The three major 'dimensions' of this scheme are, first, the spatio-temporal patterning of traditions, that in the case of the great apes includes the continental-scale, multiple-tradition patterns described above for chimpanzees and orangutans; second, the particular behavioral content of the traditions; and third, the social learning processes that handle cultural transmission. Each of these three major dimensions can be further dissected into subcomponents (Whiten et al. 2003; Whiten 2005, 2011). Different species vary much, and in different ways, across these dimensions.

By establishing what is shared in these domains between our own species and our closest primate relatives, we can make inferences about the scope of culture in the ancestors we share, from which the common characteristics have descended.

Multiple-Tradition Cultures

Key aspects of this dimension have been alluded to above: chimpanzees and orangutans display multiple-tradition cultures in the wild that to date are more extensive than those of monkeys and span a diverse range of technical and social forms of behavior. Numerous diffusion experiments have demonstrated apes' capacity to transmit and sustain multiple traditions of different kinds (Whiten et al. 2005, 2007; Dindo et al. 2011). In wild chimpanzees, each of the six most long-term study sites revealed a unique profile of traditions, such that with sufficient data on any chimpanzee's behavioral repertoire, it can be assigned to its geographical location on the basis of its cultural profile, a phenomenon well-known to us in the human case. The inference drawn is that our common ape ancestors were already considerably cultural. This level of cultural complexity would likely have been an important aspect of their way of life and provided a platform for the elaboration of culture that we can first discern in the lithic artifacts of Oldowan and Acheulean tool makers (Goren-Inbar 2011; Stout 2011).

Additionally, Whiten et al. (2009b) noted that the collated wild chimpanzee data indicate that there is a drop-off of 50 % in shared cultural traits over distances of around 700 km. Applying this scenario to Oldowan sites leads to such expectations as that for the example of Koobi Fora, as many as seven other sites, from Fejej at 49 km to Melka Kunture at 489 km, would be expected to share around half their cultural traits, with more distant sites sharing less. It is hoped that such models may stimulate more systematic comparisons of potential cultural patterning in the relevant lithic assemblages.

Cultural Content: Percussive Technology

The cultural repertoires of closely related species may also share similarities in their particular behavioral content. One graphic example of this is that large numbers of related species of passerine birds share the culturally-transmitted content that is birdsong – quite unlike other taxa such as primates, where vocal learning is minimal at best.

Perhaps of most interest to readers of the present, archaeologically-focused volume is that one of the major cultural content similarities between chimpanzees and ourselves lies in tool use, including percussive and other 'power' tool use (Whiten 2015). Chimpanzees fashion and use a greater diversity of tools (or forms of material technology, in the broadest sense) than any other non-human animal (McGrew 1992; Whiten 2011). Some are used for cleaning the body, such as leaves and other vegetation used to wipe off blood, feces or semen, or stems to probe the nostrils; some are used for comfort, such as leaf mats on wet ground; and some are used in courtship, like leaf-clipping with the teeth to make a distinctive sound. Use of hammer stones or pieces of wood to crack a variety of kinds of nut are therefore just one among a variety of other forms of tool use that represent local traditions. Nut-cracking is of particular significance in relation to the origins of stone tool manufacture, the first signs of which date to about 2.6 Ma in the fossil record. This is because the content of both stone tool making and nut-cracking using a stone hammer involve a suite of shared features: typically the body is truncally erect, and the agent is sitting; they then bring a rock down on another object they are holding or have placed on an anvil object in a precisely controlled and targeted fashion, because it is vital for success to strike either the nut (in nut-cracking) or the other stone (in knapping) at just the right place and angle and with controlled force. This necessarily requires a good appreciation of the use and control of such targeted force. If our ape ancestors used percussive tools for nut-cracking in this way, this would have provided an important preadaptation for stone tool knapping.

Nut-cracking, however, is not an isolated oddity in the chimpanzee repertoire but instead is part of a family of other percussive forms of tool use. These include clubbing, pounding (in a variety of contexts, from breaking into bees' nests, to the mashing of palm growing points that is 'pestle pounding') and stabbing into tree holes that might contain prey like bushbabies. These percussive variants are in turn elements in a broader family of 'power' tool usages that include levering and forceful puncturing (e.g., of the ground, to form tunnels down to subterranean termite nests, then exploited using long fresh stems as fishing tools). This variety in power-dependent tool use, which is described in more detail by Whiten et al. (2009b), entails a quite elaborate appreciation of the effective use of force, including controlled levels of power and the implications of different striking or levering angles. Again, the sharing of such characteristics between chimpanzees and humans suggests they are derived from a common ancestor. The emergence of the earliest stone knapping thus did not have to emerge out of the blue, but instead would have been able to exploit a matrix of relevant and useful psychological attributes concerned with power-dependent tool use.

Social Learning Processes

Earlier in this chapter, I noted that social learning of some kind is widespread in the animal kingdom. However, there are many different kinds and grades of social learning, which vary in their psychological complexity and in the information that gets transmitted from model to learner (Whiten et al. 2004). Stimulus enhancement, in which what a model does merely channels or focuses a social learner's attention, appears to account for much of the social learning that is increasingly recorded as ubiquitous in animals. Through this relatively simple process, animals can learn such things as what items to eat and where to find them, and this can sustain traditions that operate at this level of information.

When it comes to learning how to do things, however, more sophisticated forms of social learning are required. Learning by observation how to use a stone to crack a nut, or to shape another stone, for example, requires some kind of copying process. One well-known mechanism to achieve this is imitation, defined by Whiten and Ham (1992) as copying the form of an action. However, students of social learning have distinguished another process, emulation, that can also involve copying, but with a focus on the environmental results of actions rather than the actions per se. Consider as an example, a diffusion experiment in which one capuchin monkey in each of two groups was taught to open an artificial fruit, but using a different technique: one learned to grasp a knob on a flap-door and lifted the door, whereas the other learned to grasp the same knob but slid the door and its frame to one side. After these initial models were reintroduced to their groups, the two methods spread differentially, forming incipient 'lift' and 'slide' traditions in their respective groups (Dindo et al. 2009). Because both models grasped the handle on the door, stimulus enhancement could not explain the difference in what their companions learned: instead, learning by imitation or emulation was implied. In this example, imitation would involve copying the different actions involved in lifting versus sliding, whereas emulation would involve recreating the lifting or sliding action of the door, however achieved. Further experiments would be necessary to distinguish these, and relatively few of such studies exist as yet.

Much evidence converges on the conclusion that whatever copying occurs in non-human primates, it is typically of low fidelity by human standards, even in chimpanzees (Whiten 2011). This has suggested to some that what is happening is in fact only emulation, rather than imitation, and that this is what explains the paucity of evidence for cumulative culture in all but our own species, because cumulation requires good fidelity copying of each level of achievement attained, before the next innovation ratchets the behavior up a notch.

However Whiten et al. (2003) were skeptical of this conclusion. One reason is that in chimpanzees, the quality of copying has been demonstrated to be of sufficient fidelity to maintain group differences in action patterns that are quite complex. This has been particularly apparent in suites of recent diffusion experiments in which some multiple-action sequences have not only spread within groups, but have been transmitted to neighboring groups (Whiten et al. 2007). Moreover, there is evidence of imitation, as well as emulation, being used selectively, with switching between these approaches to social learning according to the learning context (Horner and Whiten 2005). In 'do-as-I-do' 'Simon-says' games of arbitrary copying that chimpanzees and orangutans (unlike monkeys) can be trained to play, both ape genera have demonstrated they are capable of bodily imitation, and there is evidence, too, of imitation of the sequential structure of more complex actions by chimpanzees (reviewed more fully in Whiten et al. 2009; Whiten 2011). Finally, in a study in which naive young East African chimpanzees (who do not crack nuts in the wild) were shown to learn nut-cracking from a skilled older individual, they were observed on occasion to mimic the ongoing nut-cracking blows of the model they were watching alongside them (Marshall-Pescini et al. 2008a; see Fig. 4.3). The fact that they had no stone in their hands and no nut in front of them means it is difficult to see this as any form of emulation, but rather as a form of imitation in which there is a bodily identification with the other individual of the kind one might expect from the operation of mirror neurons (Hopper et al. 2012; Whiten 2013b; Furhmann et al. 2014), that have been associated with imitation in humans (Iacoboni 2012). Thus, while our comparative research leads us to concur with Tennie et al. (2009; and see Tennie et al. 2016) that human imitation can be of significantly higher fidelity than in chimpanzees, the above array of evidence suggests that chimpanzees have sufficient copying ability to sustain different traditions (including tool-based ones) and draw on imitative as well as emulative processes.

Fig. 4.3 Novice juvenile chimpanzee observing proficient older individual using a stone hammer to crack nuts. Drawing from frame of video showing observing chimpanzee to imitatively match actions of the expert model (after Marshall-Pescini and Whiten 2008a: video viewable in electronic supplementary information at: http://dx.doi.org/10.1037/0735-7036.122.2.186.supp). Drawing by Jason Zampol

Imitation, Emulation, and the Cultures of the Oldowan and Acheulean

The debate about the imitative and emulative capacities of apes and their implications for the occurrence of cumulative culture bear rather directly on the nature of culture as it came to be expressed in the Oldowan and Acheulean phases of hominin cultural evolution. The transition from the Oldowan to the Acheulean itself appears as one of the first manifestations of cumulative culture in the hominin archaeological record, with the latter phase (Goren-Inbar 2011; and see Alperson-Afil and Goren-Inbar 2016) building on the cruder achievements of the earlier one (Stout 2011). Stout (2011) argues that subtle signs of cumulative progress are already discernible within the Oldowan.

One explanation for what made these cultural evolutionary steps possible is that early humans developed special new social learning capacities, in particular, imitative as opposed to emulative ones (Tomasello 1999). By contrast, our own research demonstrating chimpanzees' capacities to sustain material traditions, coupled with evidence of a degree of imitative fidelity, suggests that similar social learning powers would already have existed in the forebears of early hominin social tool makers, such that a step up in social learning would likely not be the key explanation for the emergence of cumulative lithic culture. Whiten et al. (2003), in first developing this argument, were led to propose that the psychological advances may have lain instead in those more directly underlying the cognitive requisites intrinsic to technological innovations such as those of the knapping process. As neural and cognitive advances were made in this respect, the imitative capacities were likely in place to assimilate the technological innovations of the most gifted hominin tool makers.

Some support for this argument comes from within archaeology itself, deriving from the fact that although one can discern cumulative progress in the Oldowan and Acheulean phases, it was inordinately slow. Indeed, over some periods of hundreds of thousands of years (and thus, many thousands of generations) the principal picture is one of stability of cultural inheritance, even across much of the Acheulean (Goren-Inbar 2011). However, as Mithen (1999) was perhaps the first to suggest, the transmission of skills as sophisticated as knapping Acheulean blades must have drawn on imitative capacities. Accordingly, in this phase of hominin evolution one has imitation, yet barely discernible cumulative cultural progress: so the emergence of imitation was not the magic factor responsible for the emergence of cumulative culture. Advances in the non-social cognitive requisites of biface knapping may have been more crucial in the emergence of the Acheulean lithic achievements, as well as the more elaborate forms of lithic cumulative culture that followed (Whiten et al. 2003; Faisal et al. 2010).

The Evolution of Human Culture, Deep Social Mind and the Socio-Cognitive Niche

In this final section I address the broader cognitive and ecological contexts of the evolution of human culture, referring to an analysis recently published at much greater length, that readers are encouraged to consult for more depth and detail (Whiten and Erdal 2012).

The starting point is essentially the puzzle of how our puny ape ancestors evolved into a hunting-gathering niche that extended to the hunting of large game, in direct competition with a large guild of formidable African predators including the big cats. How was this possible? An influential answer was expounded by Tooby and DeVore (1987), who proposed that evolving humans triumphed in this competition by creating a 'cognitive niche' that gave its exponents a significantly superior edge in hunting through the applications of greater intelligence.

Archaeology provides historical evidence of ancient origins to such hunting of big game, including sophisticated spears from 300 ka (Thieme 1997) as well as cut marks on large bones from much earlier times. Building on this framework, studies of peoples who have, until recently, pursued a hunting and gathering way of life (and in some cases continue to do so), have become important sources of inference about human ways of life before the very recent ($\sim$10 ka) origins of agriculture. These and other studies led Whiten (1999, 2006; Whiten and Erdal 2012) to suggest that the 'cognitive niche' concept, whilst eminently plausible as far as it goes, needs to be extended to the broader concept of a 'socio-cognitive niche', underwritten by a complex of cognitive features that together can be characterized as a uniquely human 'deep social mind'. Deep social mind refers specifically to a complex of cognitive processes, while the socio-cognitive niche (like all 'niche' concepts) refers to the ways in which the organism engages with a subset of environmental features.

Whiten and Erdal distinguish five main pillars that characterize both concepts. Human cumulative culture represents just one of these pillars. The others are language, theory of mind (aka mindreading or mentalism), egalitarianism and unprecedented levels and forms of cooperation (Fig. 4.4). The significance of this picture is that the nature of human culture must be understood as part of this larger set of socio-cognitive adaptations, between which numerous positive feedback loops can be inferred (Fig. 4.4). The crucial outcome of the operation of this complex is that a hunter-gatherer band functions in the manner of an extremely efficient group-level predator, beyond anything seen in competing social carnivores. Key features that distinguish it from the latter include organized division of labor, planned

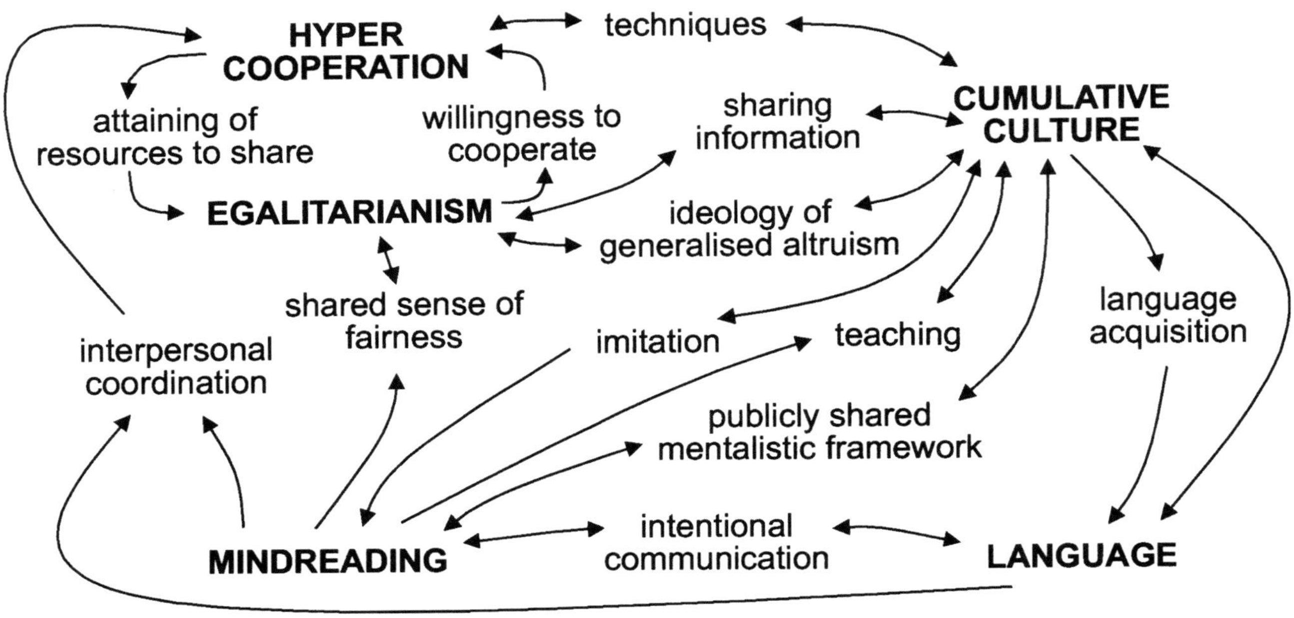

Fig. 4.4 Principal classes of social cognition (in *bold capitals*) in hunter-gatherer bands and, inferred reinforcing relationships between them (after Whiten and Erdal 2012). Note that the latter cannot be exhaustively illustrated in a legible single figure; those shown are indicative only. For explanation and discussion see text

cooperation and a material culture that has cumulatively created sophisticated weaponry.

To flesh these ideas out a little more (see Whiten and Erdal 2012 for a much extended treatment) consider Fig. 4.4. Cumulative culture and language are shown separately here, because instances of each exist independently of the other: for example, observational learning can contribute to culture without recourse to language; and language fulfills many functions daily, such as joint planning of forthcoming activities, other than specifically cultural ones. However, the two are bonded by positive feedback loops that include the use of language to transmit culture, and conversely the role of culture in the fundamental process of language acquisition. Other authors have recently sought in independent but complementary papers to extend the concept of the cognitive niche, particularly concerning culture (Boyd et al. 2011) and language (Pinker 2010) respectively; see also Sterelny (2012).

In turn mindreading – the attribution to others (and oneself) of states of mind such as ignorance and belief – has reciprocal links with both culture and language. For example on the one hand, mindreading supports explicit teaching through the correct attribution of states of knowledge and ignorance in the pupil; and conversely, one's culture provides an explicit 'folk psychology' that specifies such mental states as true and false beliefs. Egalitarianism refers to the apparently universal occurrence in hunter-gatherer bands of egalitarian behavior reflected in a range of functional contexts that include resource-sharing according to need, and a lack of explicit leadership roles (Erdal and Whiten 1996). This appears highly reciprocal, with extensive cooperation such as in foraging parties, because egalitarianism makes it functional to work together to gain the resources later shared. The links with culture include the ways in which egalitarianism is embodied in public ideologies that are transmitted from generation to generation, and the importance of content such as collaborative hunting and gathering technologies in supporting the importance of cumulative cultural capacities.

Figure 4.4 might be thought to be complicated enough, but it is important to recognize that its purpose is to outline principal pillars of the socio-cognitive niche in particular. In reality, this expansion of the concept of the cognitive niche must encompass much else that is cognitive, including the features discussed by Tooby and DeVore (1987). This should in principle expand Fig. 4.4 to show such features as tool making, and further multiple linkages of such features with those already in the diagram, but the figure would then rapidly become unwieldy. Ranging yet more widely to consider the features that make up the human adaptive complex as a whole, requires incorporation of such major changes to great ape patterns as significantly extended immaturity (long childhood and extended dependence on parental resourcing), bi-parental investment, massive encephalization, and perhaps neoteny. All of these characters, together with those outlined in Fig. 4.4, constitute an important matrix shaping and being shaped by the nature of culture in our species.

However, the five core characteristics highlighted did not spring out of the blue. All of them represent elaborations on ancestral ape foundations that can be inferred from characteristics shared with other extant great apes (Whiten and Erdal 2012). Perhaps the least strongest element of this claim concerns language, although much has been learned in recent years about the natural vocal and gestural communication of wild apes, that offers more optimism in bridging the non-linguistic/linguistic gulf than hitherto (Whiten and Erdal 2012; Slocombe 2012). In the case of culture, a much more substantial analysis of ancestral ape origins is possible, as outlined earlier and described in more detail in Whiten et al. (2009b) and Whiten (2011). In the case of the other main pillars of the socio-cognitive niche shown in Fig. 4.4, elementary forms of mindreading (notably recognition of 'seeing', 'knowing' and 'intending') have been identified over the last decade (Call and Santos 2012; Whiten 2013c); and apes are not egalitarian, yet their complex social alliances undermine simple dominance based on brute force; and various forms of cooperation have been identified in social coalitions as well as hunting and raiding parties (Whiten and Erdal 2012).

Concluding Remarks

We live at an exciting time in which it has become possible to delineate a very substantial analysis of the evolution of culture, from very ancient times indeed (if we are to include ancestral nodes with the likes of drosophila: Battesti et al. 2012), to the recent and zoologically extraordinary manifestations of cumulative human cultures. In this chapter, I have first introduced a simple hierarchical model (Fig. 4.1) that portrays the evolutionary changes that have taken place in the broadest terms and which is elaborated on further by Haidle and others in this volume (see Davidson 2016; Haidle 2016; Haidle et al. 2016; Tennie et al. 2016). I have then further dissected the nature of culture into three 'dimensions' that can be used to compare the cultural phenomena that characterize different contemporary and chrono-species, using this to reconstruct the cultural profile of our ape ancestors (and see Jordan (2015) for further application to human cultural evolution); these dimensions are the spatio-temporal patterning of traditions (particularly rich and varying multi-tradition cultures); cultural contents (including the most elaborated non-human technologies) and social learning processes (a portfolio that includes emulation and imitation). I have ended by outlining the additional and zoologically unique aspects of social cognition that together with cultural evolution itself have marked the later stages of hominin evolution.

Developing such analyses further will require deeply interdisciplinary efforts that include such disparate fields as evolutionary biology, archaeology, ethology, primatology, anthropology (from the biological and evolutionary to social and cultural), developmental psychology and the cognitive sciences generally. For some of these sources, we have to pinch ourselves to recognize what a special time window we now live in, when it has been possible to directly study other living apes and people still living by hunting and gathering, unearth undreamt-of fossil records of our past and make sense of all this and more, following the evolutionary inspirations of Darwin, Wallace, and many of their most gifted followers. The present volume represents a significant new perspective in such interdisciplinary endeavors. Like the others, this chapter represents just one segment of the larger and grander enterprise of understanding the nature and evolution of culture: and even here, many major developments and discoveries in this rapidly expanding field have had to be only briefly sketched. I have therefore striven to at least provide signposts to exciting entry points to the now vast literature on my topics.

References

Allen, J. M., Weinrich, W., Hoppitt, W., & Rendell, L. (2013). Network-based diffusion analysis reveals cultural diffusion of lobtail feeding in humpback whales. *Science, 340*, 485–488.

Alperson-Afil, N., & Goren-Inbar, N. (2016). Scarce but significant: The limestone component of the Acheulean site of Gesher Benot Ya'aqov, Israel. In M. N. Haidle, N. J. Conard, & M. Bolus (Eds.), *The nature of culture* (pp. 41–56). Dordrecht: Springer.

Battesti, M., Moreno, C., Joly, D., & Mery, F. (2012). Spread of social information and dynamics of social transmission within drosophila groups. *Current Biology, 22*, 309–313.

Boesch, C., Marchesi, P., Marchesi, M., Fruth, B., & Joulian, F. (1994). Is nut cracking in wild chimpanzees a cultural behaviour? *Journal of Human Evolution, 26*, 325–338.

Bonnie, K. E., & de Waal, F. B. M. (2006). Affiliation promotes the transmission of a social custom; Handclasp grooming among captive chimpanzees. *Primates, 47*, 27–34.

Bonnie, K. E., Horner, V., Whiten, A., & de Waal, F. B. M. (2007). Spread of arbitrary customs among chimpanzees: a controlled experiment. *Proceedings of the Royal Society of London, B, 274*, 367–372.

Boyd, R., Richerson, P. J., & Henrich, J. (2011). The cultural niche: why social learning is essential for human adaptation. *Proceedings of the National Academy of Sciences of the USA, 108*, 10918–10925.

Byers, B. E., Belinsky, K. L., & Bentley, R. A. (2010). Independent cultural evolution of two song traditions in the chestnut sided warbler. *The American Naturalist, 176*, 476–489.

Call, J., & Santos, L. R. (2012). Understanding other minds. In J. C. Mitani, J. Call, P. M. Kappeler, R. A. Palombit, & J. B. Silk (Eds.), *The evolution of primate societies* (pp. 664–681). Chicago: Chicago University Press.

Catchpole, C. K., & Slater, P. J. B. (1995). *Bird song: Biological themes and variations*. Cambridge: Cambridge University Press.

Curio, E., Ulrich, E., & Vieth, W. (1978). Cultural transmission of enemy recognition: One function of avian mobbing. *Science, 202*, 899–901.

Davidson, I. (2016). Stone tools: Evidence of something in between culture and cumulative culture? In M. N. Haidle, N. J. Conard, &

M. Bolus (Eds.), *The nature of culture* (pp. 99–120). Dordrecht: Springer.

Dean, L. G., Kendal, R. L., Schapiro, S. J., Thierry, B., & Laland, K. N. (2012). Identification of the social and cognitive processes underlying human cumulative culture. *Science, 335*, 1114–1118.

Dindo, M., de Waal, F. B. M., & Whiten, A. (2009). In-group conformity sustains different foraging traditions in capuchin monkeys (*Cebus apella*). *PLoS ONE, 4*, e7858.

Dindo, M., Stoinski, T., & Whiten, A. (2011). Observational learning along cultural transmission chains in orangutans. *Biology Letters, 7*, 181–183.

Dugatkin, L. A., & Godin, J.-G. J. (1992). Reversal of mate choice by copying in the guppy (*Poecilia reticulata*). *Proceedings of the Royal Society, B, 249*, 179–184.

Erdal, D., & Whiten, A. (1996). Egalitarianism and Machiavellian Intelligence in human evolution. In P. Mellars & K. Gibson (Eds.), *Modelling the early human mind* (pp. 139–150). Cambridge: McDonald Institute Monographs.

Faisal, A., Stout, D., Apel, J., & Bradley, B. (2010). The manipulative complexity of lower paleolithic stone toolmaking. *PLoS ONE, 5* (11), e13718.

Fragaszy, D. M., & Perry, S. (Eds.). (2003). *The biology of traditions: Models and evidence*. Cambridge: Cambridge University Press.

Fuhrmann, D., Ravignani, A., Marshall-Pescini, S., & Whiten, A. (2014). Synchrony and motor mimicking in chimpanzee observational learning. *Scientific Reports, 4*, 5283.

Galef, B. G, Jr. (1992). The question of animal culture. *Human Nature, 3*, 157–178.

Galef, B. G, Jr, & Stein, M. (1985). Demonstrator influence on observer diet preference – analysis of critical social interactions and olfactory signals. *Learning and Behavior, 13*, 31–38.

Garland, E. C., Goldizen, A. W., Rekdahl, M. L., Constantine, R., Garrigue, C., Daescher Hauser, N., et al. (2011). Dynamic horizontal cultural transmission of humpback whale song at the ocean basin scale. *Current Biology, 21*, 687–691.

Goren-Inbar, N. (2011). Culture and cognition in the Acheulian industry – a case study from Gesher Benot Ya'aqov. *Philosophical Transactions of the Royal Society, B, 366*, 1038–1049.

Griffin, A. S., & Haythorpe, K. (2011). Learning from watching alarmed demonstrators: Does the cause of alarm matter? *Animal Behaviour, 81*, 1163–1169.

Haidle, M. N. (2016). Lessons from Tasmania—cultural performance versus cultural capacity. In M. N. Haidle, N. J. Conard, & M. Bolus (Eds.), *The nature of culture* (pp. 7–17). Dordrecht: Springer.

Haidle, M. N., Conard, N. J., & Bolus, M. (2016). The nature of culture: Research goals and new directions. In M. N. Haidle, N. J. Conard, & M. Bolus (Eds.), *The nature of culture* (pp. 1–6). Dordrecht: Springer.

Haun, D. M. B., Rekers, Y., & Tomasello, M. (2012). Majority-biased transmission in chimpanzees and human children, but not orangutans. *Current Biology, 22*, 727–731.

Hill, K. (2009). Animal "culture"? In K. N. Laland & B. G. Galef (Eds.), *The question of animal culture* (pp. 269–287). Cambridge, MA: Harvard University Press.

Hopper, L. M., & Whiten, A. (2012). The comparative and evolutionary psychology of social learning and culture. In J. Vonk & T. Shackelford (Eds.), *The Oxford handbook of comparative evolutionary psychology* (pp. 451–473). New York: Oxford University Press.

Hopper, L. M., Marshall-Pescini, S., & Whiten, A. (2012). Social learning and culture in child and chimpanzee. In F. B. M. de Waal & P. F. Ferrari (Eds.), *The primate mind: Built to connect with other minds* (pp. 99–118). Cambridge, MA: Harvard University Press.

Horner, V. K., & Whiten, A. (2005). Causal knowledge and imitation/emulation switching in chimpanzees (*Pan troglodytes*) and children. *Animal Cognition, 8*, 164–181.

Horner, V., Whiten, A., Flynn, E., & de Waal, F. B. M. (2006). Faithful replication of foraging techniques along cultural transmission chains by chimpanzees and children. *Proceedings of the National Academy of Sciences of the USA, 103*, 13878–13883.

Iacoboni, M. (2012). The human mirror system and its role in imitation and empathy. In F. B. M. de Waal & P. F. Ferrari (Eds.), *The primate mind: Built to connect with other minds* (pp. 32–47). Cambridge, MA: Harvard University Press.

Jordan, P. (2015). *Technology as human social tradition: Cultural transmission among hunter-gatherers*. Oakland, CA: University of California Press.

Krützen, M., Willems, E. P., & van Schaik, C. P. (2011). Culture and geographic variation in orangutan behavior. *Current Biology, 21*, 1808–1812.

Laland, K. N., & Galef, B. G. (Eds.). (2009). *The question of animal culture*. Cambridge, MA: Harvard University Press.

Laland, K. N., & Janik, V. (2006). The animal cultures debate. *Trends in Ecology & Evolution, 21*, 542–547.

Laland, K. N., & Williams, K. (1997). Shoaling generates social learning of foraging information in guppies. *Animal Behaviour, 53*, 1161–1169.

Laland, K. N., Atton, N., & Webster, M. M. (2011). From fish to fashion: Experimental and theoretical insights into the evolution of culture. *Philosophical Transactions of the Royal Society, B, 366*, 958–968.

Langergraber, K. E., Boesch, C., Inoue, E., Inoue-Marayama, M., Mitani, J. C., Nishsida, T., et al. (2010). Genetic and 'cultural' similiarity in chimpanzees. *Proceedings of the Royal Society, B, 277*, 408–416.

Leadbeater, E., & Chittka, L. (2007). Social learning in insects—from miniature brains to consensus building. *Current Biology, 17*, R703–R713.

Leca, J.-B., Gunst, N., & Huffman, M. A. (2007). Japanese macaque cultures: Inter- and intra-troop behavioral variability of stone-handling patterns across 10 groups. *Behaviour, 144*, 251–281.

Luncz, L. V., & Boesch, C. (2014). Tradition over trend: Neighboring chimpanzee communities maintain differences in cultural behaviour despite frequent immigration of adult females. *American Journal of Primatology, 76*, 649–657.

Luncz, L. V., Mundry, R., & Boesch, C. (2012). Evidence for cultural differences between neighbouring chimpanzee communities. *Current Biology, 22*, 922–926.

Marshall-Pescini, S., & Whiten, A. (2008a). Social learning of nut-cracking behaviour in East African sanctuary-living chimpanzees (*Pan troglodytes schweinfurthii*). *Journal of Comparative Psychology, 122*, 186–194.

Marshall-Pescini, S., & Whiten, A. (2008b). Chimpanzees (*Pan troglodytes*) and the question of cumulative culture: An experimental approach. *Animal Cognition, 11*, 449–456.

McGrew, W. C. (1992). *Chimpanzee material culture: Implications for human evolution*. Cambridge: Cambridge University Press.

Mesoudi, A. (2011). *Cultural evolution: How darwinian evolutionary theory can explain human culture and synthesize the social sciences*. Chicago, IL: University of Chicago Press.

Mithen, S. (1999). Imitation and cultural change: A view from the Stone Age, with specific reference to the manufacture of hand axes. In H. O. Box & K. R. Gibson (Eds.), *Mammalian social learning: Comparative and ecological perspectives* (pp. 389–399). Cambridge: Cambridge University Press.

Müller, C. A., & Cant, M. A. (2010). Imitation and traditions in wild banded mongooses. *Current Biology, 20*, 1171–1175.

Perry, S. (2009). Are non-human primates likely to exhibit cultural capacities like those of humans, and if so, how can this be demonstrated? In K. N. Laland & B. G. Galef (Eds.), *The question of animal culture* (pp. 247–268). Cambridge, MA: Harvard University Press.

Pinker, S. (2010). The cognitive niche: Coevolution of intelligence, sociality and language. *Proceedings of the National Academy of Sciences of the USA, 107*, 8993–8999.

Price, E. E., & Whiten, A. (2012). Social learning and culture in primates: Evidence from free-ranging and captive populations. In E. A. Wasserman & T. R. Zentall (Eds.), *Handbook of comparative cognition* (pp. 862–878). New York: Oxford University Press.

Price, E. E., Lambeth, S. P., Schapiro, S. J., & Whiten, A. (2009). A potent effect of observational learning on chimpanzee tool construction. *Proceedings of the Royal Society, B, 276*, 3377–3383.

Santorelli, C. J., Schaffner, C. M., Campbell, C. J., Notman, H., Pavelka, M. S., Weghorst, J. A., et al. (2011). Traditions in wild spider monkeys are biased towards the social domain. *PLoS ONE, 6*, e16863.

Sanz, C., Call, J., & Morgan, D. (2009). Design complexity in termite-fishing tools of chimpanzees (*Pan troglodytes*). *Biology Letters, 5*, 293–296.

Shennan, S. (Ed.). (2009). *Pattern and process in cultural evolution*. London: University of California Press.

Slagsvold, T., & Wiebe, K. L. (2011). Social learning in birds and its role in shaping foraging niches. *Philosophical Transactions of the Royal Society, B, 366*, 969–977.

Slocombe, K. E. (2012). Have we underestimated great ape vocal capacities? In M. Tallerman & K. Gibson (Eds.), *Handbook of language evolution* (pp. 90–95). Oxford: Oxford University Press.

Sterelny, K. (2012). *The evolved apprentice: How evolution made humans unique*. Cambridge, MA: MIT Press.

Stout, D. (2011). Stone toolmaking and the evolution of human culture and cognition. *Philosophical Transactions of the Royal Society, B, 366*, 1050–1059.

Tennie, C., Call, J., & Tomasello, M. (2009). Ratcheting up the ratchet: On the evolution of cumulative culture. *Philosophical Transactions of the Royal Society, B, 364*, 2405–2415.

Tennie, C., Braun, D. R., Premo, L. S., & McPherron, S. (2016). The Island test for cumulative culture in the paleolithic. In M. N. Haidle, N. J. Conard, & M. Bolus (Eds.), *The nature of culture* (pp. 121–131). Dordrecht: Springer.

Thieme, H. (1997). Lower paleolithic hunting spears from Germany. *Nature, 385*, 807–810.

Thornton, A., & Clutton-Brock, T. (2011). Social learning and the development of individual and group behaviour in mammal societies. *Philosophical Transactions of the Royal Society, B, 366*, 978–987.

Tomasello, M. (1999). *The cultural origins of human cognition*. Cambridge, MA: Harvard University Press.

Tooby, J., & DeVore, I. (1987). The reconstruction of hominid behavioral evolution through strategic modelling. In W. G. Kinzey (Ed.), *The evolution of human behavior: Primate models* (pp. 183–227). New York: SUNY Press.

van Schaik, C. P., Ancrenaz, M., Borgen, G., Galdikas, B., Knott, C. D., Singleton, I., et al. (2003). Orangutan cultures and the evolution of material culture. *Science, 299*, 102–105.

Whitehead, H., & Rendell, L. (2015). *The cultural lives of whales and dolphins*. Chicago: Chicago University Press.

Whiten, A. (1999). The evolution of deep social mind in humans. In M. Corballis & S. E. G. Lea (Eds.), *The descent of mind* (pp. 155–175). Oxford: Oxford University Press.

Whiten, A. (2005). The second inheritance system of chimpanzees and humans. *Nature, 437*, 52–55.

Whiten, A. (2006). The place of "deep social mind" in the evolution of human nature. In M. A. Jeeves (Ed.), *Human nature* (pp. 207–222). Edinburgh: Royal Society of Edinburgh.

Whiten, A. (2011). The scope of culture in chimpanzees, humans and ancestral apes. *Philosophical Transactions of the Royal Society, B, 366*, 997–1007.

Whiten, A. (2012). Primate social learning, traditions and culture. In J. C. Mitani, J. Call, P. M. Kappeler, R. A. Palombit, & J. B. Silk (Eds.), *The evolution of primate societies* (pp. 681–699). Chicago: Chicago University Press.

Whiten, A. (2013a). Archaeology meets primate technology. *Nature, 498*, 303–305.

Whiten, A. (2013b). Monkeys, apes, imitation and mirror neurons. *Cortex, 49*, 2941–2943.

Whiten, A. (2013c). Humans are not alone in computing how others see the world. *Animal Behaviour, 86*, 213–221.

Whiten, A. (2015). Experimental studies illuminate the cultural transmission of percussive technologies in *Homo* and *Pan*. *Philosophical Transactions of the Royal Society, B, 370*, 20140359.

Whiten, A., & Erdal, D. (2012). The human socio-cognitive niche and its evolutionary origins. 2119–2129.

Whiten, A., & Ham, R. (1992). On the nature and evolution of imitation in the animal kingdom: Reappraisal of a century of research. *Advances in the Study of Behaviour, 21*, 239–283.

Whiten, A., & Mesoudi, A. (2008). Establishing an experimental science of culture: Animal social diffusion experiments. *Philosophical Transactions of the Royal Society, B, 363*, 3477–3488.

Whiten, A., & van Schaik, C. P. (2007). The evolution of animal 'cultures' and social intelligence. *Philosophical Transactions of the Royal Society, B, 362*, 603–620.

Whiten, A., Goodall, J., McGrew, W. C., Nishida, T., Reynolds, V., Sugiyama, Y., et al. (1999). Cultures in chimpanzees. *Nature, 399*, 682–685.

Whiten, A., Goodall, J., McGrew, W. C., Nishida, T., Reynolds, V., Sugiyama, Y., et al. (2001). Charting Cultural variation in chimpanzees. *Behaviour, 138*, 1481–1516.

Whiten, A., Horner, V., & Marshall-Pescini, S. R. J. (2003). Cultural panthropology. *Evolutionary Anthropology, 12*, 92–105.

Whiten, A., Horner, V., Litchfield, C. A., & Marshall-Pescini, S. (2004). How do apes ape? *Learning and Behavior, 32*, 36–52.

Whiten, A., Horner, V., & de Waal, F. B. M. (2005). Conformity to cultural norms of tool use in chimpanzees. *Nature, 437*, 737–740.

Whiten, A., Spiteri, A., Horner, V., Bonnie, K. E., Lambeth, S. P., Schapiro, S. J., et al. (2007). Transmission of multiple traditions within and between chimpanzee groups. *Current Biology, 17*, 1038–1043.

Whiten, A., McGuigan, H., Hopper, L. M., & Marshall-Pescini, S. (2009a). Imitation, over-imitation, emulation and the scope of culture for child and chimpanzee. *Philosophical Transactions of the Royal Society, B, 364*, 2417–2428.

Whiten, A., Schick, K., & Toth, N. (2009b). The evolution and cultural transmission of percussive technology: Integrating evidence from paleoanthropology and primatology. *Journal of Human Evolution, 57*, 420–435.

Whiten, A., Hinde, R. A., Stringer, C. B., & Laland, K. N. (2011). Culture evolves. *Philosophical Transactions of the Royal Society, B, 366*, 938–948.

Whiten, A., Hinde, R. A., Stringer, C. B., & Laland, K. N. (Eds.). (2012). *Culture evolves*. Oxford: Oxford University Press.

Zentall, T. (2012). Perspectives on observational learning in animals. *Journal of Comparative Psychology, 126*, 114–128.

Chapter 5
Scarce but Significant: The Limestone Component of the Acheulean Site of Gesher Benot Ya'aqov, Israel

Nira Alperson-Afil and Naama Goren-Inbar

Abstract In-depth study of Acheulean limestone artifacts from Gesher Benot Ya'aqov (0.79 Ma) has revealed that limestone nodules procured from fluvial deposits were transported to the lake margin and exploited throughout the occupational sequence (ca. 50 ka). Analyses of the limestone assemblages illustrate that individual artifacts go through several use-stages or complex life-histories within a single reduction sequence. This reduction sequence began with the targeting of nodules suitable for use as percussors. Use of the percussors sometimes resulted in breakage that produced flakes typical of working accidents. Broken percussors were shaped into a second morphotype, chopping tools, while cores comprise a third morphotype. These morphotypes are viewed as interrelated consecutive options. Once a morphotype was inadequate for use it was transformed into another, resulting in gradual reduction of dimensions from one type to the next. The ability to renovate/recycle implies flexibility and contingency. The consistent homogeneity of the limestone assemblages demonstrates conservatism of knowledge, transmission of the *chaîne opératoire*, specific raw materials, and flexible variations within them – all typical of a "complex" culture.

Keywords Lower Paleolithic • Cognitive abilities • Chopping tool • Core • Core-tool • Flake-tool • Percussor • *Percuteur de concassage* • Reduction sequence

Introduction

Located in the Great African Rift system, the 0.79 Ma site of Gesher Benot Ya'aqov (GBY) is bedded in the Benot Ya'aqov Formation, deposited by the paleo-Lake Hula during the Early–Middle Pleistocene. The stratigraphic sequence includes lake and lake-margin deposits in which evidence of human activities is provided by a series of 15 archaeological horizons rich in paleontological, paleobotanical, and archaeological assemblages, all assigned to MIS 18 (Goren-Inbar et al. 2000; Feibel 2004; Alperson-Afil 2008; Alperson-Afil et al. 2009; Rabinovich and Biton 2011; Sharon et al. 2011; Spiro et al. 2011; Zohar and Biton 2011). Since the beginning of excavations at GBY (in four major areas of excavation; see Goren-Inbar et al. 2000: Fig. 1), the remains of its Acheulean material culture have been continuously studied, demonstrating affinities with the African Large Flake Acheulean tradition and revealing various technological and behavioral traits of the ancient hominins.

The Acheulean techno-complex emerged in East Africa at ca. 1.76 Ma (Lepre et al. 2011) and persisted until 0.3–0.25 Ma over a wide geographical range (e.g., Kleindienst 1962; Roe 2001; Sharon 2007 and references therein). During the existence of this long cultural complex the earliest human migrations occurred, involving the Levantine Corridor as a migration route out of Africa and into Eurasia. Acheulean assemblages are commonly identified by their characteristic large cutting tools (i.e., handaxes and cleavers), which are the subject of most discussions of the Acheulean lithic repertoire and its implications for hominin technological and cognitive capacities. At the site of GBY, the Acheulean lithic assemblages incorporate three different types of raw material – basalt, flint, and limestone. Since each is characterized by particular qualities of durability, elasticity, and fragmentation, they correspondingly exhibit different technological characteristics and comprise diverse typological products.

In comparison with the other raw materials, limestone artifacts occur throughout the occupational sequence of GBY

N. Alperson-Afil (✉)
The Martin (Szusz) Department of Land of Israel Studies and Archaeology, Bar Ilan University, 5290002 Ramat Gan, Israel
e-mail: nira.alperson-afil@biu.ac.il

N. Goren-Inbar
Institute of Archaeology, Hebrew University of Jerusalem, 91905 Jerusalem, Israel
e-mail: goren@cc.huji.ac.il

Miriam N. Haidle, Nicholas J. Conard and Michael Bolus (eds.), *The Nature of Culture: Based on an Interdisciplinary Symposium 'The Nature of Culture', Tübingen, Germany*, Vertebrate Paleobiology and Paleoanthropology, DOI 10.1007/978-94-017-7426-0_5

Table 5.1 Counts of Flint, Basalt, and Limestone Artifacts throughout the GBY Sequence

Area	Layer	Flint[a]	Basalt[a]	Limestone Artifacts FFT	CCT	HX	CL	Total	Microartifacts
C	V-3	7	2	–	–	–	–	–	4
	V-4	21	6	–	–	–	–	–	8
	V-4/5	27	7	–	–	–	–	–	9
	V-5	341	94	2	2	–	–	4	333
	V-6	302	81	2	–	–	–	2	96
Jordan bank	V-5 and V-6	172	85	8	2	–	–	10	4
Unconformity C		28	20	–	5	–	–	5	–
A	I-4	70	11	–	–	–	–	–	103
	I-5	31	5	–	1	–	–	1	135
Trench II	II	66	138	1	9	1	–	11	–
B	II-2	116	17	10	2	–	–	12	170
	II-2/3	120	56	3	–	–	–	3	85
	II-4	9	5	1	–	–	–	1	–
	II-5	242	113	7	7	–	–	14	212
	II-5/6	157	49	2	2	–	–	4	1,055
	II-6/L1	1,127	1,095	22	14	5	–	41	1,506
	II-6/L2	465	990	17	6	–	–	23	2,154
	II-6/L3	647	807	24	9	–	–	33	1,570
	II-6/L4	737	1,112	21	21	–	–	42	4,727
	II-6/L4b	158	616	20	5	2	–	27	145
	II-6/L5	203	214	26	3	–	–	29	1,996
	II-6/L6	346	391	25	3	–	1	29	331
	II-6/L7	882	618	77	13	1	–	91	1,602
	II-6/7	30	58	5	–	–	–	5	37
	II-14	39	90	2	–	–	–	2	–
Unconformity B		213	282	11	10	–	–	21	5
Total		6,556	6,962	286	114	9	1	410	16,287

FFT Flakes and flake-tools; *CCT* Cores and core-tools; *HX* Handaxes; *CL* Cleavers
[a]Counts for flint and basalt take account of the entire assemblage (CCT, FFT, and bifaces; microartifacts are excluded)

in particularly low frequencies (Table 5.1). As a result, the small limestone assemblages often do not allow in-depth analysis for each of the archaeological layers, and hence only layers that yielded a minimal number of 10 limestone artifacts (larger than 2 cm) are included in the following analysis. These layers display varying frequencies of limestone product categories – cores and core-tools (CCT), flakes and flake-tools (FFT), and only rarely handaxes (HX) or cleavers (CL). Microartifacts (smaller than 2 cm) occur in higher frequencies than all of the above categories (Table 5.1).

As for the origin of the limestone nodules, they were apparently transported into the site by Acheulean occupants. Although there are sedimentary limestone rocks in the vicinity of the site (the Pliocene Gadot Formation), these are usually marly (soft) and characterized by a hard crust on the exposures' upper surfaces. They are unsuitable for knapping, and hence the limestone artifacts of GBY are foreign to the lake margin sediments (Feibel 2001), therefore the origin of the limestone raw material should be searched elsewhere. According to Belitzky (2002), four Early-Middle Pleistocene fluvial systems drained the sedimentary rocks of the Galilee into the Benot Ya'aqov Embayment, depositing limestone clasts suitable in morphology and size for knapping. Such limestone components were documented during a survey of the terraces of the southernmost fluvial system.

Typology of Limestone Artifacts

Cores and Core-Tools

The typological classification of the cores and core-tools (CCTs) shows that percussors (hammerstones), modified artifacts, and chopping tools form the bulk of the assemblage, while cores occur in very low frequencies (Table 5.2).

Percussors

The typological classification of percussors is based on their general morphology, i.e., cortical cobbles ranging from flat to globular in form, and on the presence of impact signs on their

Table 5.2 Typology of limestone cores and core-tools (counts) throughout the GBY sequence (excluding unconformities)

Typology	V-5	JB	I-5	II-2	II-5	II-5/6	II-6 L1	II-6 L2	II-6 L3	II-6 L4	II-6 L4b	II-6 L5	II-6 L6	II-6 L7	Trench II	Total
End-notched piece					2											2
Heavy duty scraper							1									1
Biface preform															1	1
Core waste										1						1
Cores						2	1		1	1				1	1	7
Angular fragment									1	2	1	1		1		6
Chopping tool		1					2	2	1		2				4	12
Percussor	2	1	1	2	1		4	2		8	2	1	3	7		34
Modified					4		6	2	6	9		1		4	3	35
Total	2	2	1	2	7	2	14	6	9	21	5	3	3	13	9	99

Fig. 5.1 Limestone knapping percussor (#2971 Layer II-6 Level 2)

surfaces and/or edges. Two distinct groups can be identified within the assemblage of percussors originating in the excavated archaeological horizons (N = 34). Percussion marks resulting from impact on hard materials (e.g., during knapping and retouching) occur on 32 artifacts. Items of this group of knapping percussors, *percuteurs de taille* (de Beaune 2000), are often flat and rounded (6–9 in the roundness scale of 1–9: Krumbein 1941). Their flat morphology ranges from flat discs (44 % of the cases) to plano-convex (sub-discoid) cobbles (52 %). The impact signs on these knapping percussors vary in intensity from minimal pecking traces to extensive weathering of the surface and/or edges. They are located mainly on the narrow edges (in 61.54 % of the cases) and to a lesser extent (38.46 %) on both the edges and the large semi-flat surfaces of the artifacts (Fig. 5.1).

The second group of percussors is very small and differs distinctly from the first. It comprises five artifacts, two originating in the excavated archaeological horizons (Layer II-6 Level 7) and the rest in the unconformity of Area B (N = 2) and in the excavation of Trench II (N = 1). These percussors are characterized by signs of heavy wear (battering, crushing, flaking, pounding), which damaged the original morphology and clearly resulted from heavy and repeated blows inflicted on a very hard surface. In addition, the damaged area is associated with intensive scars, typical of spontaneous flake removals and shattering. Such scars generally exhibit very shallow surfaces that on the one hand lack the distinctive depressions left by bulbs of percussion, but on the other hand are very rugged and unevenly broken (Fig. 5.2). It is evident that both features (battering and

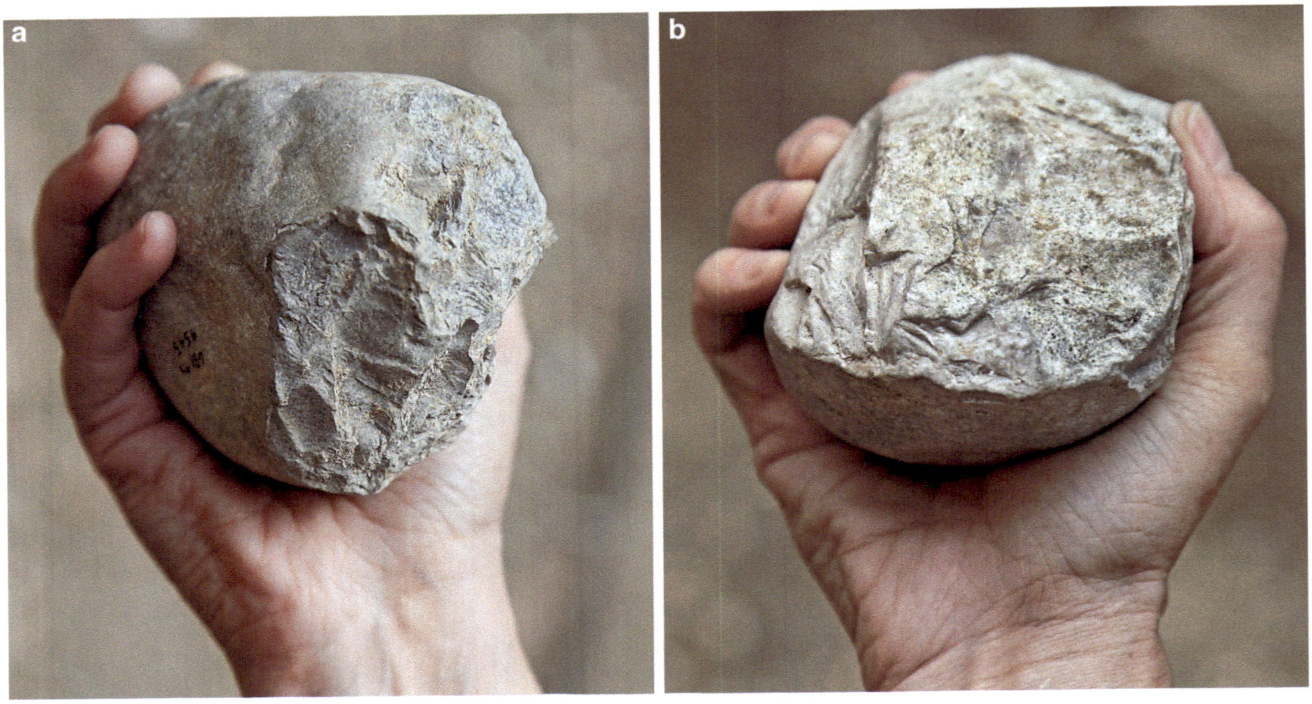

Fig. 5.2 Limestone *percuteur de concassage* (**a** #9545 Layer II-6 Level 7; **b** #7330 Layer II-6 Level 1)

shattering) derived from actions that used an extensive, violent force, most probably blows repeatedly inflicted on the same hard surface. These artifacts are classified as *percuteurs de concassage* (de Beaune 2000; see also *percuteurs de fracturation* in Soressi et al. 2011) rather than *percuteurs de taille*, as no knapper would use a percussor with such an uneven surface. We refrain from using the terminology of Mora and de la Torre (2005) or de la Torre (2011) "active hammerstones with fracture angles", but rather find de Beaune's (2000) terminology more precise. According to de Beaune, *percuteurs de concassage* are usually larger ("*volumineux*") and exhibit pronounced impact signs resulting from far stronger actions than those of regular percussors used for knapping.

Analysis of the few limestone *percuteurs de concassage* from GBY demonstrates that, like the knapping percussors, they exhibit a relatively rounded-flat morphology (either flat discs or plano-convex). The damage signs, however, occur in all cases on both the edges and the surfaces of the percussors and result in far more scars than those observed on the knapping percussors (mean number of scars: 11.6 vs. 2.9). Despite the general morphological similarity of the *percuteurs de concassage* to the knapping percussors, they are slightly larger (mean dimensions in cm – length: 9.78 vs. 9.03, width: 8.72 vs. 7.36, thickness: 6.24 vs. 5.06, circumference: 30.7 vs. 27.02).

As for their cultural affinities, there seems to be no connection between the occurrence of *percuteurs de concassage* and specific cultural entities. These artifacts are recorded in the Oldowan and the Developed Oldowan (where they are termed "active hammerstones with fracture angles": Mora and de la Torre 2005), in the Acheulean (de la Torre 2011), and in non-Acheulean Lower Paleolithic assemblages (e.g., Rufo et al. 2009). In terms of function, the unique damage signs of the *percuteurs de concassage*, in which they do not resemble any other percussion tool, enable the proposition of a variety of possible functions. In addition, unlike Mora and de la Torre (2005), we can be more precise in addressing the function of these percussors because of the variety of finds found in association with each other on the archaeological horizons of GBY:

(1) Knapping: considering their rugged and uneven surfaces, it is unlikely that they were used for knapping, which requires a surface that is convex but smooth rather than uneven. Furthermore, the powerful pounding that eventually changed the artifacts' morphology contrasts with the limited impact typical of lithic knapping.
(2) Nut cracking: the presence of pitted anvils at GBY has enabled a better understanding of the cracking of different types of nuts, also recovered from the archaeological levels (Goren-Inbar et al. 2002a). The pits that were formed on the anvils are usually small, shallow and rounded in form, and no damage signs are exhibited elsewhere. *Percuteurs de concassage*, however, demonstrate the employment of great force that is not focused on a small, limited surface. The characteristics of

the working edge of the *percuteurs de concassage* are not suitable for the focal pounding required for nut cracking.

(3) Wood working: Mora and de la Torre (2005) suggested that *percuteurs de concassage* were used for wood working. The GBY excavations did indeed uncover a preserved wood assemblage, including a worked (polished) wooden artifact (Belitzky et al. 1991) suggesting that wood working was carried out at the site. However, the use of *percuteurs de concassage* for wood working is not feasible in the context of GBY. The wood identification at GBY provided a list of 26 Mediterranean species of trees, bushes, and climbers (Goren-Inbar et al. 2002b). Of these, the oaks (*Quercus* sp.) are the hardest species but are not massive enough to produce such an extensive damage pattern as that observed on the limestone percussors. Furthermore, it is worth mentioning that no pounding damage signs have been identified on the logs and wood fragments from the site.

(4) Bone cracking: de Beaune (2000) noted that percussors are often identified as flint knapping percussors, while they were actually used for bone fragmentation. The globular morphology of the *percuteurs de concassage* and the dullness of their edges suggest they were indeed used for recurrent battering during prolonged sessions of bone fragmentation (Célérier and Kervazo 1988 in de Beaune 2000:61). At GBY, there is great variability in the frequencies of faunal remains in the different archaeological horizons, two of which are extremely rich and display the highest quantities, density, and taxonomic diversity at the site (Rabinovich et al. 2008, 2012). A separate faunal occurrence is an elephant skull with extensive pounding damage marks around its nasal foramen. The skull bears signs of pounding and cracking for marrow extraction (e.g., Goren-Inbar et al. 1994, pl. II, 102) and is associated with numerous small skull fragments (Goren-Inbar et al. 1994). The presence of fragmented bones of large mammals (e.g., elephants, rhinoceros, bovids, hippos) at GBY suggests that *percuteurs de concassage* were instrumental in the process of bone cracking.

While it is clear that the two groups of percussors – knapping and battering – differ distinctly from one another, they are both classified as percussors. The typological category of percussors at GBY includes two artifacts exhibiting signs of both knapping and battering damage on their edges and surfaces. These two cases illustrate that certain percussors were used for knapping as well as battering during their life history. The location of the damage signs, however, does not allow us to determine the order of percussion functions for these artifacts.

The Typological "Life History" of Percussors

In addition to the two percussors discussed above, the GBY limestone assemblage includes artifacts that were classified under two typological categories, being defined as a knapping percussor and assigned to an additional category as well. The necessity of such dual typological classification derives from the occurrence of percussion marks (i.e., damaged edges/surfaces) on *other* artifacts. Of the entire limestone CCT assemblage (N = 114), 24 items that were not typologically defined as percussors exhibit percussion marks. These include 15 items classified as "modified", six chopping tools, two *percuteurs de concassage* and a discoid. The presence of these artifacts in the limestone assemblage illustrates their exploitation as knapping percussors in earlier stages of their life histories. These items are remnants of the reduction sequence in which percussors were transformed into other core-tools, and it is clear that percussion marks could not have been preserved on items (e.g., cores) that had undergone further modification, removing most of their original outer surface.

Modified Artifacts

Modified artifacts (cobbles with a maximum of three flake scars and no working edges and/or platforms) are the largest category of CCTs in the limestone assemblage (N = 35; Table 5.2). As discussed above, 15 modified artifacts exhibit percussion damage on their edges and/or surfaces, suggesting their use as percussors (Fig. 5.3). As some 42 % of the modified artifacts exhibit percussion (knapping) damage, it is likely that the remaining scarred modified artifacts had

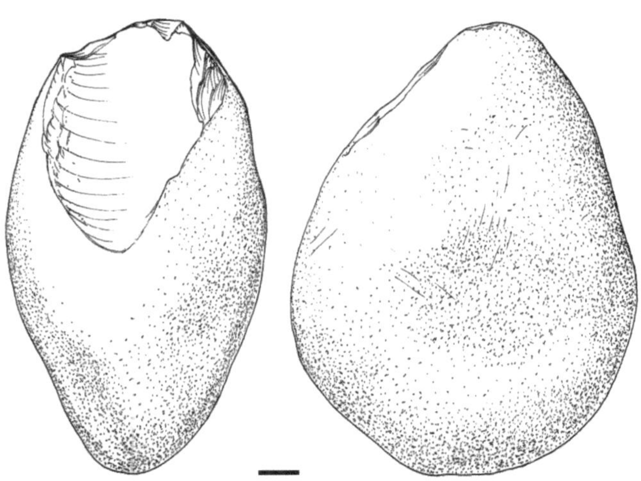

Fig. 5.3 Limestone modified artifact, scar resulting from spontaneous shattering (#7354 Trench II; scale = 1 cm)

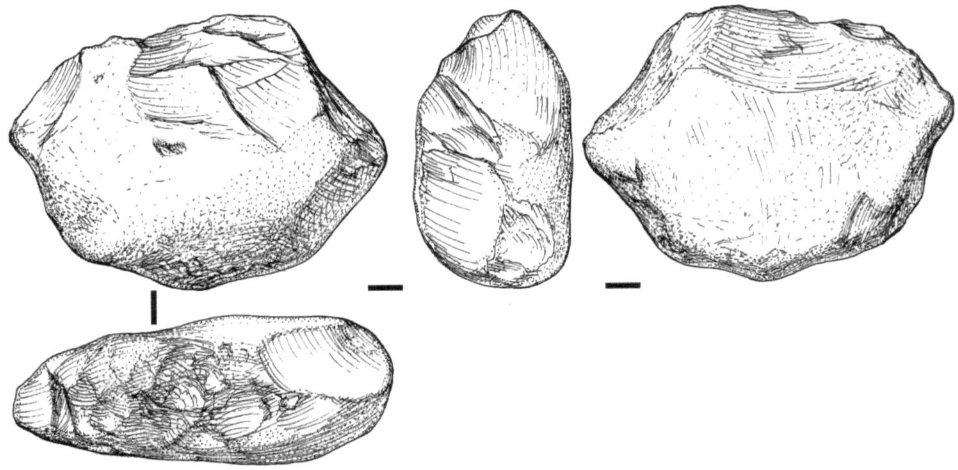

Fig. 5.4 Limestone chopping tool; note pecking/battering signs on the proximal end (#6038 Layer II-6 Level 4b; scale = 1 cm)

also been used as percussors. It is possible that spontaneous shattering of flakes occurred during use of the percussors, erasing the percussion damage signature (see below). In addition, several artifacts that are split percussors are included within the FFT category.

Chopping Tools

Chopping tools comprise some 12 % of the limestone CCT assemblage (N = 12; Table 5.2). Only a single unifacial chopping tool (i.e., a chopper) was recorded, while the remaining chopping tools exhibit a classic active edge with two platforms (Fig. 5.4). Six of the 12 chopping tools (discussed above) exhibit percussion marks.

Pitted Anvils

Pitted anvils (which are commonly recorded on basalt) occur on limestone artifacts and are identified on both FFTs (N = 3) and CCTs (N = 20). Pitted anvils of the CCT category are all complete and occur mostly on percussors (N = 16), though they were also found on a single "modified" item, a chopping tool, an amorphous core, and a heavy duty scraper. They are mostly cortical, with only one item bearing no cortex, and the majority of items (N = 14) exhibit cortical coverage on 76–100 % of their surface. Their size and shape fall within the relatively large category of percussors (mean dimensions in mm – length: 89.95, width: 76.25, thickness: 53.00, circumference: 271.11).

Cores and Other Core-Tools

In the small sample of seven cores, six were classified as "core varia" and a single one as "amorphous core". The category of core-tools includes two end-notched pieces, a heavy duty scraper, a cleaver, nine handaxes, and a biface preform (Table 5.2). In addition, the only remnant of a limestone core-waste artifact was recorded within the CCT category. Angular fragments, which occur in very low frequencies, were also included in the CCT category as they cannot be considered flakes.

Flakes and Flake-Tools

Among limestone flakes and flake-tools the large majority of artifacts, over 70 % in each of the layers, consists of unretouched flakes. The few retouched items (N = 19) include side-scrapers, notches, and a variety of retouched flakes (Table 5.3). The small number of each of the tool types precludes a meaningful typological description. Core-trimming elements are absent.

Technology of Limestone Artifacts

Cores and Core-Tools

Though recorded throughout the occupational sequence of GBY, limestone CCTs occur in very low frequencies, precluding separate in-depth analysis of each archaeological layer. Thus, in order to obtain a meaningful technological description of the limestone component of GBY, we have assembled all limestone CCTs according to major typological groups.

The general characteristics of limestone CCTs suggest that the majority of artifacts are complete, patinated, and abraded or heavily abraded (Table 5.4). By and large, chopping tools and cores exhibit better preservation than percussors and modified artifacts (Table 5.4). This could be correlated with

Table 5.3 Typology of limestone flakes and flake-tools[a]; Percentages within layer/level

Typology	II-2	II-5	II-6 L1	II-6 L2	II-6 L3	II-6 L4	II-6 L4b	II-6 L5	II-6 L6	II-6 L7	Total
Side-scrapers			4.55			4.76	5.00				3
Typical end-scraper					4.17						1
Atypical borer										1.30	1
Notch		14.29	4.55			4.76	5.00	3.85			5
Denticulate							5.00				1
Alternately retouched beaks								3.85			1
Retouched flake			9.09				5.00		8.00	1.30	6
Naturally backed knife									4.00		1
Flake	100.00	85.71	81.82	76.47	87.50	71.43	70.00	88.46	84.00	93.51	213
Blade					8.33	4.76				1.30	4
Split percussors				23.53		14.29	10.00	3.85	4.00	2.60	13
Total	10	7	22	17	24	21	20	26	25	77	249

[a]Assemblages which consist of a minimum number of 10 limestone artifacts (larger than 2 cm)

Table 5.4 General characteristics of limestone cores and core-tools core-tool

Preservation	Percussors		[a]Split percussors		Modified		Chopping tools		Cores	
	N = 34		N = 13		N = 35		N = 12		N = 7	
	N	%	N	%	N	%	N	%	N	%
Fresh					1	2.86			1	14.29
Slightly abraded	11	32.35	2	15.38	8	22.86	7	58.33	2	28.57
Abraded	21	61.76	4	30.77	24	68.57	5	41.67	4	57.14
Heavily abraded	1	2.94	5	38.46	2	5.71				
Exfoliated/decayed	1	2.94	2	15.38						
Patina										
Patina	31	91.18	12	92.31	35	100.00	10	83.33	5	71.43
Double patina	3	8.82	1	7.69			2	16.67	2	28.57
Breakage										
Complete	31	91.18			32	91.43	12	100.00	7	100.00
Lateral	1	2.94			3	8.57				
Distal	1	2.94								
Distal and proximal	1	2.94								
Split			13	100.00						
Cortex	N = 32		N = 13		N = 33		N = 12		N = 7	
None					2	6.06			2	28.57
1–25 %					3	9.09			2	28.57
26–50 %	1	2.94	1	7.69	4	12.12	2	16.67	1	14.29
51–75 %	3	9.38	1	7.69	5	15.15	6	50.00	1	14.29
76–100 %	28	87.50	11	84.62	19	57.58	4	33.33	1	14.29

[a]*Note* split percussors, included in the FFT category, are presented here due to their technological nature

cortical coverage, where abrasion predominates and which covers larger areas in the two latter categories (the "cortex" is but the outer face of the limestone cobbles).

An interesting pattern emerges from the characteristics of different categories of cores and core-tools. First, as expected, percussors exhibit the highest frequencies of cortical coverage, which decreases gradually from modified artifacts to chopping tools and then to cores (Table 5.4). As for breakage patterns, chopping tools and cores are all complete, while breakage is seen on 8.82 % of the percussors. Interestingly, double-patinated items also gradually increase from 8.82 % of percussors, through 16.67 % of chopping tools, to 28.57 % of cores (Table 5.4). Such "gradual" changes are also seen in the metric measurements of these categories: percussors are generally longer, wider, and thicker than chopping tools, which are generally longer, wider, and thicker than cores (Table 5.5). The circumference of these categories decreases similarly (Table 5.5).

Flakes and Flake-Tools

Flakes and flake-tools are extensively patinated, though the frequency of double patination is markedly low (Table 5.6). While this taphonomic situation depends on exposure to atmospheric condition, the preservation of the FFTs at GBY is mainly determined by in situ chemical processes that affect the limestone and cause very high frequencies of abraded surfaces, though with internal variation, as shown in Table 5.6.

A trait shared by most of the limestone flaked pieces is the presence of cortex on the dorsal face (Table 5.6). Examination of the extent of cortical coverage suggests that most artifacts exhibit cortical coverage of 75 % on their dorsal face (Table 5.7). Correspondingly, when the number of dorsal scars on limestone flakes is examined, it is evident that the preceding knapping involved very few removals, with an average of no more than four flake removals (Table 5.8).

Table 5.5 Descriptive statistics of limestone cores and core-tools

	Percussors	Split percussors	Modified	Chopping tools	Cores
Maximum	(N = 34)	(N = 13)	(N = 35)	(N = 12)	(N = 7)
Maximum	123	123	125	108	109
Median	96	97	65	95.50	67
Minimum	64	72	33	55	31
Mean	*95.85*	*98.61*	*70.62*	*89.92*	*69.85*
Std dev	12.50	14.85	26.66	18.68	32.01
Std err mean	2.14	4.11	4.50	5.39	12.09
Length					
Maximum	120	114	206	106	103
Median	91	89	70	88.50	64
Minimum	55	60	30	25	25
Mean	*90.76*	*83.69*	*71.65*	*81.75*	*62.42*
Std dev	15.65	16.66	35.13	24.81	31.38
Std err mean	2.68	4.62	5.93	7.16	11.86
Width					
Maximum	94	117	96	100	90
Median	76	76	46	73.50	60
Minimum	53	57	23	52	27
Mean	*75.05*	*82.92*	*53.08*	*73.25*	*56.28*
Std dev	11.44	18.57	22.33	14.12	23.97
Std err mean	1.96	5.15	3.77	4.08	9.05
Thickness					
Maximum	79	61	68	65	49
Median	52	36	29	46	25
Minimum	37	22	15	37	18
Mean	*52.70*	*37.61*	*36.00*	*48.25*	*32.00*
Std dev	11.04	10.57	16.01	10.30	12.32
Std err mean	1.89	2.93	2.70	2.97	4.65
Circumference	(N = 24)	(N = 13)	(N = 6)	(N = 9)	(N = 7)
Maximum	335	339	349	325	303
Median	276	267	237	295	205
Minimum	98	210	113	154	91
Mean	*270.13*	*272.61*	*233.83*	*264.67*	*198.00*
Std dev	50.97	34.70	82.05	60.69	85.72
Std err mean	10.40	9.62	33.49	20.23	32.40
[a]Number of scars	(N = 20)	(N = 9)	(N = 35)	(N = 12)	(N = 7)
Maximum	8	6	4	12	20
Median	2.50	2	2	7	4
Minimum	1	1	1	2	3
Mean	*2.95*	*2.66*	*2.22*	*7.08*	*6.71*
Std dev	2.03	1.41	0.84	3.32	5.99
Std err mean	0.45	0.47	0.14	0.96	2.26
Sum	59	24	78	85	47

[a]Excluded from the counts are items without scars

Table 5.6 General characteristics of limestone flakes and flake-tools; percentages (top) and counts (bottom)

Layer	Patina			Preservation				Breakage		Cortex		Total
	None	Patina	Double	Fresh	Slightly abraded	Abraded	Heavily abraded	Complete	Broken	None	Cortex	
II-2	–	80.00	20.00	–	20.00	40.00	40.00	60.00	40.00	50.00	50.00	10
		(8)	(2)		(2)	(4)	(4)	(6)	(4)	(5)	(5)	
II-5	28.57	57.14	14.28	14.28	14.28	57.14	14.28	71.43	28.57	14.28	85.71	7
	(2)	(4)	(1)	(1)	(1)	(4)	(1)	(5)	(2)	(1)	(6)	
II-6/L1	13.63	77.27	9.09	13.63	13.63	59.09	13.63	59.09	40.90	50.00	50.00	22
	(3)	(17)	(2)	(3)	(3)	(13)	(3)	(13)	(9)	(11)	(11)	
II-6/L2	5.88	88.23	5.88	5.88	23.52	52.94	17.64	52.94	47.05	35.29	64.70	17
	(1)	(15)	(1)	(1)	(4)	(9)	(3)	(9)	(8)	(6)	(11)	
II-6/L3	–	95.83	4.16	8.33	29.16	45.83	16.66	50.00	50.00	45.83	54.16	24
		(23)	(1)	(2)	(7)	(11)	(4)	(12)	(12)	(11)	(13)	
II-6/L4	4.76	90.47	4.76	19.04	28.57	38.09	14.28	52.38	47.61	28.57	71.42	21
	(1)	(19)	(1)	(4)	(6)	(8)	(3)	(11)	(10)	(6)	(15)	
II-6/L4b	5.00	85.00	10.00	–	15.00	35.00	50.00	55.00	45.00	30.00	70.00	20
	(1)	(17)	(2)		(3)	(7)	(10)	(11)	(9)	(6)	(14)	
II-6/L5	–	80.76	19.23	–	7.69	23.07	69.23	42.30	57.69	38.46	61.53	26
		(21)	(5)		(2)	(6)	(18)	(11)	(15)	(10)	(16)	
II-6/L6	12.00	80.00	8.00	–	44.00	52.00	4.00	24.00	76.00	28.00	72.00	25
	(3)	(20)	(2)		(11)	(13)	(1)	(6)	(19)	(7)	(18)	
II-6/L7	5.19	92.20	2.59	1.29	7.79	33.76	57.14	35.06	64.93	28.57	71.42	77
	(4)	(71)	(2)	(1)	(6)	(26)	(44)	(27)	(50)	(22)	(55)	

Table 5.7 Extent of cortical coverage on limestone flakes and flake-tools; percentages (top) and counts (bottom)

Layer	No cortex	1–25 %	26–50 %	51–75 %	76–100 %	Indet.	Total
II-2	50.00 (5)	–	–	–	40.00 (4)	10.00 (1)	10
II-5	14.28 (1)	–	14.28 (1)	14.28 (1)	57.14 (4)	–	7
II-6/L1	50.00 (11)	4.54 (1)	–	4.54 (1)	36.36 (8)	4.54 (1)	22
II-6/L2	35.29 (6)	17.64 (3)	–	5.88 (1)	41.17 (7)	–	17
II-6/L3	45.83 (11)	12.50 (3)	12.50 (3)	4.16 (1)	16.66 (4)	8.33 (2)	24
II-6/L4	28.57 (6)	9.52 (2)	9.52 (2)	14.28 (3)	33.33 (7)	4.76 (1)	21
II-6/L4b	30.00 (6)	5.00 (1)	20.00 (4)	–	35.00 (7)	10.00 (2)	20
II-6/L5	38.46 (10)	–	–	7.69 (2)	30.76 (8)	23.07 (6)	26
II-6/L6	28.00 (7)	4.00 (1)	8.00 (2)	4.00 (1)	4.00 (1)	52.00 (13)	25
II-6/L7	28.57 (22)	1.29 (1)	5.19 (4)	–	25.97 (20)	38.96 (30)	77
Total	(85)	(12)	(16)	(10)	(70)	(56)	249

Breakage patterns vary amongst assemblages, with a general majority of complete items in most layers (Table 5.6). Nevertheless, fragmented items and distal and proximal breaks are common (Table 5.9). The breakage patterns are not associated with the preservation or patination state of the limestone artifacts, or with their size; consequently, it is possible that these patterns result from spontaneous breakage (i.e., percussor shattering) during the knapping process (Figs. 5.3 and 5.5).

Analyses of the limestone FFTs resulted in several observations that seem to characterize all the examined assemblages and can be summarized as follows:

(1) The frequency of limestone FFTs is relatively low in comparison with other raw materials.
(2) Complete limestone FFTs characterize most layers, though fragmented items and distal and proximal breaks are common in all archaeological horizons.
(3) The vast majority of limestone FFTs consists of unretouched flakes.
(4) Limestone FFTs are small in size with very similar mean length and width (i.e., somewhat rounded, non-elongated, flakes) (Table 5.8).
(5) Limestone core trimming elements are absent from the assemblages.
(6) The number of dorsal scars and the extent of cortical coverage imply that most limestone FFT production was carried out during the early stages of decortication.

From these patterns, no standardized industry of limestone FFT production can be identified at GBY. Rather, limestone FFTs seem to exhibit characteristics of byproducts of limestone core-tool production. In order to evaluate the entire sequence of limestone knapping at the site, we should look within the CCT category for the possible contributors of the limestone flake component.

Discussion: The Limestone Reduction Sequence

In comparison to the other raw materials used by the GBY Acheuleans, the limestone assemblage exhibits a different utilization mode and approach (Fig. 5.6). While basalt and flint consistently exhibit higher frequencies of flakes/flake-tools than cores/core-tools, as well as varying ratios of these two categories throughout the archaeological sequence, the utilization of limestone seems to be fairly uniform throughout the sequence, with low frequencies of limestone artifacts and a ratio of cores/core-tools to flakes/flake-tools closer to 2 (Fig. 5.6). Even before systematic examination of the limestone component, these patterns clearly indicate a particular exploitation mode in which flakes/flake-tools are not the main target products.

Within the examined sample (i.e., originating in the excavated archaeological horizons: Layers II-2, II-5, II-6 Levels 1–7), the number of limestone FFTs (N = 247) and the number of observed scars on the CCTs (N = 261) are very similar. This similarity indicates that there is no shortfall of flakes that could have originated in the knapping of CCTs. Furthermore, it suggests that the limestone FFTs were the products of in situ knapping and utilization. If the limestone

a b c d

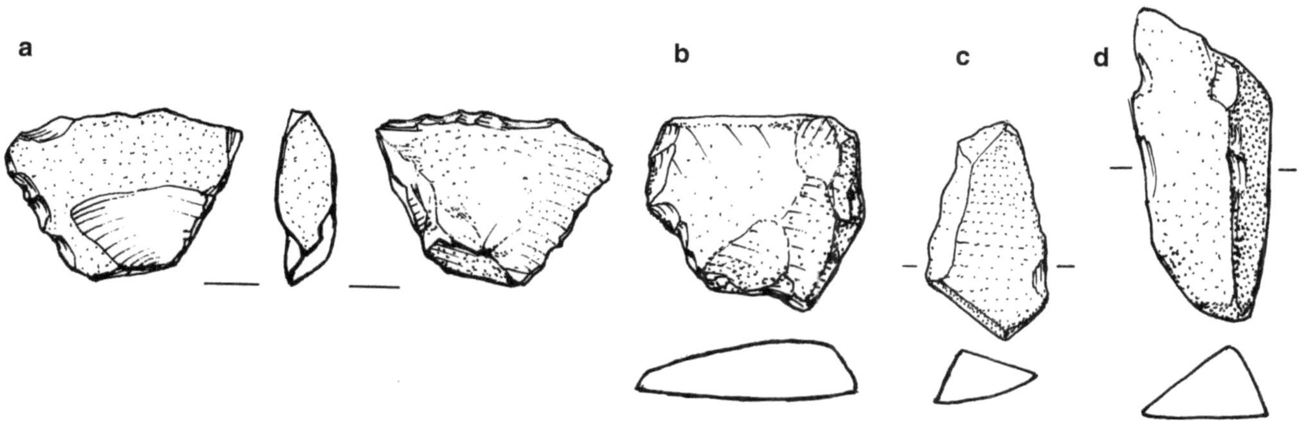

e

f

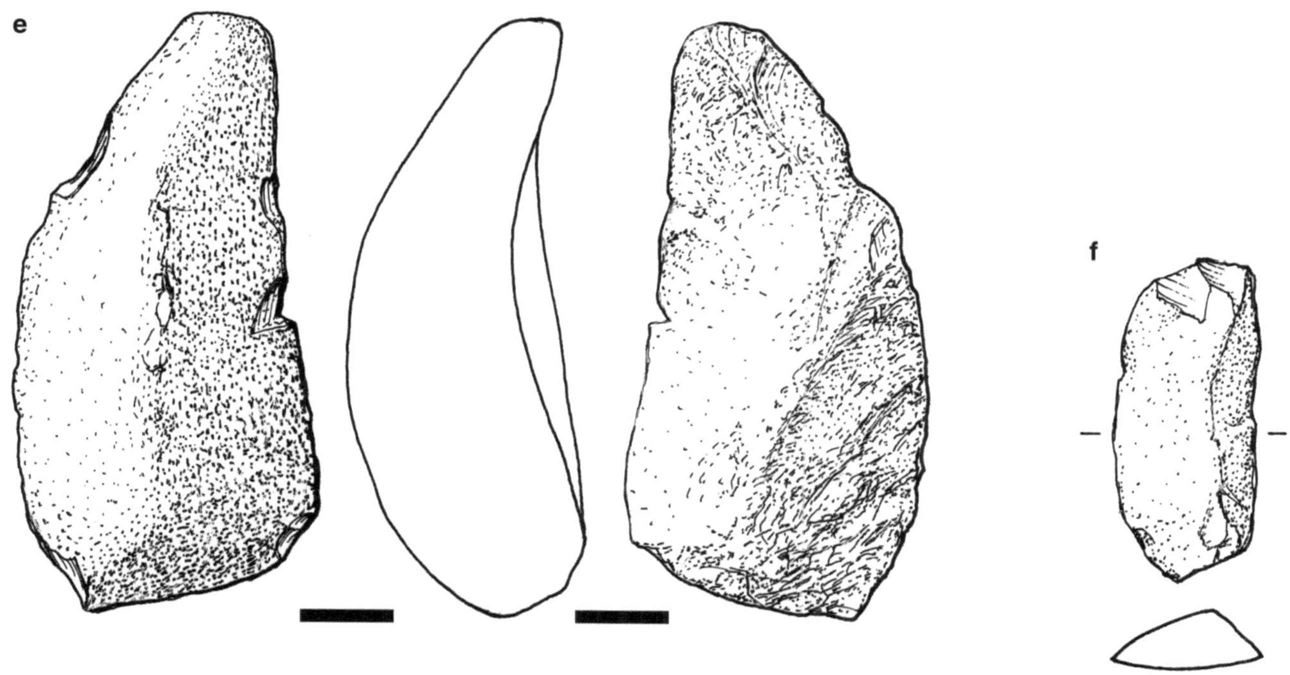

g

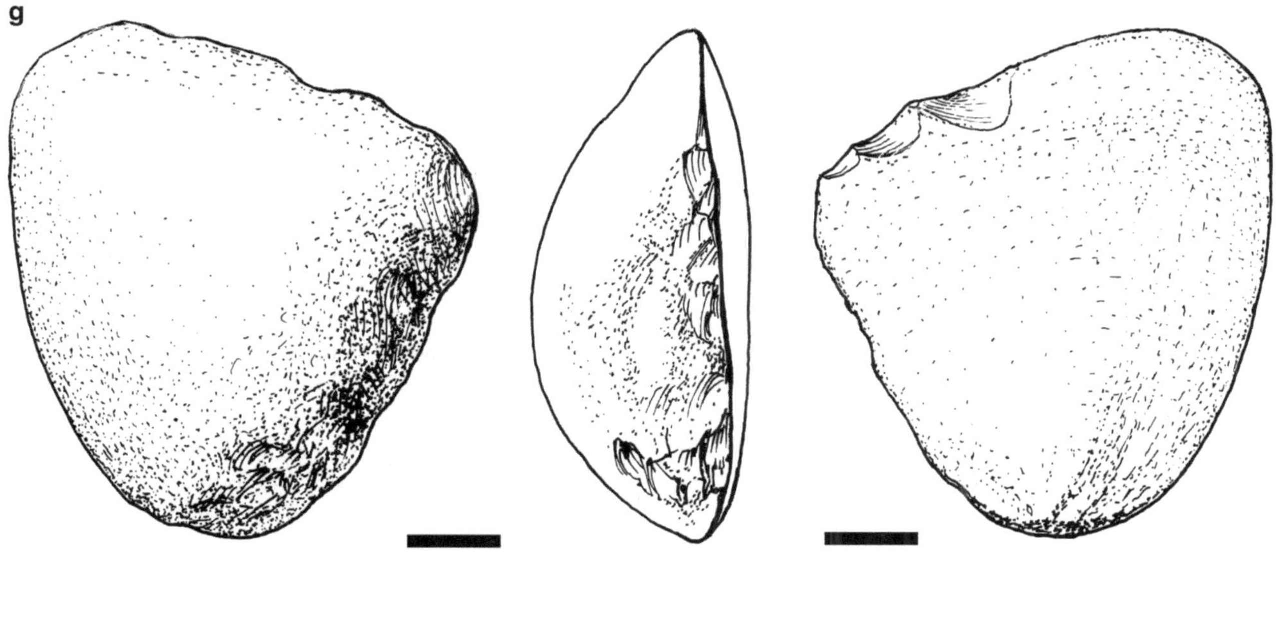

◀ **Fig. 5.5** Limestone FFTs (**a** #9633 Layer II-6 Level 5; **b** #13311 Layer II-6 Level 3; **c** #1693 Layer II-6 Level 4b; **d** #14322 Layer II-6 Level 7; **e** #4329 Layer II-6 Level 6; **f** #2100 Layer II-6 Level 4; **g** #3440 Layer II-6 Level 4b; scale = 1 cm)

Table 5.8 Descriptive statistics of limestone flakes and flake-tools

	II-2	II-5	II-6 L1	II-6 L2	II-6 L3	II-6 L4	II-6 L4b	II-6 L5	II-6 L6	II-6 L7
Maximum										
Maximum	41	82	76	117	64	144	91	113	93	110
Median	27.50	38	32	31	32.50	34	30.50	26	29	28
Minimum	21	33	23	22	22	22	20	20	21	21
Mean	*28.80*	*47.57*	*34.86*	*47.17*	*35.70*	*48.38*	*39.60*	*29.76*	*33.28*	*33.10*
Std dev	7.28	18.63	12.83	30.27	12.12	34.66	21.41	17.59	16.13	15.38
Std err mean	2.30	7.04	2.73	7.34	2.47	7.56	4.78	3.45	3.22	1.75
Flaking length										
Maximum	38	82	57	93	64	144	89	107	93	93
Median	24.50	35	24.50	28	30	31	29.50	24.50	22	25
Minimum	15	24	18	21	20	15	15	17	15	12
Mean	*26.20*	*42.57*	*28.59*	*39.47*	*32.91*	*43.28*	*36.35*	*28.03*	*28.56*	*28.74*
Std dev	8.18	19.26	9.66	21.18	11.24	32.78	21.94	16.97	16.95	13.62
Std err mean	2.58	7.27	2.06	5.13	2.29	7.15	4.90	3.32	3.39	1.57
Flaking width										
Maximum	31	61	72	117	51	106	69	96	66	107
Median	23	34	26	29	25	29	25.50	18	23	23
Minimum	19	19	17	14	14	11	14	13	15	10
Mean	*24.60*	*37.42*	*31.09*	*38.76*	*27.37*	*37.28*	*30.55*	*22.07*	*26.04*	*26.00*
Std dev	4.64	16.08	13.03	29.95	10.29	27.05	16.74	15.80	11.54	14.47
Std err mean	1.46	6.07	2.77	7.26	2.10	5.90	3.74	3.09	2.30	1.67
Thickness										
Maximum	14	39	22	43	27	61	44	53	36	45
Median	9	12	12.50	15	12	13	10.50	9.50	9	10
Minimum	8	9	7	7	5	8	7	4	6	4
Mean	*9.8*	*18.28*	*13.40*	*17.29*	*13.29*	*17.42*	*15.65*	*10.76*	*11.52*	*12.44*
Std dev	2.09	10.87	4.53	11.07	6.58	12.98	10.13	8.91	6.71	7.26
Std err mean	0.66	4.10	0.96	2.68	1.34	2.83	2.26	1.74	1.34	0.82
N	10	7	22	17	24	21	20	26	25	77
Number of scars										
Maximum	3	2	8	6	9	5	4	4	6	8
Median	1.50	2	3	2.50	2	2.50	3	3	4	2
Minimum	1	1	1	1	1	1	1	1	2	1
Mean	*1.83*	*1.60*	*3.91*	*2.66*	*3.12*	*2.64*	*2.38*	*2.83*	*3.71*	*2.68*
Std dev	0.98	0.54	2.46	1.66	2.60	1.27	1.12	1.16	1.49	1.56
Std err mean	0.40	0.24	0.71	0.48	0.65	0.34	0.31	0.47	0.56	0.28
[a]N	6	5	12	12	16	14	13	6	7	29

[a]Counts exclude cortical flakes with no scars

reduction sequence was oriented towards the production of flakes and flake-tools, one would expect their number to exceed the observed number of scars on the CCTs. However, considering the great similarity of the two values, the characteristics of the FFTs (i.e., cortical, non-retouched, small-sized), and the dominance of percussors in these assemblages, it is most probable that the limestone flakes originate primarily from battering and accidental shattering as a result of the use of limestone percussors. In each of the archaeological horizons, the pattern is more variable. In some assemblages the number of observed scars on CCTs is smaller than the number of FFTs and in other layers it is larger. The presence and frequencies of limestone microartifacts in each of these layers extend the observed variability.

Thus, percussors and modified artifacts form the majority of limestone core-tools, and these categories display similar patterns of breakage and cortical coverage, with a majority of complete items and a particularly high cortical coverage (Table 5.4). In terms of metric dimensions, however, there is a general similarity in size between percussors and chopping tools. This similarity may suggest that limestone pebbles were selected to fit a particular size and form (i.e., fist size). The

Table 5.9 Breakage patterns of limestone flakes and flake-tools; percentages (top) and counts (bottom)

Layer	Complete	Distal	Proximal	Distal and proximal	Distal and lateral	Proximal and lateral	Lateral	Fragment	Indet.	Split	Total
II-2	60.00 (6)	–	–	10.00 (1)	20.00 (2)	–	10.00 (1)	–	–	–	10
II-5	71.43 (5)	28.57 (2)	–	–	–	–	–	–	–	–	7
II-6/L1	59.09 (13)	27.27 (6)	–	4.54 (1)	–	–	4.54 (1)	4.54 (1)	–	–	22
II-6/L2	52.94 (9)	–	5.88 (1)	–	–	5.88 (1)	–	11.76 (2)	–	23.52 (4)	17
II-6/L3	50.00 (12)	4.16 (1)	20.83 (5)	–	4.16 (1)	4.16 (1)	4.16 (1)	12.50 (3)	–	–	24
II-6/L4	52.38 (11)	4.76 (1)	4.76 (1)	–	–	4.76 (1)	–	14.28 (3)	4.76 (1)	14.28 (3)	21
II-6/L4b	55.00 (11)	10.00 (2)	10.00 (2)	–	–	–	5.00 (1)	10.00 (2)		10.00 (2)	20
II-6/L5	42.30 (11)	15.38 (4)	3.84 (1)					30.76 (8)	3.84 (1)	3.84 (1)	26
II-6/L6	24.00 (6)	8.00 (2)	16.00 (4)	4.00 (1)			16.00 (4)	28.00 (7)	–	4.00 (1)	25
II-6/L7	35.06 (27)	12.98 (10)	9.09 (7)	2.59 (2)	10.39 (8)	2.59 (2)	7.79 (6)	16.88 (13)	–	2.59 (2)	77
Total	(111)	(28)	(21)	(5)	(11)	(5)	(14)	(39)	(2)	(13)	249

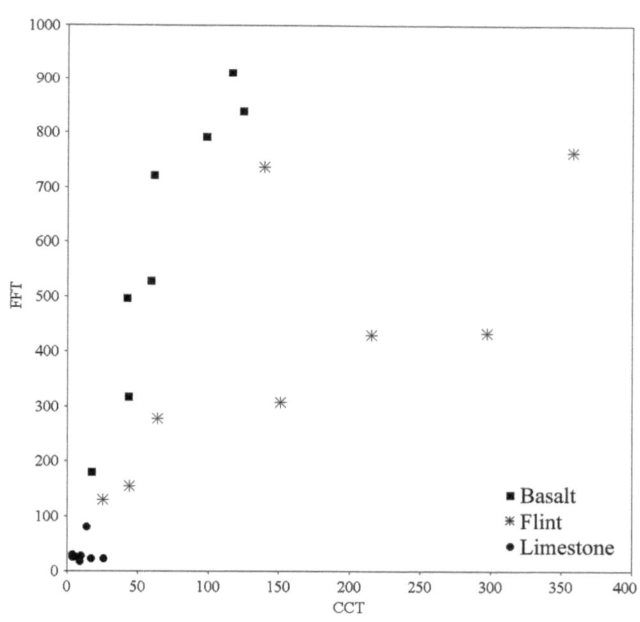

Fig. 5.6 Counts of cores and core-cools (excluding bifaces) and flakes and flake-tools from the 8 levels of Layer II-6 (Levels 1–7) by different raw materials

slight size gradient in several metric characteristics of these two categories enables a general reconstruction of the "lifetime" of the limestone pebble – first as a percussor (larger, more cortical, fewer scars), and then, possibly after spontaneous breakage of the utilized percussor, modification into a chopping tool (slightly smaller, less cortical, more scars, more frequently double patinated). Finally, the concluding stage in this limestone *chaîne opératoire* may possibly be the transformation of a chopping tool into a core – smallest, less cortical, and even more frequently double patinated.

A classic reduction sequence is determined by a succession of procedures, which generally begin with procurement of the raw material and terminate with production of the desired end product, its use, and finally its discard. The GBY limestone component, however, presents a dynamic reduction sequence in which the desired end product (i.e., a percussor) is actually the starting point of the sequence (Fig. 5.7).

Such a flexible and dynamic operational sequence is an indication of long-term planning and hence advanced cognitive abilities (see below), expressed in an intentional selection of the limestone raw material that involved decisions about the future function and therefore dictated the size/dimensions of the selected pebble/nodule. From the very first stages of the limestone reduction sequence, the mobility of artifacts from their sedimentary source (fluvial terraces) necessitated means of transporting the limestone pebbles/nodules – means that were different from those used for the basalt slabs (Goren-Inbar 2011b).

The hominins of GBY apparently had a deep understanding of the association between the characteristics of raw materials and their particular mode of exploitation in the different reduction sequences. Since basalt percussors occur in GBY at even higher frequencies than those of limestone, one may ask what were the specific qualities of limestone that made its use desirable. Clearly, the basalt percussors differ in qualities from those of limestone, the latter being a softer stone. Studies concerned with the qualities of limestone percussors illustrate the fact that

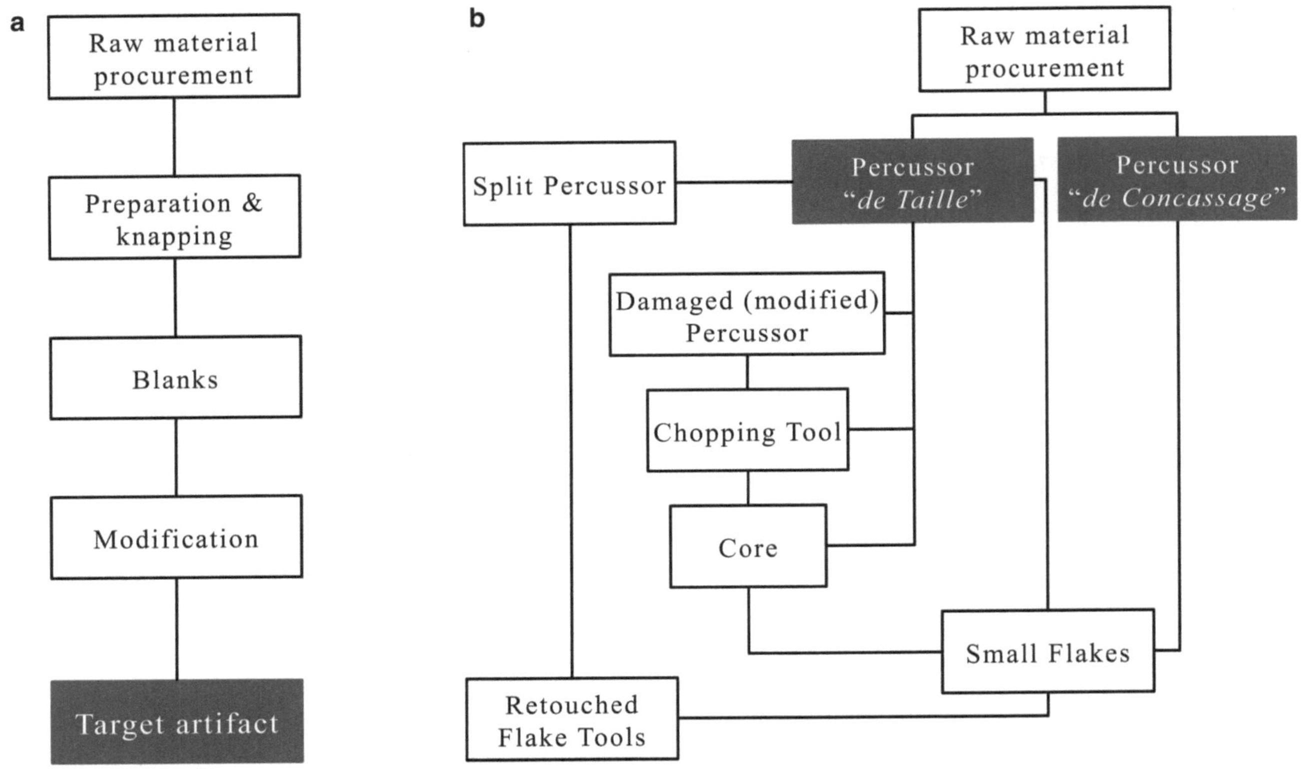

Fig. 5.7 Schematic flow charts of reduction sequences: **a** A generic model; **b** The GBY limestone model

their use resembles "soft hammer" techniques (Pelegrin 2000); this technique is indeed observed at GBY, as demonstrated by the presence of its traits on basalt, and even more clearly on flint, flakes (Sharon and Goren-Inbar 1999; Goren-Inbar and Sharon 2006). An antler fragment with evident marks of its use as a percussor was discovered at GBY (Goren-Inbar 2011a). Both limestone percussors and the antler one were most probably instrumental in the final stages of biface modification. At present we lack experimental data comparing the application of different percussors (antler, limestone, basalt) and their byproducts. Until such data are available, our understanding of the particular selection modes and their reasoning cannot encompass detailed reconstruction of the full processes of decision-making regarding the use of the different percussors.

Cognitive Abilities

A previous study of the basalt reduction sequence at GBY demonstrated that the Acheuleans' cognitive abilities were developed and encompassed a variety of cognitive fields (Goren-Inbar 2011b and references therein; cognitive terminology after Coolidge and Wynn 2009). The acquisition of the basalt raw material and its repetition throughout the entire Acheulean sequence of the site are indicative both of *large-scale spatial thinking* and of *long-term memory*. Exactly the same patterns characterize the acquisition of the limestone component. While we lack precise knowledge of the collecting spots of the basalt slabs and the limestone nodules, the geology of the region furnishes information on the accessibility of these rocks. Clearly, *advanced planning* was part of the reduction sequence of the limestone, but in contrast with the basalt, it is impossible to suggest *cooperative provisioning* for the former. The expression of *contingency* also differs between the two raw materials. While the basalt reduction sequence demonstrates a variety of modes for extracting large flakes (for the production of bifaces; see Goren-Inbar 2011b), *contingency* in limestone modification is expressed by the hominins' ability to transform one type into another, a behavior that clearly illustrate the *technical and procedural know-how (procedural cognition)*. The contribution that emerges from the analysis of the limestone component, and its comparison to the other results from GBY, is extremely instructive about cognitive abilities. The reduction sequences for basalt and limestone described here illustrate the multi-facetted abilities of the hominins. Moreover, both behaviors are documented along a time trajectory of 50 ka at the site. They have clear implications for the *communication abilities* that form a segment of the general *social cognition* of the Acheulean.

The Acheulean hominins of GBY repeatedly reoccupied the paleo-Lake Hula margin, where they habitually used fire

(Goren-Inbar et al. 2004; Alperson-Afil 2008; Alperson-Afil and Goren-Inbar 2010), systematically butchered animals (Rabinovich et al. 2008, 2012), exploited aquatic resources (Alperson-Afil et al. 2009), collected plant food and cracked nuts (Goren-Inbar et al. 2002a), and transported artifacts of different raw materials to and from the site (Sharon and Goren-Inbar 1999; Goren-Inbar and Sharon 2006; Goren-Inbar et al. 2011). These activities were carried out within a formalized, modern conceptualization of their living space (Alperson-Afil et al. 2009). Despite its scarcity, analyses of the limestone component have contributed significant information on the behavior of the GBY hominins. Their primary goal in obtaining limestone nodules was their use as percussors, so that other limestone artifacts can be considered byproducts. The fact that these byproducts are integrated within the limestone assemblages in a formal, consistent manner is evidence of contingency. In addition, the ability of the GBY hominins to transform one type into another while diverging from the original plan implies cultural complexity as well as flexibility. As with other aspects of the lithic assemblages of GBY, the limestone component exhibits consistent typological and technological homogeneity along the time trajectory for millennia. This conservatism of the transmission of knowledge about specific raw materials, their reduction sequences, and the flexible variations within them is typical of a "complex" culture.

Acknowledgements This study was carried out with the support of an ongoing grant awarded by the Israel Science Foundation (Grant No. 300/06) to the Center of Excellence Project Title: "The Effect of Climate Change on the Environment and Hominins of the Upper Jordan Valley between ca. 800 ka and 700 ka ago as a Basis for Prediction of Future Scenarios." N. Alperson-Afil is grateful for the support of the Martin Buber Society of Fellows, Hebrew University of Jerusalem. The authors wish to thank Gadi Herzlinger, Arik Malinsky-Buller, Gonen Sharon, Gaby Laron (photography), and Leonid Zeiger (artifact drawings). The authors are also grateful for the contributions of S. Gorodetsky who edited the manuscript with her usual professionalism.

References

Alperson-Afil, N. (2008). Continual fire-making by hominins at Gesher Benot Ya'aqov, Israel. *Quaternary Science Reviews, 27*, 1733–1739.

Alperson-Afil, N., & Goren-Inbar, N. (2010). *The Acheulian site of Gesher Benot Ya'aqov: Ancient flames and controlled use of fire* (Vol. II). Dordrecht: Springer.

Alperson, N., Sharon, G., Zohar, I., Biton, R., Melamed, Y., Kislev, M. E., et al. (2009). Spatial organization of hominins activities at Gesher Benot Ya'aqov, Israel. *Science, 326*, 1677–1680.

Belitzky, S. (2002). The structure and morphotectonics of the Gesher Benot Ya'aqov area, northern dead sea rift, Israel. *Quaternary Research, 58*, 372–380.

Belitzky, S., Goren-Inbar, N., & Werker, E. (1991). A Middle Pleistocene wooden plank with man-made polish. *Journal of Human Evolution, 20*, 349–353.

Célérier, G., & Kervazo, B. (1988). Livret-guide pour la visite du gisement de Pont d'Ambon, Bourdeilles. In *Le peuplement magdalénien, Centenaire de la découverte de l'Homme de Chnncelade, Actes du Colloque International, Chancelade*, 10–14 October 1988.

Coolidge, F. L., & Wynn, T. (2009). *The Rise of Homo sapiens – The evolution of modern thinking*. Chichester, UK: Wiley-Blackwell.

de Beaune, S. A. (2000). *Pour une Archéologie du Geste*. Paris: Editions CNRS.

de la Torre, I. (2011). The Early Stone Age lithic assemblages of Gadeb (Ethiopia) and the Developed Oldowan/early Acheulean in East Africa. *Journal of Human Evolution, 60*, 768–812.

Feibel, C. S. (2001). Archaeological sediments in lake margin environments. In J. K. Stein & W. R. Farrand (Eds.), *Sediments in archaeological contexts* (pp. 127–148). Salt Lake City: University of Utah Press.

Feibel, C. S. (2004). Quaternary lake margins of the Levant Rift Valley. In N. Goren-Inbar & J. D. Speth (Eds.), *Human paleoecology in the Levantine corridor* (pp. 21–36). Oxford: Oxbow Books.

Goren-Inbar, N. (2011a). Behavioral and cultural origins of Neanderthals: A Levantine perspective. In S. Condemi & G.-C. Weniger (Eds.), *150 years of Neanderthal discoveries: Continuity and discontinuity* (pp. 89–100). Dordrecht: Springer.

Goren-Inbar, N. (2011b). Culture and cognition in the Acheulian industry – A case study from Gesher Benot Ya'aqov. *Philosophical Transactions of the Royal Society B, 366*, 1038–1049.

Goren-Inbar, N., & Sharon, G. (2006). Invisible handaxes and visible Acheulian biface technology at Gesher Benot Ya'aqov, Israel. In N. Goren-Inbar & G. Sharon (Eds.), *Axe age: Acheulian tool-making from quarry to discard* (pp. 111–135). London: Equinox.

Goren-Inbar, N., Lister, A., Werker, E., & Chech, M. (1994). A butchered elephant skull and associated artifacts from the Acheulian site of Gesher Benot Ya'aqov, *Israel. Paléorient, 20*(1), 99–112.

Goren-Inbar, N., Feibel, C. S., Verosub, K. L., Melamed, Y., Kislev, M. E., & Tchernov, E., et al. (2000). Pleistocene milestones on the out-of-Africa corridor at Gesher Benot Ya'aqov, Israel. *Science, 289*, 944–974.

Goren-Inbar, N., Sharon, G., Melamed, Y., & Kislev, M. (2002a). Nuts, nut cracking, and pitted stones at Gesher Benot Ya'aqov, Israel. *Proceedings of the National Academy of Sciences of the U.S.A, 99*, 2455–2460.

Goren-Inbar, N., Werker, E., & Feibel, C. S. (2002b). *The Acheulian site of Gesher Benot Ya'aqov, Israel: The wood assemblage* (Vol. I). Oxford: Oxbow Books.

Goren-Inbar, N., Alperson, N., Kislev, M. E., Simchoni, O., Melamed, Y., & Ben-Nun, A., et al. (2004). Evidence of hominin control of fire at Gesher Benot Ya'aqov, Israel. *Science, 304*, 725–727.

Goren-Inbar, N., Grosman, L., & Sharon, G. (2011). The record, technology and significance of the Acheulian giant cores of Gesher Benot Ya'aqov, Israel. *Journal of Archaeological Sciences, 38*, 1901–1917.

Kleindienst, M. R. (1962). Components of the East African Acheulian assemblage: An analytic approach. In G. Mortelmans & J. Nenquin (Eds.), *Actes du IVème Congrès Panafricain de Préhistoire et de l'Etude du Quaternaire* (pp. 81–111). Tervuren: Musée Royal de l'Afrique Centrale.

Krumbein, W. C. (1941). Measurement and geological significance of shape and roundness of sedimentary particles. *Journal of Sedimentary Research, 11*.

Lepre, C. J., Roche, H., Kent, D. V., Harmand, S., Quinn, R. L., & Brugal, J.-P., et al. (2011). An earlier origin for the Acheulian. *Nature, 477*, 82–85.

Mora, R., & de la Torre, I. (2005). Percussion tools in Olduvai Beds I and II (Tanzania): Implications for early human activities. *Journal of Anthropological Archaeology, 24*, 179–192.

Pelegrin, J. (2000). Les techniques de débitage laminaire au Tardiglaciaire: Critères de diagnose et quelques réflexions. In B. Valentin, P. Bodu, & M. Christensen (Eds.), *L'Europe Centrale et Septentrionale au Tardiglaciare* (Vol. 7, pp. 73–86). Paris: Mémoire du Musée de Préhistoire de l'Ile-de-France.

Rabinovich, R., & Biton, R. (2011). The Early-Middle Pleistocene faunal assemblages of Gesher Benot Ya'aqov – Taphonomy and paleoenvironment. *Journal of Human Evolution, 60*, 357–374.

Rabinovich, R., Gaudzinski, S., & Goren-Inbar, N. (2008). Systematic butchering of Fallow deer (*Dama*) at the Early Middle Pleistocene Acheulian site of Gesher Benot Ya'aqov, (Israel). *Journal of Human Evolution, 54*, 134–149.

Rabinovich, R., Gaudzinski-Windheuser, S., Kindler, L., & Goren-Inbar, N. (2012). *The Acheulian site of Gesher Benot Ya'aqov: Mammalian taphonomy – The assemblages of layers V-5 and V-6* (Vol. III). Dordrecht: Springer.

Roe, D. A. (2001). The Kalambo Falls large cutting tools: A comparative metrical and statistical analysis. In J. D. Clark (Ed.), *Kalambo falls prehistoric site* (pp. 429–599). Cambridge: Cambridge University Press.

Rufo, M. A., Minelli, A., & Peretto, C. (2009). The limestone industry of the Paleolithic site of Isernia La Pineta: An interpretative model of the behavioural strategies. *L'Anthropologie, 113*, 78–95.

Sharon, G. (2007). *Acheulian large flake industries: Technology, chronology, and significance*. BAR International Series 1701. Oxford: Archaeopress.

Sharon, G., & Goren-Inbar, N. (1999). Soft percussor use at the Gesher Benot Ya'aqov Acheulian site? *Mitekufat Haeven, 28*, 55–79.

Sharon, G., Alperson-Afil, N., & Goren-Inbar, N. (2011). Cultural conservatism against variability in the continual Acheulian sequence of Gesher Benot Ya'aqov, Israel. *Jounal of Human Evolution, 60*, 387–397.

Soressi, M., Rendu, W., Texier, J.-P., Émilie Claud, L. D., d'Errico, F., & Laroulandie, V., et al. (2011). Pech-de-l'Azé I (Dordogne, France): nouveau regard sur un gisement moustérien de tradition acheuléenne connu depuis le XIXe siècle. *Mémoire XLVII de la Société Préhistorique Française*, 95–132.

Spiro, B., Ashkenazi, S., Starinsky, A., & Katz, A. (2011). Strontium isotopes in *Melanopsis* sp. as indicators of variation in hydrology and climate in the Upper Jordan Valley during the Early-Middle Pleistocene, and wider implications. *Journal of Human Evolution, 60*, 407–416.

Zohar, I., & Biton, R. (2011). Land, lake, and fish: Investigation of fish remains from Gesher Benot Ya'aqov (paleo-lake Hula). *Journal of Human Evolution, 60*, 343–356.

Chapter 6
Technological Transformations Imply Cultural Transformations and Complex Cognition

Lyn Wadley

Abstract As archaeologists, we cannot access culture or cognition directly; we can only interpret levels of cultural or cognitive complexity from circumstantial evidence or from technological evidence. Some technologies cannot be achieved without complex cognition. Interpreting technological, cognitive and cultural complexity requires carefully constructed bridging theory between archaeologically-recovered data and interpretations about behavior and human capacity. Some of the technologies that imply complex cognition also involve permanent transformations of their ingredients. Such technologies imply that the artisans were capable of conceptualizing cultural transformations. Iron Age material culture items clearly demonstrate the link between technological and cultural transformations. Metal furnaces are symbolically linked to human fertility and motifs on a number of artifacts are reminiscent of human rites of passage. I suggest here that material culture and motifs in the deeper past may similarly connect to cultural transformations, even though it is unlikely that the meaning of these can be decoded.

Keywords Technology • Culture • Behavior • Symbolism • Bridging theory • Middle Stone Age • Iron Age • Rites of passage

Introduction

What is special about the culture of *Homo sapiens*? A simplistic answer is that modern humans are capable of using an entire suite of attributes that we interpret as cultural, for example, altruism, symbolic thought and burial of the dead. This seems not to be the case for animals; they are more likely to display one or two attributes that overlap with those associated with human cultural behavior. It is not my intention here to make a distinction between human and animal culture, but rather to explore some aspects of culture that can be inferred for people who were physically like us from about 100 ka. The potential for complex culture is inextricably linked to the level of cognitive complexity attained by people. For this reason my study of cultural transformations is linked to interpretations of cognitive development as well as to the concept of technological transformations. Archaeologists can never access culture or cognition directly; we can only interpret the level of cultural or cognitive complexity indirectly from circumstantial evidence. This might make the interpretation of cultural behavior seem speculative. Yet, interpreting technology or subsistence is no different, even though archaeologists often feel more comfortable about doing this. We do not excavate technology or subsistence directly; we infer a level of technology from artifacts and their contexts, and we infer subsistence strategies from artifacts and identified organic remains. Interpreting technological, cognitive and cultural complexity requires carefully constructed bridging theory between archaeologically-recovered data and interpretations of human behavior. Before we attribute abilities or intention to early human subjects, we need to be sure that they possessed the requisite mental architecture and cognitive capacity. When examining technology and inferring cognitive capacity from this, we need to ask whether certain tasks are reliant on particular cognitive capacities and could not be achieved without them.

I mention a few of the previously published viewpoints about cultural attributes that are supposed to overlap with those of people today, then I move to examine material culture and its behavioral correlates from a different perspective. The first of the published arguments, put forward by Davidson and Noble (1992) and others, is that the colonization of Australia implies exceptional planning and

L. Wadley (✉)
Evolutionary Studies Institute, University of the Witwatersrand, Johannesburg, PO WITS 2050, South Africa
e-mail: Lyn.Wadley@wits.ac.za

Miriam N. Haidle, Nicholas J. Conard and Michael Bolus (eds.), *The Nature of Culture: Based on an Interdisciplinary Symposium 'The Nature of Culture', Tübingen, Germany*, Vertebrate Paleobiology and Paleoanthropology, DOI 10.1007/978-94-017-7426-0_6

risk-taking of the kind associated with people like ourselves. This seems compelling evidence that, after about 60 ka, people had cognition similar to that of our own.

Perhaps one of the most advertised of the attributes thought to signal behavior like that of our own is symbolism. Many artifacts are considered to be symbolic; amongst the most convincing are those which may have functioned as personal ornaments, either stitched to clothing or strung and worn directly on the body. Perforated marine or estuarine shells, with wear patterns suggesting that they were strung, occur at widely spaced sites from the Near East to southern Africa. Those from the Near East are 100 ka (Bar-Yosef Mayer et al. 2009), while North African perforated shells are older than 80 ka (Bouzouggar et al. 2007), and the South African examples from Blombos (Fig. 6.1b) and Sibudu are ca. 72 ka (Henshilwood et al. 2004; d'Errico et al. 2005, 2008). The perforated shells from Blombos and Sibudu are from different genera, but they look similar. Some Neanderthal sites in Europe have also yielded putative beads. If, as many archaeologists believe, beads imply that their makers were expressing group or individual identity, then Neanderthals as well as modern humans practised symbolic behavior. The engraved ochre from Blombos (Fig. 6.1c) is frequently cited as further evidence for early symbolic material culture. The most celebrated of the pieces, with a crosshatched design between boundary lines, is between 78 and 72 ka, from the Still Bay Industry, but older pieces have ages of about 100 ka (Henshilwood et al. 2009). Klein Kliphuis, in the Western Cape, South Africa, also has a piece of engraved ochre; this one is from the Howiesons Poort Industry and is probably about 60 ka (Mackay and Welz 2008). The extraordinarily large collection of 270 engraved ostrich eggshell fragments from the Howiesons Poort Industry of Diepkloof Rock Shelter in the Western Cape (Fig. 6.1a) is certainly older than 60 ka (Texier et al. 2010). The ladderlike motifs on the majority of them seem

Fig. 6.1 Middle Stone Age ornaments from South Africa. **a** Engraved ostrich eggshell from Diepkloof; **b** perforated *Nassarius* shells from Blombos; **c** engraved ochre slab from Blombos. Images reproduced courtesy of P.-J. Texier (**a**) and C. Henshilwood (**b** and **c**)

convincing evidence for a cultural tradition in design, albeit that the traditions were more protracted than those in the modern sense. The eggshell is from water bottles used, like those in the Kalahari, for everyday subsistence tasks. At the very least, the appearance of perforated shells, and engraved ochre and eggshell, represent new behavior and significant cultural markers within the Middle Stone Age (MSA). We can reach this conclusion whether or not we interpret the artifacts as evidence for early symbolic activities.

Culture must extend beyond a kinship group, indeed material culture can be an accessory to forming and maintaining cultural groups outside of kin bonds. Material culture items not linked to subsistence are often exchanged as part of periodic group aggregations in historically recorded hunter-gatherer societies (Wadley 1987). There are many advantages to the maintenance of large groups even if the aggregations are brief and last only a few weeks. The subsistence benefits of wide-ranging social links are obvious in environments where resources are unevenly distributed seasonally. The social benefits are even greater. This applies not just to obtaining mates and for carrying out ritual requiring large groups, but also for an exchange of information. As Wilson and others point out, cooperation is most beneficial for tasks where the difficulty exceeds the capacity of individuals. Wilson and colleagues tested this prediction using the game of Twenty Questions and found that groups outperformed individuals, especially when task difficulty increased (Wilson et al. 2004). The average of many individuals' estimates can be surprisingly close to the truth, even when their separate values lie remarkably far from it. This phenomenon can be called the "wisdom of crowd effect" (Lorenz et al. 2011). Under the right circumstances one can expect that innovation would be part of such interaction because social need is sometimes the stimulus for innovation.

We do not yet know when in the past aggregations and dispersals became part of hunter-gatherer fluid band membership. Nor do we know when material culture became an important facilitator of this process. Material culture may have been used as an accessory to forming cultural groups outside of kin bonds. This could be the anticipated role of items such as shell beads and engraved ochre or ostrich eggshell. The study is worth pursuing, but here I concentrate on other aspects of culture.

Some technological strategies that we infer from MSA data seem to have potential to inform us about the cognitive abilities achieved by people in the past. One of these is snaring, which requires complex executive functions of the brain. Making a snare obliges a person to hold in mind actions that are out-of-sight and to strategize accordingly. Several new social circumstances are enabled through the invention of the snare, which can be set and tended by people of all ages. Snares empower women as well as young and old band members and snares are easy for individuals (as opposed to groups) to create and manage. Snares may have been used at Sibudu in the Howiesons Poort, ~65 ka (Wadley 2010a), because the fauna is dominated by tiny blue duiker that dwell in forests (Clark and Plug 2008). In addition, the Howiesons Poort layers alone contain the remains of many small carnivores that are susceptible to being caught in snares as opposed to nets. What we do presently know is whether technology, such as snares, was driven by economic or cultural needs; either is possible.

I suggest that another novel way of looking for ancient links between cognition and culture is to search for archaeological evidence of deliberate transformations.

Technological Transformations

Technological transformations are a good place to start because they are likely to signal capacity for the concepts of cultural transformations that I shall examine shortly. I need to explain the connection. Transformative technologies bring together disparate raw materials and change them for all time. This process requires executive functions of the brain to facilitate goal-directed actions, anticipation of problems, analogical reasoning, and planning over long distances or time (Coolidge and Wynn 2001). Such technologies are different from those that use technical expertise acquired through apprenticeship (Wynn and Coolidge 2007). Alloying metals and the production of fired ceramics are examples of the type of innovative and transforming technologies requiring enhanced executive functions (Wynn and Coolidge 2007). An older example of transformative technology is the deliberate heat-treatment of rock to improve its qualities for stone tool knapping through changing its crystalline structure (Brown et al. 2009; Mourre et al. 2010). The manufacture of compound adhesives in the past is another way in which multi-faceted technology requiring multi-tasking can provide a clue to complex cognition possessed by artisans in the past. A variety of compound adhesive recipes was used in Sibudu Cave by at least 70 ka and at Rose Cottage in South Africa by about 65 ka (Wadley et al. 2009). Neanderthals used bitumen as adhesive in Syria (Boëda et al. 2008), and early glues are also known in Europe (Mazza et al. 2006) where it is possible that the adhesive-making process was also complicated, involving the use of compound ingredients. Some birds weave intricate nests, and others even make adhesives, but these actions are instinctive, and they use simply coded operational sequences in which, as Haidle (2010) would explain, the distance between problem and solution is far smaller than that demonstrated by the human action of making glues for composite hunting weapons.

Replication of compound adhesives (Wadley et al. 2009) demonstrates the complexity of the process. Making reliable adhesive involves a calculated manipulation of disparate ingredients such as plant gum and iron oxide ground from naturally occurring nodules. After lithic inserts are attached to their hafts with moist adhesive, the composite tools must be dried near a fire, using controlled heat and vigilance to prevent burning or leaking adhesive. The plasticity of plant gum, the aggregate-size of ground iron oxides and the heat of fires are inconsistent natural features that require a slightly different recipe and processing procedure for each adhesive manufacturing event. Intense concentration and mental flexibility are a necessary part of the process. Making compound adhesives is a complex task that seems to require modern working memory capacity (Wadley 2010b). Thus, the making of compound adhesives (as opposed to simple glues) provides circumstantial evidence for complex cognition. Some compound adhesives irreversibly transform their ingredients which cannot subsequently be separated into their original, individual components. I argue that people who practised such technical transformations are likely also to have understood the concepts of cultural transformation. I return to this issue later.

Effective glue allows flexible placement of lithic inserts and this is one of the main reasons for using glue (Rots 2002). At Sibudu, Lombard (2007, 2011) has shown that Howiesons Poort segments were rotated, for example, transversely, longitudinally or diagonally – resulting in a variety of composite tools designed for disparate tasks. Mental rotation of a single shape to achieve worthwhile end-products is a different process from rotating cores during knapping. It involves using a suite of similarly shaped segments to create a diverse package of composite tools, each having a separate purpose. It involves a complex span task of the kind that psychologists use today to measure Working Memory Capacity (Kane et al. 2004), that is, the ability to control attention to keep relevant matter in active memory or to enable relevant matter to be easily retrieved from inactive memory.

Cultural Transformations

Transformation as a cultural theme crosscuts gender, ethnicity and place. Only its time depth is uncertain and, as archaeologists, we must resolve this issue because dealing with time is our speciality. Cultural transformation irreversibly shapes identity. The transformations from girl to woman and boy to man are permanent and irreversible. Such changes are often socially marked by rites of passage that invoke symbolism to reinforce people's altered status. It is symbolically appropriate that permanent alterations are made to the initiate's body (though this does not happen in all societies today); these may include tattoos, circumcision, finger joint amputation or dental modification (Mitchell and Plug 1997). In order to provide some appropriate examples of technical and social transformations, I first explore evidence from the recent past because here there are some clear links between material culture and social transformations. However, I am not advocating the use of ethnography to interpret behavior in the deep past.

Technical and social transformations can be metaphors for each other. In Bantu-speaking Africa, iron-using probably began somewhere near Lake Victoria (Phillipson 1985:171) and the use of iron was widespread by the first millennium. Iron smelting is a good example of irreversibly transforming a natural product. Not unexpectedly, the process became an analogy for life and some stages of metal production in particular mirror behavior during human transformation ceremonies. Iron-smelting is generally conducted in seclusion (Collett 1993), just as initiations are carried out at some distance from a village and smelting is ritually performed by men. The symbolism associated with furnaces concerns fertility (Collett 1993). Some African iron smelting furnaces, which are moulded clay, have modelled clay breasts, vaginal openings, testicle-like bellows and penis-like blow pipes (Schmidt 2009). Furnaces become human bodies and reproductive systems. Shona furnaces in Zimbabwe replicate women's bodies, even mimicking their bodily scarifications used to mark adulthood (Collett 1993). The furnaces are sometimes decorated with clay replicas of protective, beaded waistbands that women wear during childbirth (Schmidt 2009). The same motifs appear on items such as ceramics, walls of houses, beaded clothing, and wooden headrests.

As in the case of iron production, ceramic manufacture transforms natural products: clay, tempering materials and water are mixed, kneaded and changed irreversibly through firing. The motifs on ceramics may communicate their user's gender, age and status (Armstrong et al. 2008). Southeast Bantu-speakers conceptually link women's reproductive capability, fecundity of the land, ceramic manufacture and reproduction (Boeyens et al. 2009). Consequently, pots can be metaphors for the transformations in women's lives (Whitelaw 1993). In Zimbabwe, pots used by Karanga-speakers symbolize wombs, and an unfired pot is appropriately a simile for a girl who has not reached puberty (Whitelaw 1993).

The elaborately decorated ceramic human heads from Lydenburg, South Africa (Inskeep and Maggs 1975), probably dating to the ninth or tenth century, have been linked to rites of passage and they may have been instructive models. Markings on these heads seem to advertise adulthood: four of them look as if they have dental mutilations, while facial markings resemble adult scarifications. During initiation ceremonies for Bantu-speakers, several icons were used for male and female instruction (Prins and Hall 1994). Ceramic figurines used during initiation lessons are commonly found in excavated courtyards. Often these are broken, presumably intentionally, as a sign of people's life changes. At Schroda, the tenth century Zhizo capital in northern Limpopo Province,

Fig. 6.2 A broken female figurine from the Iron Age settlement at Schroda, South Africa. Note the phalliform head, the emphasized navel, and the bodily scarification. Reproduced with the permission of J. A. Van Schalkwyk and the Ditsong Museum, Pretoria, South Africa

South Africa, approximately 2000 figurine fragments were recovered (Van Schalkwyk 2002:71), some with obvious genitalia and sometimes marked with lines that seem to be scarification. Part of one Schroda figurine is illustrated here (Fig. 6.2). Venda mothers in South Africa traditionally gave clay figurines like these to their daughters. On marriage, these figurines of women would be taken by the young wife to her new home. It was said that the head of the clay figurine (and by implication the girl's spirit) belonged to her father who conceived her and this explains the phallic form of the figurine's head (Wood 2002:84, 90).

It would clearly be ill-advised to use such African ethnography to interpret behaviour either in the MSA or in the European Paleolithic. Yet I cannot resist the suggestion that some form of cultural transformation may also be depicted in selected examples of the ancient material culture. There are some similarities, for instance, between the Schroda figurines and some of the 'Venus' figurines from Europe, even though the differences in age and geography make the comparison imprudent and inappropriate. Though it would be naive to interpret both types of figurine in the same way, the ivory figurine from Hohle Fels, Germany, which is possibly 35 ka (Conard 2009), has some features that are tantalizingly like those from African initiation figurines. In the first instance, none of the figurines is truly representational; the exaggerated form of the figures deliberately conveys a cultural message. The emphasis is on breasts and genitalia. The unrealistically large and upstanding breasts on the Hohle Fels figurine are intended to be those of a young, fertile woman. Since the figure appears to be naked, the markings below the breasts might be tattoos of the kind that symbolically mark adulthood. Finally, the vestigial head and limbs leave the viewer in no doubt that his or her gaze must be directed to the young woman's reproductive capability. The lack of facial features suggests, moreover, that the artist had no intention of depicting an individual woman, but rather fertile womankind. Many animal figurines from European Paleolithic sites also bear incised geometric designs and these markings need to be part of a separate study. The ca. 40 ka 'Lion man' figurine from Hohlenstein-Stadel, Germany (Kind et al. 2014) (see Nowell 2016: Fig. 9.1), has parallel lines marking his upper arms, and these suggest that scarification may have marked men as well as women (I make an assumption here because the part-human, part-animal figure is neither clearly male nor female). Another important feature embodied in the figurine from Hohlenstein-Stadel is the evidence it provides that people of the time were able to abstract salient attributes of humans and animals and then conceive of transformations of humans and/or animals into 'fantasy creatures' comprising part animal and part human. Abstract concepts and executive functions may have appeared only late on the evolutionary scene (Wynn et al. 2009). Executive functions are the behavioral manifestation of enhanced working memory and the Hohlenstein-Stadel 'Lion-man' is convincing evidence for modern executive functions (Wynn et al. 2009). In Namibia, the painted slab from Apollo 11, which may date to ca. 28 ka, also contains a representation of a creature that is part feline, part human (Wendt 1976), so the concept is pan-continental.

Farther back in time there is a greater level of interpretive difficulty. In southern Africa, geometric designs were applied, not to human representations as in the recent examples described, but to ostrich eggshell water bottles, pieces of ground ochre or to pebbles. The engraved ostrich eggshell water bottles from Diepkloof Rock Shelter that are ca. 60–65 ka are of special interest. The motifs occurring on close to 300 eggshell fragments seem to come from standardized templates. Ladder-like designs are most common, though sub-parallel lines, curved lines and crosshatching also occur (Texier et al. 2010). The water bottle designs were repeated; they were intentional and not accidental. Water bottles are functional items today and probably were in the past, too, though functionality need not exclude other kinds of meaning. Today, in the Kalahari, ostrich eggshell water bottles are owned by married women, and while we have no means of knowing what their ownership was in the past, the shell containers are easily broken and are therefore likely to have been made and used by adults. As with pots in Iron Age societies, the use of shell water bottles in the domestic sphere does not necessarily exclude them from bearing symbolism or cultural meaning. Since the engraved designs do not improve the functionality of the bottles, a cultural reason for the engraving must therefore be sought. Decorated eggshell water bottles seem to have been common, at least in the Diepkloof area before 60 ka, so the Texier et al. (2010) interpretation of a symbolic, cultural tradition lasting several thousand years is entirely reasonable. Since the ladder-like motif is particularly common at Diepkloof it seems unlikely that it, or the other designs, were intended to denote private ownership of bottles.

It is similarly tempting to interpret the cross-hatched incisions on ochre from Blombos (Henshilwood et al. 2009) and on pebbles from other MSA sites in South Africa, like Klein Kliphuis (Mackay and Welz 2008), and Palmenhorst/Rössing, in Namibia (Wendt 1975), as designs of the kind that mark some form of human social transformation. The most spectacular of the Blombos ochre engravings, mentioned briefly earlier, is from the M1 phase (78 and 72 ka) and it depicts a smoothly ground ochre plaque with cross-hatched incisions that lie between sub-parallel boundary lines (Henshilwood et al. 2009) (Fig. 6.1c). It is the only one of its kind at Blombos, even though several other ochre pieces have motifs or fragments of motifs on them, even in occupations with ages of approximately 100 ka. Henshilwood and d'Errico (2011) suggest that the symbolic codes on the Blombos engraved ochre are 'almost secret'. The rarity of engraved ochre and pebbles suggests that this expression of symbolism was different from the seemingly 'more public' display of symbols on engraved eggshell water bottles. We cannot decode the meanings embodied in the ancient incisions on shell or stone, but we are able to place these symbolic expressions firmly within the context of the technological transformations that were practised at the time – deliberate heating of stone or ochre to improve their desired qualities, and the elaborate use of pyrotechnology for a variety of purposes, including the manufacture of compound adhesives.

Closing Remarks on the Nature of *Homo sapiens* Culture

Homo sapiens culture, unlike that of animals, is multifaceted, multi-component and flexible. Ancient *Homo sapiens* culture must be inferred from material culture, with the use of bridging theory as the link between it and archaeological interpretations. From a theoretical viewpoint, the ability to effect permanent transformations on products from nature is symptomatic of complex cognition and it provides circumstantial evidence that people were simultaneously capable of thinking about transformations of all kinds, including social ones. Cultural transformations expressed through material culture may have included rites of passage to mark the achievement of adulthood or entry into a spirit world. Such rites are most likely to have taken place during group aggregations when many people were present, and the possibility of group fluidity in the deep past needs exploration. Material culture items that appear alongside or as part of the first evidence for transformative technologies may incorporate an increasing desire by people to control and mark, not only the natural world, but also human life cycles. In cultural contexts, transformative events, such as rites of passage, irreversibly shape people's identities. Transformed material culture provides suitable analogies for such human behavior.

Acknowledgements I acknowledge the financial support of the National Research Foundation (NRF). The opinions expressed in this paper are not necessarily those of the NRF. I also acknowledge the ongoing support of the University of the Witwatersrand. I thank Miriam Haidle for inviting me to Tübingen to take part in the stimulating "Nature of Culture" conference. I thank all the conference organizers for their outstanding kindness and hospitality.

References

Armstrong, J., Whitelaw, G., & Reusch, D. (2008). Pots that Talk, *Izinkamba Ezikhulumayo*. *Southern African Humanities, 20*, 513–548.

Bar-Yosef Mayer, D., Vandermeersch, B., & Bar-Yosef, O. (2009). Shells and ochre in Middle Paleolithic Qafzeh cave, Israel: Indications for modern behavior. *Journal of Human Evolution, 56*, 307–314.

Boëda, É., Bonilauri, S., Connan, J., Jarvie, D., Mercier, N., Tobey, M., et al. (2008). New evidence for significant use of bitumen in Middle Palaeolithic technical systems at Umm El Tlel (Syria) around 70,000 BP. *Paléorient, 34*, 67–83.

Bouzouggar, A., Barton, N., Vanhaeren, M., d'Errico, F., Collcutt, S., Higham, T., et al. (2007). 82,000-year-old shell beads from North Africa and implications for the origins of modern human behavior. *Proceedings of the National Academy of Sciences of the USA, 104*, 9964–9969.

Boeyens, J., van der Ryst, M., Coetzee, F., Steyn, M., & Loots, M. (2009). From uterus to jar: The significance of an infant pot burial from Melora Saddle, an early nineteenth-century African farmer site on the Waterberg Plateau. *Southern African Humanities, 21*, 213–238.

Brown, K. S., Marean, C. W., Herries, A. I. R., Jacobs, Z., Tribolo, C., Braun, D., et al. (2009). Fire as an engineering tool of early modern humans. *Science, 325*, 859–862.

Clark, J. L., & Plug, I. (2008). Animal exploitation strategies during the South African Middle Stone Age: Howiesons Poort and post-Howiesons Poort fauna from Sibudu Cave. *Journal of Human Evolution, 54*, 886–898.

Collett, D. P. (1993). Metaphors and representations associated with precolonial iron-smelting in eastern and southern Africa. In T. Shaw, P. Sinclair, B. Andah, & A. Okpoko (Eds.), *The Archaeology of Africa: Food, metals and towns* (pp. 499–511). London: Routledge.

Conard, N. (2009). A female figurine from the basal Aurignacian of Hohle Fels in southwestern Germany. *Nature, 459*, 248–252.

Coolidge, F. L., & Wynn, T. (2001). Executive functions of the frontal lobes and the evolutionary ascendancy of *Homo sapiens*. *Cambridge Archaeological Journal, 11*, 255–260.

Davidson, I., & Noble, W. (1992). Why the first colonization of the Australian region is the earliest evidence of modern human behavior. *Archaeology in Oceania, 27*, 135–142.

d'Errico, F., Henshilwood, C., Vanhaeren, M., & van Niekerk, K. (2005). *Nassarius kraussianus* shell beads from Blombos Cave: Evidence for symbolic behaviour in the Middle Stone Age. *Journal of Human Evolution, 48*, 2–14.

d'Errico, F., Vanhaeren, M., & Wadley, L. (2008). Possible shell beads from the Middle Stone Age layers of Sibudu Cave, South Africa. *Journal of Archaeological Science, 35*, 2675–2685.

Haidle, M. N. (2010). Working-memory capacity and the evolution of modern cognitive potential: Implications from animal and early human tool use. *Current Anthropology, 51*, S149–S166.

Henshilwood, C. S., & d'Errico, F. (2011). Middle Stone Age engravings and their significance to the debate on the emergence of symbolic material culture. In C. S. Henshilwood & F. d'Errico (Eds.), *Homo symbolicus: The dawn of language, imagination and spirituality* (pp. 75–96). Amsterdam: John Benjamins.

Henshilwood, C. S., d'Errico, F., & Vanhaeren, M. (2004). Middle Stone Age shell beads from South Africa. *Science, 304*, 404.

Henshilwood, C. S., d'Errico, F., & Watts, I. (2009). Engraved ochres from the Middle Stone Age levels at Blombos Cave, South Africa. *Journal of Human Evolution, 57*, 27–47.

Inskeep, R. R., & Maggs, T. M. O'C. (1975). Unique art objects in the Iron Age of the Transvaal, South Africa. *South African Archaeological Bulletin, 30*, 114–138.

Kane, M. J., Hambrick, D. Z., Tuholski, S. W., Wilhelm, O., Payne, T. W., & Engle, R. W. (2004). The generality of working memory capacity: A latent-variable approach to verbal and visuospatial memory span and reasoning. *Journal of Experimental Psychology: General, 133*, 189–217.

Kind, C.-J., Ebinger-Rist, N., Wolf, S., Beutelspacher, T., & Wehrberger, K. (2014). The smile of the lion man. Recent excavations in Stadel Cave (Baden-Württemberg, south-western Germany) and the restoration of the famous Upper Palaeolithic figurine. *Quartär, 61*, 129–145.

Lombard, M. (2007). The gripping nature of ochre: The association of ochre with Howiesons Poort adhesives and Later Stone Age mastics from South Africa. *Journal of Human Evolution, 53*, 406–419.

Lombard, M. (2011). Quartz-tipped arrows older than 60 ka: Further use-trace evidence from Sibudu, KwaZulu-Natal, South Africa. *Journal of Archaeological Science, 38*, 1918–1930.

Lorenz, J., Rauhut, H., Schweitzer, F., & Helbing, D. (2011). How social influence can undermine the wisdom of crowd effect. *Proceedings of the National Academy of Sciences of the USA, 108*, 9020–9025.

Mackay, A., & Welz, A. (2008). Engraved ochre from a Middle Stone Age context at Klein Kliphuis in Western Cape of South Africa. *Journal of Archaeological Science, 35*, 1521–1532.

Mazza, P. P. A., Martini, F., Sala, B., Magi, M., Colombini, M. P., Giachi, G., et al. (2006). A new Palaeolithic discovery: Tar-hafted stone tools in a European Mid-Pleistocene bone-bearing bed. *Journal of Archaeological Science, 33*, 1310–1318.

Mitchell, P. J., & Plug, I. (1997). Ritual mutilation in Southern Africa: Gender and ethnic identities and the possibilities of archaeological recognition. In L. Wadley (Ed.), *Our gendered past: Archaeological studies of gender in Southern Africa* (pp. 135–166). Johannesburg: Witwatersrand University Press.

Mourre, V., Villa, P., & Henshilwood, C. S. (2010). Early use of pressure flaking on lithic artifacts at Blombos Cave, South Africa. *Science, 330*, 659–662.

Nowell, A. (2016). Childhood, play and the evolution of cultural capacity in Neanderthals and modern humans. In M. N. Haidle, N. J. Conard, & M. Bolus (Eds.), *The nature of culture* (pp. 87–97). Dordrecht: Springer.

Phillipson, D. W. (1985). *African archaeology* (1st ed.). Cambridge: Cambridge University Press.

Prins, F., & Hall, S. (1994). Expressions of fertility in the rock art of Bantu-speaking agriculturists. *African Archaeological Review, 12*, 171–203.

Rots, V. (2002). *Hafting traces on flint tools: Possibilities and limitations of macro- and microscopic approaches*. Ph.D. dissertation, Katholieke Universiteit Leuven.

Schmidt, P. R. (2009). Tropes, materiality, and ritual embodiment of African iron smelting furnaces as human figures. *Journal of Archaeological Method and Theory, 16*, 262–282.

Texier, P.-J., Porraz, G., Parkington, J., Rigaud, J.-P., Poggenpoel, C., Miller, C., et al. (2010). A Howiesons Poort tradition of engraving ostrich eggshell containers dated to 60,000 years ago at Diepkloof Rock Shelter, South Africa. *Proceedings of the National Academy of Sciences of the USA, 107*, 6180–6185.

Van Schalkwyk, J. A. (2002). Metaphors and meanings: Conceptualising the Schroda clay figurines. In J. A. Van Schalkwyk & E. O. M. Hanisch (Eds.), *Sculptured in clay: Iron Age figurines from Schroda, Limpopo Province, South Africa* (pp. 69–79). Pretoria: National Cultural History Museum.

Wadley, L. (1987). *Hunters and gatherers of the Later Stone Age, Southern Transvaal.* Oxford: British Archaeological Reports International Series 380.

Wadley, L. (2010a). Were snares and traps used in the Middle Stone Age and does it matter? A review and a case study from Sibudu, South Africa. *Journal of Human Evolution, 58*, 179–192.

Wadley, L. (2010b). Compound-adhesive manufacture as a behavioral proxy for complex cognition in the Middle Stone Age. *Current Anthropology, 51*, S111–S119.

Wadley, L., Hodgskiss, T., & Grant, M. (2009). Implications for complex cognition from the hafting of tools with compound adhesives in the Middle Stone Age, South Africa. *Proceedings of the National Academy of Sciences of the USA, 106*, 9590–9594.

Wendt, W. E. (1975). Notes on some unusual artifacts from south West Africa. *Cimbebasia Ser. B, 2*(6), 179–186.

Wendt, W. E. (1976). "Art mobilier" from the Apollo 11 Cave, South West Africa: Africa's oldest dated works of art. *South African Archaeological Bulletin, 31*, 5–11.

Whitelaw, G. (1993). Customs and settlement patterns in the first millennium AD: Evidence from Nanda, an Early Iron Age site in the Mngeni Valley, Natal. *Natal Museum Journal of Humanities, 5*, 47–81.

Wilson, D. S., Timmel, J. J., & Miller, R. R. (2004). Cognitive cooperation: when the going gets tough, think as a group. *Human Nature, 15*, 225–250.

Wood, M. (2002). Poupée de Fertilité: an interview revisited. In J. A. Van Schalkwyk & E. O. M. Hanisch (Eds.), *Sculptured in clay: Iron Age figurines from Schroda, Limpopo Province, South Africa* (pp. 81–93). Pretoria: National Cultural History Museum.

Wynn, T., & Coolidge, F. L. (2007). A Stone Age meeting of minds. *American Scientist, 96*, 44–51.

Wynn, T., Coolidge, F., & Bright, M. (2009). Hohlenstein-stadel and the evolution of human conceptual thought. *Cambridge Archaeological Journal, 19*, 73–83.

Chapter 7
Neanderthal Utilitarian Equipment and Group Identity: The Social Context of Bifacial Tool Manufacture and Use

Thorsten Uthmeier

Abstract Since personal ornamentation is quite rare prior to the onset of the Upper Paleolithic and, speaking in quantitative terms, lacks the widespread occurrence that would allow for conclusions regarding standardization, this study focuses on stone artifacts as possible social markers. The article explores the role of bifaces as signals for social identity and, at the same time, tries to take into account the temporal and spatial dynamics of operational chains. As a case study, the two main complexes of the European Late Middle Paleolithic with bifaces – the Mousterian of Acheulean Tradition (MtA) and the Micoquian – are investigated. In general, it is assumed that the contemporaneity of these industries, combined with similar environments and land use patterns, reduces the influence of functional factors. Bifaces are identified as spatially and chronologically stable elements, while concepts of core reduction vary. The importance of bifaces in MtA and Micoquian lithic systems is explained by their potential for resharpening. In both the MtA and the Micoquian, bifaces are means providing partial independence from raw material sources. Qualitative comparisons of the operational chains show marked differences, especially in advanced stages of resharpening. As surrogates of their respective operational chains, bifacial tools are considered social makers. The entire operational chain is seen as reducing social insecurity by materially reinforcing intimate social ties in regular face-to-face-contacts, whereas the tools alone signal social identity in contexts of less frequent interaction with socially distant individuals or even random contact with members of other collectives.

T. Uthmeier (✉)
Institut Für Ur- und Frühgeschichte,
Friedrich-Alexander-Universität Erlangen-Nürnberg, Kochstraße 4/18, 91054 Erlangen, Germany
e-mail: thorsten.uthmeier@ufg.phil.uni-erlangen.de

Keywords European Late Middle Paleolithic • Mousterian of Acheulean Tradition • Micoquian • Handaxes • *Keilmesser* • Resharpening • Symbolic interaction

Introduction

In the ongoing discussion of major changes in the behaviors of Paleolithic humans, extrasomatic storage of information is of central importance (Mellars 1973, 1996; McBearty and Brooks 2000; Bar-Yosef 2002; cf. Gamble 2007:37). Objects transmitting information dissociated from their immediate material properties are referred to as symbols, their meaning being understood by both the sender and the recipient though shared norms, values and a common code (Chase 1991:195). These are "social makers" (Chase 1991:197), and one may add that the repeated transcription of social relations into different, yet durable and visible materials both creates and maintains social ties in the memories of the individuals that make reasonable use of them (for the theory of transcriptions see Jäger 2002, 2003). In contrast to intimate face-to-face relations, secure decoding in extended networks requires uniform or at least similar repetition (Chase and Dibble 1987:44). Since the communicative function of objects is unfolded through its embedment into social networks, the scale of uniformity (or standardization) depends on the size of the network (e.g., Gamble 1999: Table 2.8, 2007: Table 5.5). Several isolated, often not artificially altered objects from the early to late Middle Paleolithic (e.g., Chase and Dibble 1987) include fossilized marine animals and plant remains found at considerable distances from their original source among the lithic assemblages. As each is unique, these may well be understood as having a symbolic meaning (about the depth of time, or, more trivial, the memory of a certain situation: Leroi-Gourhan 1984:451–453), but only for the individual. Due to their diversity, they lack a function as medium to communicate. Since their discovery, personal

Miriam N. Haidle, Nicholas J. Conard and Michael Bolus (eds.), *The Nature of Culture: Based on an Interdisciplinary Symposium 'The Nature of Culture', Tübingen, Germany*,
Vertebrate Paleobiology and Paleoanthropology, DOI 10.1007/978-94-017-7426-0_7

ornaments found in the Grotte du Renne at Arcy-sur-Cure suggest that Neanderthals were indeed the first to use symbols in the strict sense (for a summary see Caron et al. 2011). Many hypotheses have been proposed to explain this phenomenon, ranging from independent – and primal – Neanderthal innovation to acculturation and/or exchange after initial contact with anatomically modern humans. While the case of Grotte du Renne is complicated by both the contemporaneous presence of modern humans in Europe and discussions about the integrity of site formation processes (Higham et al. 2010; Mellars 2010; but see Caron et al. 2011 for a reply), further data for independent Neanderthal use of symbols comes from Spain. At two cave sites which only contained Middle Paleolithic layers, Cueva de los Aviones and Cueva Antón (Zilhao et al. 2010), marine shells were collected approximately 50 to 45 ka from the then distant seashores. However, these objects are nearer to fossils collected than symbolic artifacts because holes in them could not be proven to be artificial perforations and the collection is quite diverse in species and size. Perhaps the diversity mirrors differences in meaning on the level of intimate networks, but the lack of standardization hinders a function as commonly understood symbols.

The Social Context of Artifact Manufacture

Middle Paleolithic objects discussed as "non-utilitarian" artifacts (e.g., Bednarik 1992:31–32) are rare. From a qualitative point of view, these are restricted to "personal ornaments" or "suspended objects of adornment" (Álvarez Fernández and Jöris 2007:32). In general, such pendants are "final objects" (in the sense of Weißmüller 1995:20): after manufacture, they are only rarely repaired, and often the operational chains result in one or a limited number of objects. When attached to the body or clothes, information stored in pendants like those from the Grotte du Renne or the Cuevas Antón and Aviones is ascribed to the individual. Pendants therefore also refer to the social position, status and prestige of their owners. If so, to receive one would almost certainly be combined with some sort of property rights, whether of the pendant itself or access to some kind of social value. Such objects may have been exchanged, but I expect that this would be more the exception than the rule. On the contrary, Middle Paleolithic "utilitarian" objects, which in the preserved material culture with few exceptions are lithics, are "transient objects" (Weißmüller 1995:20). These result from a process of manufacture and resharpening, which produces not one standardized form, but a variety of forms depending on the intensity of use and resharpening. The discard of one and the same reduction sequence may or may not be dispersed in time and distributed to different places or even sites (e.g., Uthmeier 2004a). Although normally conducted by individuals, the material outcome of lithic manufacture and use is likely to be shared between group members, even more so as not only totally exhausted items, but also those with potential for further use are to be found in the artifact scatters of a site. Apart from the marked differences in quantitative aspects during the Middle Paleolithic, it is generally questionable if the distinction between "non-utilitarian" artifacts as objects loaded with symbolic meaning and "utilitarian" artifacts as objects manufactured and used without much reflection is a meaningful one. Following W. Weißmüller (2003:173: "*Zeichencharakter*"), A. M. Byers (1994:370) or, more recently, C. Gamble (2007:67–69, 162–163), I would instead argue that utilitarian objects may also endow social identity (much like the "passive style" suggested by Sackett 1982, 1986; cf. Chase 1991:197; for an overview of an interpretation of artifacts as symbols, see also Porr 1998; Hahn 2003; Kienlin 2005). Things used day-to-day and classified as functional equipment are interwoven with knowledge and gestures of production and use. Like "habits" (Weber 1964:21; cf. Weede 1992:54), the use of equipment becomes second-nature and the interdependent cultural relation between them and the user itself is no longer conscious. Often, the cultural imprint only becomes apparent when the user is directly confronted with equipment from other cultural contexts (Mauss 1979:100; cf. Apel 2001:23). To conclude, "utilitarian artifacts" may well function as objects that confer cultural identity, even more so as these items are (in contrast to "non-utilitarian" body ornaments) potentially shared after primal manufacture among group members. While W. Weißmüller (2003) has suggested that this only accounts for "final objects" (e.g., tanged projectiles), it is argued here that curated "transient objects" – despite their variable form – also store information beyond their primary properties. Middle Paleolithic tool manufacture and use is generally oriented towards such reduction sequences. However, processes of resharpening alter lithic tools not accidentally, but in predetermined ways (for scrapers: Dibble 1988; for bifaces: Richter 1997:203–206 and Fig. 163). Therefore, it is assumed that transformations of lithic raw materials, whether the reduction of cores or the resharpening of modified pieces, are equally suitable for the creation and reproduction of group identity. Due to the transient nature of objects, the unit of lithic analysis in this regard cannot be specific static objects, but must be the operational chain (Boëda et al. 1990; see also Porr 2005:72). It is this unit, and not the single artifact, that has stored the overall technological and practical knowledge. Equally, it is the operational chain that is repeatedly conducted in public, admittedly by individuals, but visible to every member of the group, both the flaking activity itself and the products produced. As cultural information is transcribed to the lithic products, informed group members at least (but to a certain degree also members of other groups) must have been able to draw conclusions about their social relations with the makers by observing the artifacts, depending on the degree of familiarity with the production process. In the words of W. Weißmüller (2003:181, my translation), sites become "technological libraries". This is where cultural identity comes into play. If Neanderthals were more driven by rationality than by instinct, then they were also confronted with emotional and social insecurity based on subjective expectations and evaluation of the behavior of the others. George Herbert Mead (1934) has described this as the gap in between one's own view of the ego, which he named the "I", and the way ego is looked upon by the others, termed "me" (Mead 1934; for a brief summary see Mikl-Horke 2001:192–199; Münch 2002a:265–290). Insecurity results from the possibility that these two views are not congruent. "Symbolic interactionism" (Blumer 1937, 1969; cf. Münch 2002b:257–282) is a social means to minimize situational insecurity not only by oral communication, but also by the use of objects. Apart from facial expression, gestures, clothes and other habits, identical or similar manufacture of lithics is another way to confirm that all group members share the same worldview.

For what kind of social groups is this kind of interaction relevant (Fig. 7.1)? Symbolic interaction is essential for the socialization of individuals and the maintenance of "intimate social networks" (Gamble 1999: Table 2.8). Whereas the notion that tight social bonds are connected to insecurity may be, at first glance, surprising, it is common sense that the nature of social interactions between socially distant individuals is difficult to predict. The social unit of rare or even accidental face-to-face contacts would be "extended networks" of 100 to 400 actors, and "global networks" with 2,500 individuals (network classification taken from Gamble

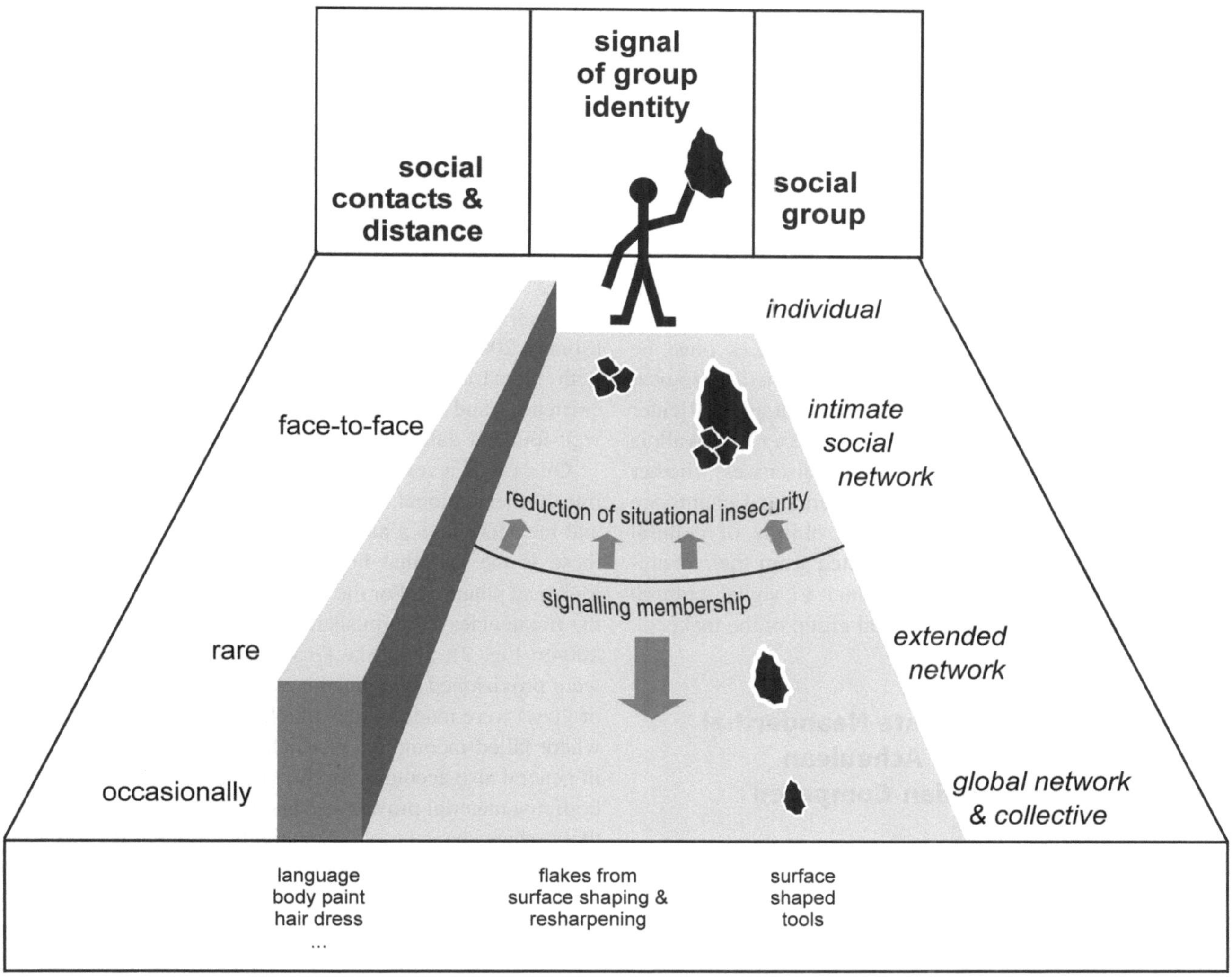

Fig. 7.1 The social context of bifacial tool manufacture and use. Schematic explanation of the key terminology used. Different sizes of signals of group identity reflect their expected importance in different social contexts

1999: Table 2.8). To operationalize global networks, one may prefer the definition of technological "collectives" (Uthmeier 2004b:16; see also Weißmüller 2003:178 for a distinction between "individual" and "collective concept reservoir"): a collective is a social group that is held together by shared values, but at the same time so large that not all members know one another personally. Being of secondary importance for the daily life of individuals, other group members are mainly recognized in specific situations where their behavior and/or material equipment testifies to membership.

Coming back to the transient nature of lithic artifacts, the question is whether entire sequences are meaningful for signaling cultural identities. The key lies in the situational differences of the construction of cultural identity within a specific group, and occasionally presenting it to non-members of this group. While in egalitarian societies members of a group would witness the entire manufacturing process and recognize minor differences (in core preparation, detaching of lateral sharpening flakes, etc.), those of other groups would mainly be confronted with only isolated segments of this process. The latter could be either represented by single pieces belonging to the active equipment of individuals met intentionally or randomly, or by the unrelated waste of different flaking events at inhabited and/or abandoned sites (see also Tostevin 2007). Even if social interaction takes place, it is highly improbable that the complete procedure of artifact manufacture would be made explicit to non-members of the group. Therefore, objects symbolizing social identity on an inter-group level not only have to be standardized, non-interchangeable representatives for the production process, but at the same time, their distinctiveness must be comprehensible without explanation. From a methodological point of view, these objects are reference samples (Richter 1997: Fig. 161; "*Belegstücke*", e.g., *Keilmesser* or Levallois cores). The next section of this article discusses whether contemporaneous Middle Paleolithic operational chains are complex enough to be understood as objects of cultural identity, and whether objects dissociated from their manufacture context are sufficiently distinct to signal cultural identity to people foreign to the social group of the makers.

The Role of Bifaces in Late Neanderthal Groups: Mousterian of Acheulean Tradition and Micoquian Compared

Here, European industries with surface-shaped tools from the first half of MIS 3 (Richter 1997:70 "*formüberarbeitete Werkzeuge*"; Boëda et al. 1990:45 "*façonnage*"), e.g., lithics covered by negatives from soft hammer retouch on one or – more often – both surfaces, are investigated as a case study to elucidate Neanderthal utilitarian equipment as symbols for group identity (in the following, the terms "bifacially surface-shaped tools" and "bifaces" are used synonymously). The presence of surface-shaped tools characterizes the Micoquian (for a summary see Richter 1997:219–254, 2002; Chabai 2004; Jöris 2005), sometimes also termed "*Keilmessergruppen*" (Veil et al. 1994), and the Mousterian of Acheulean Tradition (in the following: "MtA"; Soressi 2002, 2004a, b; Soressi and Hays 2003; Wragg Sykes 2010). Environmental and absolute dates from many Micoquian sites (Richter 1997: Table 9.10; Uthmeier and Chabai 2010: Table 1) and the MtA in the strict sense ("*MTA à bifaces cordiformes*"; Soressi 2002: Table 1) support the interpretation that these industries were contemporaneous between 60 and 35 ka. This absolute chronological framework also seems to hold true for Mousterian assemblages with bifaces from Great Britain (White and Jacobi 2002; Soressi 2004b: Fig. 1 "*MTA à bout-coupés anglais*"; Wragg Sykes 2010), which show strong affinities to the French MtA and, at the same time, are markedly different from the Micoquian (Wragg Sykes 2010:30). Whether these industries evolved after MIS 4 or came into existence earlier is less secure. For the Micoquian in Central Europe, both a short (Richter 1997) and a long (Jöris 2005) chronology exists, whereas further east, the maximal ages for assemblages of the Micoquian (from Zaskalnaya V and Kabazi II) obtained so far by ESR dates and environmental studies point to the very beginning of the last glaciation (Uthmeier and Chabai 2010: Table 1). Finally, the relation between the MtA from Southwestern France and the more northerly distributed, yet older "*Moustérien à bifaces triangulaires plats*" is not clear (Soressi 2002:7). As a minimal consensus, it can be said that both industries were contemporaneous for most of their existence, and that there is, at least at the moment, no well-founded data for an offset of their first appearance.

On a broad level, the Micoquian, which can be subdivided into further regional, chronological and/or functional facies, and the MtA have a number of features in common. One of these is the fact that both industries are characterized by logistical planning. For the MtA, Type A, this is illustrated by the frequencies of unfinished and finished handaxes (Soressi 2004b: Fig. 9). Sites like Grotte XVI and Pech de l'Azé 1 were provisioned with handaxes that (from a systemic point of view) were made at sites like Le Moustier or La Rochette, where failed incomplete products dominate. Planning depth in general also accounts for the Micoquian. Studies based on both raw material provenance and tool reduction have shown that surface-shaped tools were highly mobile artifacts carried from site to site (e.g. Richter 1997; Uthmeier and Chabai 2010:209; Uthmeier 2012). One impressive example (Fig. 7.2) is from Kabazi V in Crimea: while almost all of the waste from the manufacture of a biface was refitted (Veselsky 2008: Figs. 16.1 to 16.15), the tool itself – reconstructed by a plastic cast (Veselsky 2008: Fig. 16.16) – was absent and most probably taken to another camp. Planning depth can also be seen in intense re-sharpening and re-use of surface-shaped tools. Strategies to extend the use-life of surface-shaped tools are a major feature of both the MtA (Soressi 2002:127–134, 2004a:355–356; Wragg Sykes 2010:25–26) and the Micoquian (e.g., Richter 1997:185–206; Jöris 2001:65–67). Some bifacial tools in MtA and Micoquian assemblages are extremely thin. Examples from the upper Middle Paleolithic layer of the southern German site of Weinberghöhlen (Koenigswald et al. 1974) can be compared to a scheme developed by Jan Apel (2001:34–43) for the manufacture of Late Neolithic flint daggers. Based on actual experiments, Apel proposed five stages correlated to different ascending degrees of learning which range from imitation to long term active teaching. With a ratio between width and thickness of 1:5 (calculated according to data from 29 items; Allworth-Jones 1986: Table 3), Micoquian leafpoints from Weinberghöhlen belong to J. Apel's fourth and fifth stages of the production of half products. These stages can only be mastered after years of apprenticeship with repeated oral instructions to correct mistakes, followed by individual practical exercise. It follows that the manufacture of thin surface-shaped leafpoints was too complex to be learned by simple observation and imitation. Admittedly, most Middle Paleolithic bifacial tools are less thin, but the fact that symmetry in outline is often combined with standardized cross-sections still suggests that "technological knowledge" and, equally important, practical "know-how" (see Schiffer and Skibo 1987:597; cf. Apel 2001:27) was obtained through active teaching. The assumption that Micoquian artifact manufacture had to be learned though intense social interaction is further strengthened by the observation that in many assemblages, several concepts of blank production were combined. One example for this is the combination of Levallois and Quina concepts in Central European Micoquian assemblages ("M.M.O.-A"; Richter 1997:224–235). Similar observations were made for the MtA, where the Levallois concept, the *méthode discoïde*, and a method to produce elongated flakes have been documented (Soressi 2004a:347). From an archaeological point of view, this complicates recognition of the entire "collective concept reservoir", as W. Weißmüller (2003:178) has termed it. Depending on site

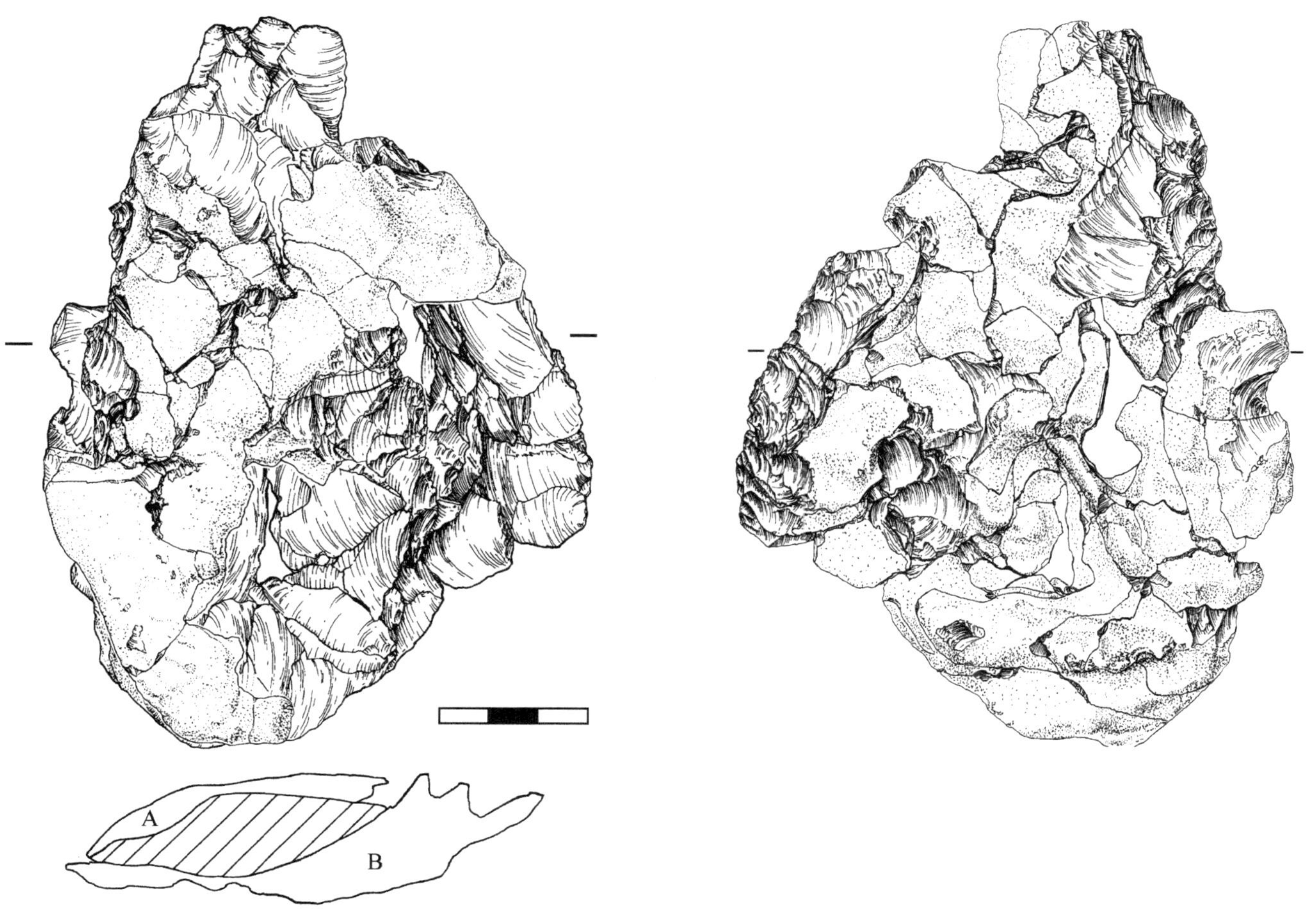

Fig. 7.2 Two views of a refitted flint plaquette surface-shaped to produce a bifacial preform at Kabazi V, Level III/4-2 (from Veselsky 2008: Figs. 16.1 and 16.2, reproduced with permission by the author)

function, raw material quality and availability, concepts may or may not be present in a given assemblage.

Detailed investigations into subsistence strategies of 50 archaeological horizons from Micoquian sites in Crimea (Uthmeier and Chabai 2010) provide an estimation of the number of individuals acting together in daily life. When high resolution data is used to calculate days with enough meat to cover average caloric needs (Uthmeier and Chabai 2010: Fig. 11), for C. Gamble's (1999: Table 2.8) effective group, food resources would have lasted only for several days. Given that this accounts for sites where fireplaces and pits support intensive site use, this argues for the presence of family-size intimate groups. Kill and butchering stations with high minimal numbers of individuals of prey do not necessarily contradict this hypothesis as they are palimpsests of different hunting episodes or – more speculative – the remains of cooperative hunting. How did these groups exploit a region over the long-term? Were they able to establish some kind of property rights for the region in which they lived? Investigating 13 Micoquian assemblages from the small cave of Sesselfelsgrotte, J. Richter (1997, 2001) reconstructed long-term occupation patterns for the Lower Altmühl Valley in Germany (Fig. 7.3). From cyclic increase and decline in the diversity of lithic raw materials (Richter 1997:144–147) combined with investigations into the intensity of tool use (Richter 1997:173–182), Richter

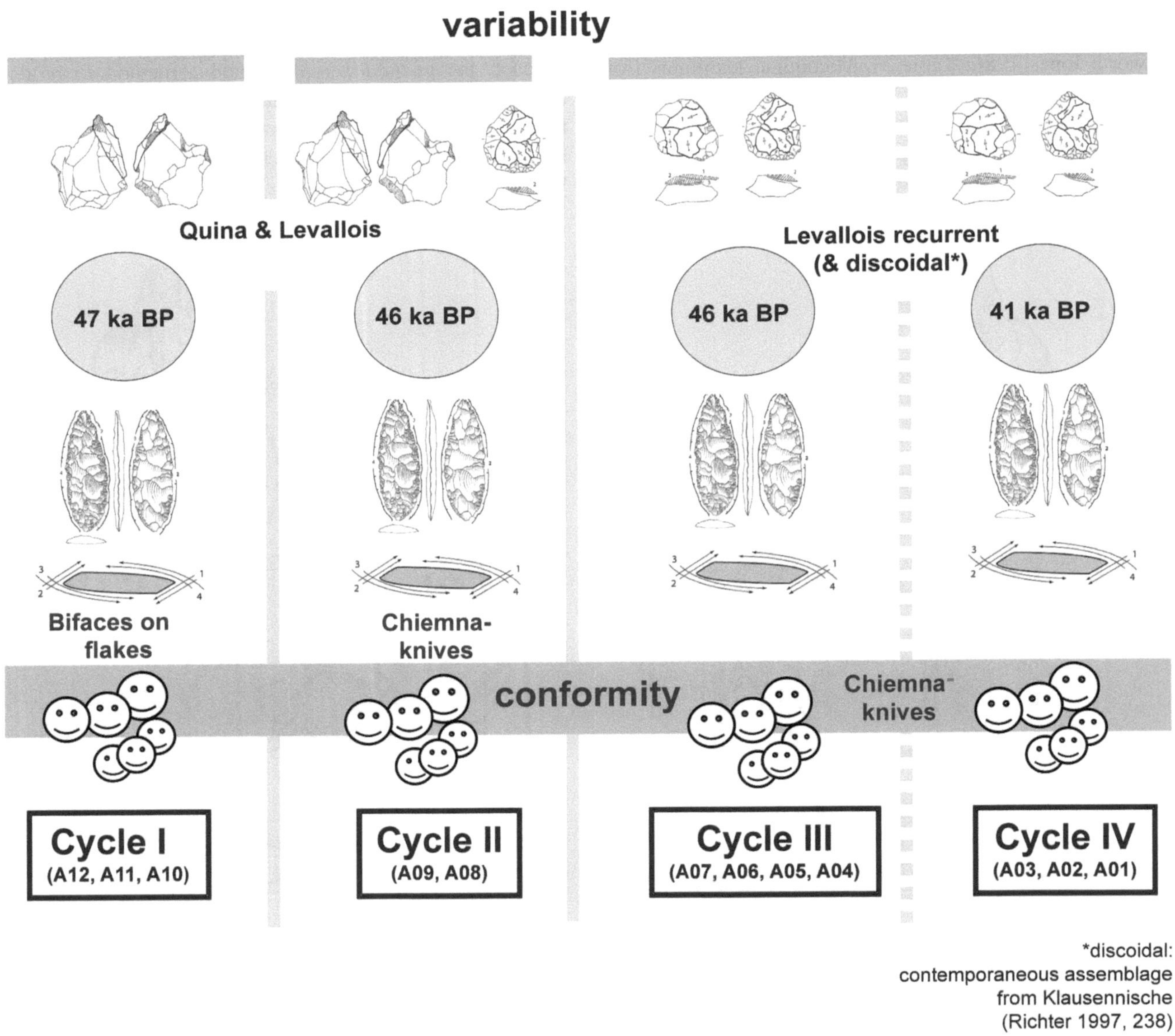

Fig. 7.3 Main attributes of cycles within the Micoquian of Sesselfelsgrotte (assemblages A12 to A01). Each cycle results from a – presumably uninterrupted – occupation of the wider region, separated by gaps on a millennial scale (compiled on the basis of information found in Richter 1997)

concluded that several assemblages in each case represent a temporal segment of a total of four occupations of the Altmühl Valley used as a seasonal territory (Richter 1997:206–209, Figs. 159, 164). At the start of each occupation cycle, stays in the Sesselfelsgrotte were short and characterized by numerous micro-moves heading for resources on an encounter basis, whereas at the end of each cycle, stays were longer and micro-moves led to procurement of specific resources. In fact, each cycle corresponds to the establishment of an optimized pattern of resource procurement. Since high quality flint from the source of Baiersdorf had to be rediscovered during the course of each cycle, Richter (2001) suggested that the transfer of knowledge was interrupted. Although relevant absolute dates (Richter 2002: Fig. 4; "*Innerer Bereich*") for the first three cycles could be similar within one-sigma level of error (Richter 2002: Fig. 5), qualitative attributes of the lithic assemblages indeed indicate gaps of considerable duration between the cycles. Whilst the Quina concept is the dominant concept of simple blank production in the first cycle, it is accompanied by the Levallois concept in the second. An additional new feature of the latter cycle is the presence of "Chiemna" knives. For the final two cycles, the predominance of the Levallois concept indicates technological continuity, but this time there is a considerable gap in absolute dates between them. With regard to regional traditions of cultural identity, it becomes clear that temporarily more or less intense occupied regions were left empty for quite long times. Combined with the quantitative data from Crimea, one has to assume a constant shift of regions used and abandoned by low overall population numbers, an assumption that is in good accordance with demographic estimations by J.-P. Bouquet-Appel and P.-Y. Demars (2000). Apart from this, the diachronic data from Sesselfelsgrotte gives insight into objects important for Middle Paleolithic cultural identity. Although there is considerable variability in concepts for core reduction under comparable environmental conditions, major features of the manufacture and reduction of surface-shaped tools remain stable. One reason for this stability might be functional in that bifaces are related to moves to gather resources. In such a critical aspect for group survival, one has to expect that decisions concerning the acceptance of innovations would tend to be conservative (for decision-making in egalitarian societies see Eisenhauer 2002). However, despite environmental stability within the G-layers of Sesselfelsgrotte, minor changes by all means do occur among bifaces, e.g., the shift to Chiemna knives, pointing to the acceptance of innovations. Compared to concepts of reduction of cores, this happens on a low level within the "*Keilmesser*" concept. Therefore, it is concluded that operational chains behind bifaces were more important for the creation of cultural identity than those for the reduction of cores.

According to Marie Soressi (2004b:13–14, Fig. 3), the most important differences between the MtA and the Micoquian with regard to bifaces are the cross-sections. In the MtA, the typical surface-shaped tool is a more or less symmetrical cordate handaxe with biconvex cross-section. To the contrary, most Micoquian surface-shaped tools have a plano-convex or plan-convex/plan-convex cross-section. It is important to understand that cross-sections represent concepts of manufacture and re-sharpening. For example, plan-convex/plan-convex cross-sections result from what Gerhard Bosinski (1967:43) termed "*wechselseitig-gleichgerichtete Kantenbearbeitung*", a mode typical for Micoquian assemblages, whereas parallel cross-sections reflect the use of plaquettes as raw material (see also Boëda 1995). Like in the case study of the Altmühl Valley, concepts of manufacture and reduction of surface-shaped tools remain remarkably stable during the duration of the MtA. Due to their assumed importance in the construction of social membership in Late Middle Paleolithic groups, further analysis focuses on this part of the tool kit.

In the following, two geographical clusters of the Micoquian, the Central European Micoquian and the Crimean Micoquian, are treated separately. Like other widespread industries with alleged conformity in material culture (e.g. the linear pottery culture: Moddermann 1988), they show some differences that help to examine "diversity in uniformity". Conversely, the MtA of southwestern France (Soressi 2002) and Great Britain (White and Jacobi 2002; Wragg Sykes 2010) are taken as an entity, because central features of bifacial production and reduction are identical (which qualitatively also accounts for MtA Types A and B; Soressi 2004a:361).

Central European Micoquian assemblages (Fig. 7.4) are characterized by a variety of typologically different surface-shaped tools, Among these, backed bifacial knives, or *Keilmesser*, are considered to be the most significant form. Working step analyses by J. Richter (1997) and O. Jöris (2001) have shown that plan-convex *Keilmesser* were the starting point of resharpening processes that led to bifacial pieces of identical cross section, but totally unrelated outlines. After a long use-life made possible by resharpening of the convex upper surface and ventral thinning, some *Keilmesser* were finally discarded as bifacial scrapers. In cases where reduction started from simple backed bifacial knives ("*Bocksteinmesser*"), the final products were bifacial scrapers with one working edge, whereas the reduction of backed bifacial knives with angular working edges ("*Klausennischemesser*") ended up as bifacial scrapers with one or two angular working edges. In rare cases, "*Bocksteinmesser*" were transformed by establishing a similar functional end at the base. To refer to the final typological variability, the many relations between primary, secondary and even tertiary forms were termed "families of surface-shaped tools"

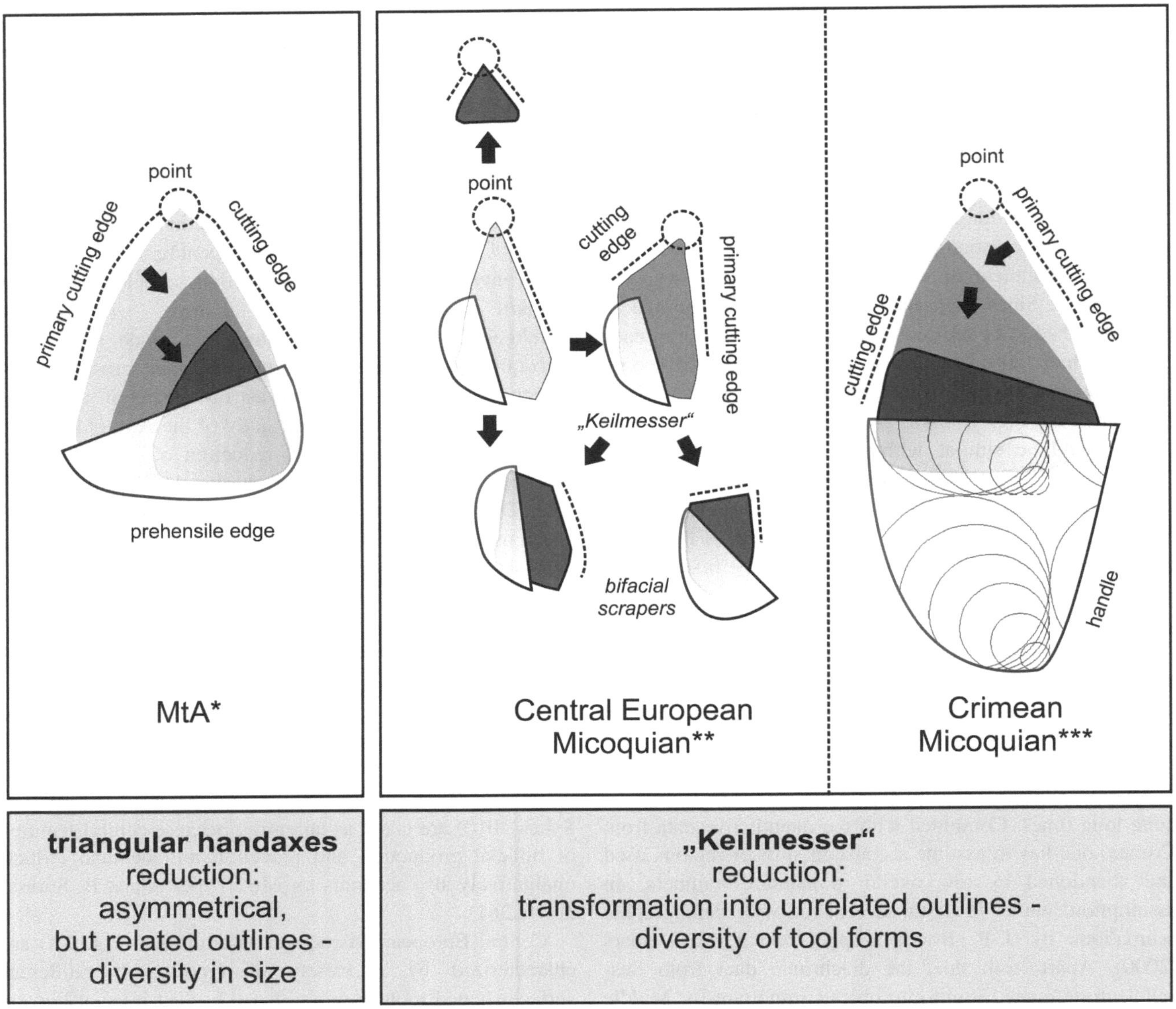

Fig. 7.4 Operational chains in the late European Middle Paleolithic (compiled after *Soressi 2002; **Richter 1997; ***Richter 2004)

(Richter 1997:204; my translation). Concepts for the production of blanks from cores varied. Quina cores are known from an early phase ("M.M.O.-A"; Richter 1997:243) while later assemblages are dominated by Levallois recurrent cores ("M.M.O.-B" or "Levallois-Micoquien"; Richter 1997:243). The *méthode discoïde* is rare, but has been securely documented in Kůlna cave by E. Boëda (1995). Interestingly, the distribution of the different concepts shows no geographical pattern, thus suggesting the mobility of both individuals and groups.

At first glance, surface-shaped tools of the Crimean Micoquian (Fig. 7.4) seem to be in good accordance with the Central European record (Chabai et al. 1995, 2002). Again, *Keilmesser* can be found amongst surface-shaped bifacial tools, and some bifaces have a plan-convex/plan-convex cross section pointing to the "*wechselseitig-gleichgerichtete Kantenbearbeitung*". However, these items are rare and more related to plaquettes as raw blocks rather than to anticipated tool forms, and not subject to intense resharpening. In general, it is possible to differentiate between three major concepts for surface-shaped bifacial tools: backed bifacial knives (or *Keilmesser*), bifacial scrapers, and bifacial points (Uthmeier 2004a). Extended use-lives made possible by recurrent resharpening are mainly found in plan-convex bifacial scrapers (Richter 2004). Working steps of resharpening were confined to the right lateral edge, while the base remained unaltered, possibly because the pieces were still hafted. With advanced reduction, bifacial side scrapers with originally one predominant straight to convex lateral working edge became more and more asymmetrical. Towards the very end, some pieces have a working edge perpendicular to the base. In addition to the transformation of the surface-shaped tools itself, there is another important feature: the Crimean Micoquian makes intensive use of flakes from surface shaping (Chabai 2004; Uthmeier 2012). In fact, apart from simple flake cores true concepts of core reduction are missing. Instead, blanks from the reduction of bifacial tools were employed as an effective strategy to provide autonomy from raw material sources. This can be seen in the correlation between mean percentages of bifaces and transportation distances to raw material sources in different facies of the Crimean Micoquian. "Ak-Kaya" assemblages were situated near raw material sources and show (on average) high numbers of – in this case large – bifaces, whereas "Starosele" and "Kiik-Koba" assemblages, which were located farther from qualitatively better raw material sources, have low percentages of small and reduced bifaces. At the latter sites, the demand for cutting edges was satisfied by modification of larger flakes from the on-site reduction of imported bifaces.

Compared to both entities of the Micoquian surveyed here, bifaces from the contemporaneous MtA in the strict sense ("*MTA à bifaces cordiformes*" and "*MTA à bout-coupés anglais*") are markedly different (Soressi 2004b:13–17). Most of them (Fig. 7.4) can be classified as thin handaxes with bi-convex cross sections and symmetrical cordate outlines. While French handaxes usually have an unworked or blunt, slightly rounded base (Soressi 2002, 2004a, b), those from Great Britain may have a straighter, often sharp base and a less pointed tip (Wragg Sykes 2010:23). True *Keilmesser* are lacking. Instead, backed knifes on elongated flakes ("*couteaux à dos retouchés*"; Soressi 2002:6–7) are another typical tool type. Not only primary tool forms but biface reduction strategies also show discrete characteristics. M. Soressi (2004a:355–356) was able to show that resharpening in the MtA aimed at the maintenance of the length of one lateral working edge. During their use-life, handaxes lost their symmetry as well as part of their bi-convex cross section, the latter because resharpening of the upper surface was prepared by ventral thinning, often longitudinal. Nevertheless, even heavily reduced items can still be identified as handaxes, and they still have two active working edges (differing in total length). As in the Crimean Micoquian, blanks struck with soft hammer technique during the production and resharpening of bifaces were used for modification mainly into scrapers and raclettes (Soressi 2002:81–87). Other tool types were made on flakes from cores of the Levallois concept, the *méthode discoïde* or from "*nucléus unipolaires semitournants*" (Soressi 2002:104–106, 173–195).

Discussion: Neanderthal Utilitarian Equipment and Group Identity

Although the terms "*débitage*" and "*façonnage*" are used as representatives for the broadest classes of lithic artifact manufacture (Boëda et al. 1990:45), it appears as if, in late Middle Paleolithic industries with bifaces, *façonnage* plays a more important role in the material construction of group identity than the reduction of cores. This is attested by the chronological and/or spatial variability in *débitage* concepts in the MtA and the Central European Micoquian under similar environmental conditions, which is in stark contrast to the chronological constancy of operational changes for bifaces in all of the three industries examined (Fig. 7.5). Between the largely contemporaneous MtA, Central European Micoquian and Crimean Micoquian, operational chains for biface reduction were found to be distinct, complex, and standardized enough to be taken as means for the construction and maintenance of group identity. On an intra-group level, e.g., intimate networks, it has to be assumed that apart from the biface itself, a set of typical flakes or even "waste" from its production may have been sufficient to signal group identity. When contacts between socially more distant individuals or intimate groups are considered, e.g., "extended networks" and "global networks" or collectives, MtA handaxes and Micoqian *Keilmesser* are to such an extent representatives of manufacturing processes that they have signaled group identity even when isolated. However, it has to be stressed that functional equipment must have been of minor importance when compared to attributes that could have been observed from larger distances, e.g., body painting, hair dress, language, etc.

Against the background of the highly distinctive bifaces, there is a surprising agreement between the MtA and the Central European Micoquian in the presence of concepts for core reduction, e.g., the Levallois concept and the *méthode discoïde*. Two different, yet not mutually exclusive explanations can be offered. First, the two concepts of blank production have a long history traced back to MIS 10 (Levallois concept; Delagnes and Meignen 2006: Table 2) and MIS 6 (*méthode discoïde*; Delagnes and Meignen 2006: Table 1). Such long periods of existence could go back to long-term, spatially widely distributed traditions, or result from repeated re-invention. Second, the flow of technological knowledge may go back to exchange between contemporaneous groups, as proposed for the presence of the Quina concept in the Micoquian of Sesselfelsgrotte (Richter 2000). The other feature that is found in more than one social entity is the use of blanks from surface shaping as blanks for tools with simple lateral modifications in both the MtA and the Crimean

	MtA*	Micoquian**	Crimean Micoquian***
object of group identity	handaxe *bout de coupe*	backend bifacial knife	backend bifacial knife bifacial convergent scraper/ bifacial point

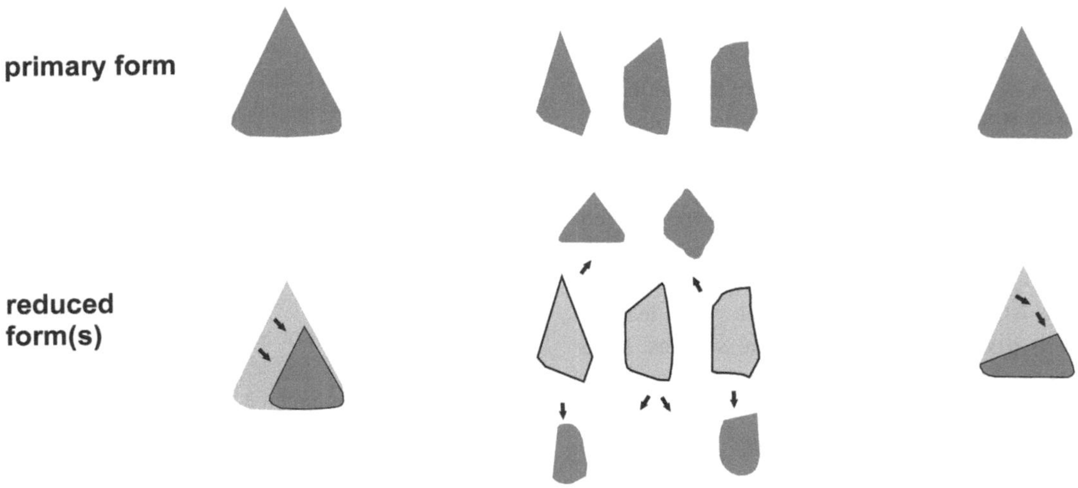

	MtA*	Micoquian**	Crimean Micoquian***
reduction	distortion of symmetry	unrelated outlines	(extremely) asymmetrical
use of flakes from surface shaping	up to 2/3 of all tools	rare	tools mainly on these
blank manufacture	Levallois *„Methode discoïde"* blanks from surface shaping *„nucléus unipolaires semitournants"*	Levallois *„Methode discoïde"* Quina	blanks from surface shaping simple flake cores

Fig. 7.5 Neanderthal utilitarian equipment and group identity indicated by operational chains for the manufacture and reduction of bifaces (compiled on the basis of information found in *Soressi 2002; **Richter 1997; ***Uthmeier 2004b)

Micoquian. However, this feature is not evenly distributed amongst assemblages of the MtA. While well-known from MtA Type A assemblages, bifaces and the number of flakes from their manufacture amongst blanks is by far less frequent in MtA assemblages of Type B (Soressi 2004a, b). This is explained by a decrease in long-term planning of moves in Type B sites (Soressi 2004a, b). In this, e.g., a high interest in independence from raw material sources due to high residential as well as logistical mobility, the Crimean Micoquian (Uthmeier and Chabai 2010) is very similar to the MtA, Type A. It becomes clear that in this case, the quantity of bifaces and the use of flakes from surface shaping as blanks is a consequence of the land use pattern employed rather than direct connection between groups. In the Central European

Micoquian, bifaces are more frequent at residential camps, but no use is made of blanks from surface shaping, indicating somehow lower degrees of planning depth and/or lower levels of mobility than in the previous industries. Arguments for a land use pattern with longer stays at some locations include considerable "times of activity" calculated on the basis of working hours deduced from actual experiments for the production and use of different types of blanks and tools (Uthmeier 2004c: Table 14.3). On a more general level, one may discuss whether the observed differences in both the operational chains and the resulting tool forms can be described in terms of "function" or "style". Following R. C. Dunnell (1978:199), "The definition of function [...] is frequently a synonym of 'use'. [...] Stylistic similarity is [...] the result of direct cultural transmission once chance similarity in a context of limited possibilities is excluded." (for an assessment of this definition from an evolutionary point of view see Shennan 2005:136). In fact, the function of bifaces typical for the MtA and the Micoquian is very similar: in both cases these are multi-functional tools which make use of a point (in most cases implemented by convergent retouched edges of a bifacial surface-shaped blank) and a primary lateral working edge. Perhaps most important, and yet equally in agreement, is the possibility for recurrent resharpening. Are, then, different realizations of similar functional demands a question of style? Since concepts of resharpening target the longevity of working edges closely intertwined with land use patterns, operational chains have a "functional" significance for the well-being of the groups. Against this background, one may ask if the differences observed are at all deliberate, e.g., group decisions brought about by discourse, or related to chance (Shennan 2005). The small size of groups (on the level of extended networks) and the low overall demographic density presumably leading to isolation for longer periods would speak for such a notion. On the other hand, with the data at hand, there is no clear chronological trend between the industries, and operational chains exclude one another, making simple "drift" less probable. Apart from this, objects develop social meaning through social interaction, which gives them a relevance in every-day life that is independent from possible evolutionary (macro-scale) developments.

Conclusions

Although egalitarian hunter-gatherers are expected to be open to cooperation and integration of individuals from socially distant groups, by comparing the manufacture and use of surface-shaped tools the MtA and the Micoquian can be understood as separated collectives. Within the Micoquian, the Central European Micoquian and the Crimean Micoquian show differences in the operational chains for both *façonnage* and *débitage*, but it is doubtful whether the resulting bifacial tools and production waste are distinct enough to be recognized as socially separate technological knowledge without participating observation. Therefore, it is more probable that both entities constitute one collective which consists of at least two extended networks with slightly different strategies of lithic curation. Transfer of knowledge, and exchange of social actors, should have been facilitated between these entities, and hindered between them and the MtA.

Acknowledgements The author would like to thank G.-C. Weniger, M. Bolus, and an anonymous reviewer who helped to improve the present article, as well as R. Miller for the revision of the English language and, last but not least, M. Haidle and M. Bolus for the possibility to participate in the "Nature of Culture" workshop.

References

Allsworth-Jones, P. (1986). *The Szeletian and the transition from Middle to Upper Palaeolithic in Central Europe*. Oxford: Oxford University Press.

Álvarez Fernández, E., & Jöris, O. (2007). Personal ornaments in the Early Upper Paleolithic of Western Eurasia: An evaluation of the record. *Eurasian Prehistory, 5*(2), 31–44.

Apel, J. (2001). *Daggers, knowledge and power. The social aspects of flint-dagger technology*. Uppsala: Uppsala University Department of Archaeology & Ancient History.

Bar-Yosef, O. (2002). The Upper Paleolithic revolution. *Annual Review of Anthropology, 31*, 363–393.

Bednarik, R. (1992). Palaeoart and archaeological myths. *Cambridge Archaeological Journal, 2*, 27–57.

Blumer, H. (1937). Social psychology. In E. P. Schmidt (Ed.), *Man and society* (pp. 144–198). New York: Prentice Hall.

Blumer, H. (1969). *Symbolic interactionism*. Englewood Cliffs: Prentice Hall.

Boëda, E. (1995). Steinartefakt-Produktionssequenzen im Micoquien der Kulna-Höhle. *Quartär, 45*(46), 75–98.

Bosinski, G. (1967). *Die mittelpaläolithischen Funde im westlichen Mitteleuropa*. Köln: Böhlau.

Bouquet-Appel, J.-P., & Demars, P.-Y. (2000). Neanderthal contraction and modern human colonization of Europe. *Antiquity, 74*, 544–552.

Byers, A. A. (1994). Symboling and the Middle-Upper Palaeolithic Transition: A theoretical and methodological critique. *Current Anthropology, 35*, 369–399.

Caron, F., d'Errico, F., del Moral, P., Santos, F., & Zilhão, J. (2011). The reality of Neandertal symbolic behavior at the Grotte du Renne, Arcy-sur-Cure, France. *PLoS ONE, 6*(6), e21545.

Chabai, V. P. (2004). *The Middle Paleolithic of Crimea: Stratigraphy, chronology, typological variability and eastern European context*. Simferopol: Shlyakh. (in Russian).

Chabai, V. P., Marks, A. E., & Yevtushenko, A. I. (1995). Views of the Crimean Middle Paleolithic: Past and present. *Préhistoire Européene, 7*, 59–79.

Chabai, V. P., Richter, J., Uthmeier, T., & Yevtushenko, A. I. (2002). Neue Forschungen zum Mittelpaläolithikum auf der Krim. *Germania, 80*, 441–473.

Chase, P. G. (1991). Symbols and Paleolithic artifacts: Style, standardization, and the imposition of arbitrary form. *Journal of Anthropological Science, 10*, 193–214.

Chase, P. G., & Dibble, H. L. (1987). Middle Paleolithic Symbolism: A review of current evidence and interpretations. *Journal of Anthropological Science, 6*, 263–296.

Delagnes, A., & Meignen, L. (2006). Diversity of Lithic Production systems during the Middle Paleolithic in France. Are there any chronological trends? In E. Hovers & S. L. Kuhn (Eds.), *Transitions before the Transition. Evolution and stability in the Middle Paleolithic and the Middle Stone Age*. (pp. 85–107). New York: Springer.

Dibble, H. (1988). Typological aspects of reduction and intensity of utilisation of lithic resources of the French Mousterian. In H. Dibble & A. Montet-White (Eds.), *Upper Pleistocene Prehistory of Western Eurasia* (pp. 181–197). Philadelphia: University of Pennsylvania.

Dunnell, R. C. (1978). Style and function: A fundamental dichotomy. *American Antiquity, 43*, 192–202.

Eisenhauer, U. (2002). *Untersuchungen zur Siedlungs- und Kulturgeschichte des Mittelneolithikums in der Wetterau*. Bonn: Habelt.

Gamble, C. (1999). *The Palaeolithic societies of Europe*. Cambridge: Cambridge University Press.

Gamble, C. (2007). *Origins and reevolutions. Human identity in earliest prehistory*. Cambridge: Cambridge University Press.

Boëda, E., Geneste, J.-M., & Meignen, K. (1990). Identification de chaînes opératoires lithiques du Paléolithique ancien et moyen. *Paléo, 2*, 43–80.

Hahn, H. P. (2003). Dinge als Zeichen – eine unscharfe Beziehung. In U. Veit, T. L. Kienlin, C. Kümmel, & S. Schmidt (Eds.), *Spuren und Botschaften. Interpretationen materieller Kultur* (pp. 29–51). Münster: Waxmann.

Higham, T., Jacobi, R., Julien, M., David, F., Basell, L., Wood, R., et al. (2010). Chronology of the Grotte du Renne (France) and implications for the context of ornaments and human remains within the Châtelperronian. *Proceedings of the National Academy of Science of the USA, 107*, 20234–20239.

Jäger, L. (2002). Transkriptivität. Zur medialen Logik der kulturellen Semantik. In L. Jäger & G. Stanitzek (Eds.), *Transkribieren. Medien/Lektüre* (pp. 19–41). München: Fink.

Jäger, L. (2003). Transkription - zu einem medialen Verfahren an den Schnittstellen des kulturellen Gedächtnisses. *TRANS. Internet-Zeitschrift für Kulturwissenschaften, No. 15/2003*. http://www.inst.at/trans/15Nr/06_2/jaeger15.htm. Accessed 1 July 2012.

Jöris, O. (2001). *Der mittelpaläolithische Fundplatz Buhlen (Grabungen 1966–1969). Stratigraphie, Steinartefakte und Fauna des Oberen Fundplatzes*. Bonn: Habelt.

Jöris, O. (2005). Zur chronostratigraphischen Stellung der spätmittelpaläolithischen Keilmessergruppen. Der Versuch einer kulturgeographischen Abgrenzung einer mittelpaläolithischen Formengruppe in ihrem europäischen Kontext. *Bericht der Römisch-germanischen Kommission, 34*, 49–153.

Kienlin, T. L. (Ed.). (2005). *Die Dinge als Zeichen. Kulturelles Wissen und materielle Kultur*. Bonn: Habelt.

Leroi-Gourhan, A. (1984). *Hand und Wort: Die Evolution von Technik, Sprache und Kunst* (3rd ed.). Suhrkamp: Frankfurt a. M.

Mauss, M. (1979). *Sociology and psychology*. London: Routledge.

McBrearty, S., & Brooks, A. S. (2000). The revolution that wasn't: A new interpretation of the origin of modern human behavior. *Journal of Human Evolution, 39*, 423–463.

Mead, G. H. (1934). *Mind, self, and society*. Chicago: University of Chicago Press.

Mellars, P. A. (1973). The character of the Middle-Upper transition in southwest France. In C. Renfrew (Ed.), *The explanation of cultural change: Models* (pp. 255–276). London: Duckwort.

Mellars, P. A. (1996). *The Neanderthal legacy. An archeological perspective from Western Europe*. New York: Princeton University Press.

Mellars, P. A. (2010). Neanderthal symbolism and ornament manufacture: The bursting of a bubble? *Proceedings of the National Academy of Science of the USA, 107*, 20147–20148.

Mikl-Horke, G. (2001). *Soziologie: Historischer Kontext und soziologische Theorie-Entwürfe* (5th ed.). München: Oldenbourg Wissenschaftsverlag.

Moddermann, P. J. R. (1988). The linear pottery culture: Diversity in uniformity. *Berichten van de Rijksdienst voor het Oudheidkundig Bodemonderzoek, 38*, 63–140.

Münch, R. (2002a). *Soziologische Theorie. Band 1: Grundlegung durch die Klassiker*. Frankfurt: Campus Verlag.

Münch, R. (2002b). *Soziologische Theorie. Band 2: Handlungstheorie*. Frankfurt: Campus Verlag.

Porr, M. (1998). Die Postmoderne Archäologie in Großbritannien. In K. H. Eggert & U. Veit (Eds.), *Theorie in der Archäologie: Zur englischsprachigen Diskussion* (pp. 183–216). Münster: Waxmann.

Porr, M. (2005). The making of bifaces and the making of the individual. In C. Gamble & M. Porr (Eds.), *The Hominid individual in context. Archaeological investigations of Lower and Middle Palaeolithic landscapes, locales and artefacts* (pp. 68–80). London & New York: Routledge.

Richter, J. (1997). *Sesselfelsgrotte III: Der G-Schichten-Komplex der Sesselfelsgrotte – Zum Verständnis des Micoquien*. Saarbrücken: Saarbrücker Druckerei und Verlag.

Richter, J. (2000). Social memory among late Neanderthals. In J. Orschiedt & G.-C. Weniger (Eds.), *Neanderthals and modern humans – Discussing the transition. Central and Eastern Europe from 50.000–30.000 B.P.* (pp. 123–132). Mettmann: Neanderthal Museum.

Richter, J. (2001). For lack of a wise old man? Late Neanderthal land use patterns in the Altmühl River Valley, Southern Germany. In N. J. Conard (Ed.), *Settlement dynamics of the Middle Paleolithic and Middle Stone Age* (Vol. 1, pp. 205–220). Tübingen: Kerns Verlag.

Richter, J. (2002). Die ^{14}C-Daten aus der Sesselfelsgrotte und die Zeitstellung des Micoquien/M.M.O. *Germania, 80*, 1–22.

Richter, J. (2004). Copies of flakes. Operational sequences of foliate pieces from Buran-Kaya III Level B1. In V. P. Chabai, K. Monigal, & A. E. Marks (Eds.), *The Middle Paleolithic and Early Upper Paleolithic of Eastern Crimea. The Paleolithic of Crimea III* (pp. 233–247). Liège: Université de Liège.

Sackett, J. R. (1982). Approaches to style in lithic archaeology. *Journal of Anthropological Archaeology, 1*, 59–112.

Sackett, J. R. (1986). Isochrestism and style: A clarification. *Journal of Anthropological Archaeology, 5*, 266–277.

Schiffer, M. B., & Skibo, J. M. (1987). Theory and experiment in the study of technological change. *Current Anthropology, 28*, 595–622.

Shennan, S. (2005). The evolutionary dynamics of cultural traditions. In K. T. Kienlin (Ed.), *Die Dinge als Zeichen: Kulturelles Wissen und materielle Kultur* (pp. 133–147). Bonn: Habelt.

Soressi, M. (2002). Le Moustérien de tradition acheuléenne du sud-ouest de la France. Discussion sur la signification du faciès à partir de l'étude comparée de quatre sites: Pech-de-l'Azé, Le Moustier, La Rochette et la Grotte XVI. Ph.D. Dissertation, Université Bordeaux I. http://www.paleoanthro.org/dissertations/marie%20soressi.pdf. Accessed 1 July 2012.

Soressi, M. (2004a). From the Mousterian of Acheulean Tradition type A to type B: A change in technical tradition, raw material, task, or settlement dynamics? In N. J. Conard (Ed.), *Settlement dynamics of the Middle Paleolithic and Middle Stone Age II* (pp. 343–366). Tübingen: Kerns Verlag.

Soressi, M. (2004b). Die Steintechnologie des Spätmoustérien. Ihre Bedeutung für die Entstehungsgeschwindigkeit modernen Verhaltens und die Beziehung zwischen modernem Verhalten und biologischer Modernität. *Mitteilungen der Gesellschaft für Urgeschichte, 13*, 9–28.

Soressi, M., & Hays, M. (2003). Manufacture, transport and use of Mousterian bifaces. A case study from the Perigord (France). In M. Soressi & H. L. Dibble (Eds.), *Multiple approaches to the study of bifacial technologies* (pp. 125–147). Philadelphia: University of Pennsylvania.

Tostevin G. B. (2007). Social intimacy, artefact visibility and acculturation models of Neanderthal-Modern Human interaction. In P. A. Mellars, K. Boyle, O. Bar-Yosef, & C. Stringer (Eds.), *Rethinking the human revolution. New behavioural and biological perspectives on the origin and dispersal of modern humans* (pp. 341–357). Cambridge: McDonald Institute for Archaeological Research.

Uthmeier, T. (2004a). Transformation analysis and the reconstruction of on-site and off-site activities: Methodological remarks. In V. P. Chabai, K. Monigal, & A. E. Marks (Eds.), *The Middle Paleolithic and Early Upper Paleolithic of Eastern Crimea. The Paleolithic of Crimea III* (pp. 175–191). Liège: Université de Liège.

Uthmeier, T. (2004b). *Landnutzungsmuster im Mittelpaläolithikum der Halbinsel Krim, Ukraine. Ein Beitrag zu Ökonomie und Soziologie der Neandertaler*. Habilitation Dissertation, University of Cologne.

Uthmeier, T. (2004c). *Micoquien, Aurignacien und Gravettien in Bayern. Eine regionale Studie zum Übergang vom Mittel- zum Jungpaläolithikum*. Bonn: Habelt.

Uthmeier, T. (2012). The Transition from Middle- to Upper Palaeolithic at Buran Kaya III, Crimea (Ukraine): A case of conceptual continuity of lithic artefact manufacture? In A. Pastoors & M. Peresani (Eds.), *Flakes not blades: The role of flake production at the onset of the Upper Palaeolithic* (pp. 239–260). Mettmann: Neanderthal Museum.

Uthmeier, T., & Chabai, V. P. (2010). Neanderthal subsistence tactics in the Crimean Micoquian. In N. J. Conard (Ed.), *Settlement dynamics of the Middle Paleolithic and Middle Stone Age* (Vol. III, pp. 195–234). Tübingen: Kerns-Verlag.

Veil, S., Breest, K., Höfle, H.-C., Meyer, H. H., Plisson, H., Urban-Küttel, B., et al. (1994). Ein mittelpaläolithischer Fundplatz aus der Weichsel-Kaltzeit bei Lichtenberg, Lkr. Lüchow-Dannenberg. Zwischenbericht über die archäologischen und geowissenschaftlichen Untersuchungen 1987–1992. *Germania, 72*, 1–66.

Veselsky, A. P. (2008). Kabazi V: Production and rejuvenation of bifacial tools. In V. P. Chabai, J. Richter, & T. Uthmeier (Eds.), *Kabazi V: Interstratification of Micoquian and Levallois-Mousterian camp sites, Part II*. The Palaeolithic Sites of Crimea 4 (pp. 455–479). Simferopol & Köln: Shlayakh.

von Koenigswald, W., Müller-Beck, H., & Pressmar, E. (1974). *Die Archäologie und Paläontologie in den Weinberghöhlen bei Mauern (Bayern), Grabungen 1937–67*. Tübingen: Verlag Archaeologica Venatoria.

Weede, E. (1992). *Mensch und Gesellschaft. Soziologie aus der Perspektive des methodologischen Individualismus*. Tübingen: Mohr Siebeck.

Weber, M. (1964). *Wirtschaft und Gesellschaft. Studienausgabe*. Köln: Kiepenreuther und Witsch.

Weißmüller, W. (1995). *Die Silexartefakte aus den Unteren Schichten der Sesselfelsgrotte. Ein Beitrag zum Problem des Moustérien*. Saarbrücken: Saarbrücker Druckerei und Verlag.

Weißmüller, W. (2003). Von Gestaltungsfreiheit und Standardisierungszwang: Zum Zeichencharakter paläolitischer Steinartefakte. In U. Veit, T. L. Kienlin, C. Kümmel, & S. Schmidt (Eds.), *Spuren und Botschaften. Interpretationen materieller Kultur* (pp. 173–186). Münster: Waxmann.

White, J., & Jacobi, R. (2002). Two sides to every story: Bout coupé handaxes revisted. *Oxford Journal of Archeology, 21*, 109–133.

Wragg Sykes, R. M. (2010). Beyond bout coupés: The dynamic role of bifaces in the British Mousterian. *Lithics: The Journal of the Lithic Studies Society, 31*, 20–32.

Zilhão, J., Angelucci, D. E., Badal-García, E., d'Errico, F., Daniel, F., Dayet, L., et al. (2010). Symbolic use of marine shells and mineral pigments by Iberian Neandertals. *Proceedings of the National Academy of Sciences of the USA, 107*, 1023–1028.

Chapter 8
Tracing Group Identity in Early Upper Paleolithic Stone and Organic Tools – Some Thoughts and Many Questions

Michael Bolus

Abstract While discussions about identity in the Upper Paleolithic usually focus on art, decorated objects, and personal ornaments, regarding style as one crucial topic, organic tools and especially stone artifacts have been considered to a much lesser degree. This paper tries to assess the significance of stone and organic tools, representing the most common archaeological record beyond art and ornaments, for establishing group identity in the early Upper Paleolithic. It starts providing a short overview of some major contributions addressing style with regard to stone artifacts and then screens the archaeological record. Problems result from the lack of an unambiguous definition of 'style' and from the lack of applicable parameters to decide whether differences between tools have to be interpreted in terms of different styles or rather in terms of different types. In both cases it is not clear if and in which way identity is conveyed. Both stone and organic tools appear to be weak indicators for group identity and even with data added by other artifact categories such as personal ornaments, decorated objects and art objects the chance of getting positive results is rated to be rather low for the early Upper Paleolithic.

Keywords Aurignacian • Protoaurignacian • Uluzzian • Artifacts • Style • Type • Cultural capacities • Cultural performances

M. Bolus (✉)
Heidelberg Academy of Sciences and Humanities, Research Center "The Role of Culture in Early Expansions of Humans", Eberhard Karls Universität Tübingen, Rümelinstraße 23, 72070 Tübingen, Germany
e-mail: michael.bolus@uni-tuebingen.de

Introduction

Discussions about identity in the Upper Paleolithic usually focus on art, decorated objects, and personal ornaments (see, e.g., Bar-Yosef 2002; Vanhaeren and d'Errico 2006; Bolus and Conard 2008; Conard 2008), while organic tools and especially stone artifacts have been considered to a much lesser degree (e.g., Close 1978; Sackett 1982; Barton 1997) since the signal given by these artifact categories seems to be less clear. When distinguishing between group or social identity and personal identity, it becomes obvious that most of the relevant papers are dealing with the former rather than with the latter.

Since it is obviously much less clear if group identity can be expressed through stone and organic tools, it is necessary to find parameters to assess the significance – if there is one – of these artifacts for establishing social/group (and personal) identity. Hence, this paper raises the question: Can we trace group identity in the most common archaeological record beyond art and ornaments and if so, how?

Style in the Archaeological Discourse

Style seems to be a crucial topic when trying to trace identity in prehistoric times. One question must therefore be if style, beyond artifacts with clear symbolic meaning, can also be expressed by seemingly profane objects such as stone and organic tools. In case this should turn out to be true, it might be possible to deduce social/group identity from the style of stone and organic tools.

While most colleagues seem to agree on the significance of style, the definitions of style vary within relevant key papers. Polly Wiessner (1983) in her study on Kalahari San projectile points distinguishes between assertive style which carries information about personal identity, separating the individual from similar group members, and emblemic style

Miriam N. Haidle, Nicholas J. Conard and Michael Bolus (eds.), *The Nature of Culture: Based on an Interdisciplinary Symposium 'The Nature of Culture', Tübingen, Germany*, Vertebrate Paleobiology and Paleoanthropology, DOI 10.1007/978-94-017-7426-0_8

which is an active means of establishing group identity, while C. Michael Barton (1997) in his study on stone tools, style, and social identity uses the terms ‘passive style’ und ‘active style’. Following Barton, both assertive style and emblemic style (*sensu* Wiessner) are representatives of active style. For him, individual social groups should be characterized by particular variants of a lithic artifact class, while passive style results from “stochastic variability [that] has the *potential* to affect a much wider range of lithic variability” and thus “makes it considerably more difficult to evaluate the capacity for lithics to convey information about social identity” (Barton 1997:144). Still another approach is suggested by James R. Sackett (1982) who uses the term ‘isochrestic style’, which refers to choices made between variants that are equivalent in use. This means that isochrestic style is an expression of a certain degree of flexibility that groups or individuals possess for choosing specific cultural performances from a common pool of cultural capacities. In this sense, “cultural performances represent the actual sets of cultural attributes expressed by an organism or a group”, which can be deduced from the archaeological record, while “cultural capacities of a defined analytical unit (species, population, or group) are theoretical constructs and express the potential range of cultural performances in different subunits at a given time.” (for more details on cultural capacities versus cultural performances see Haidle et al. 2015; quotations from that paper).

The relevance of stylistic behavior for the exchange of information is highlighted by Martin Wobst (1977) who states that style in artifacts can be symbolic. This stresses the assumption that style in general may be an indicator of group identity.

This being said and with the awareness in mind, that other definitions of style could be added, how can we trace style in stone and organic tools and determine its implications for recognizing cultural identity then?

The Archaeological Evidence

One key question raised here is, how to identify “identity markers [as] forms of expression of previous group identity materialized in the archaeological record” (Müller-Scheeßel and Burmeister 2006:18; translation from German by MB), in this case: materialized in stone and organic tools?

Following Sarah Wurz (2008), Paola Villa et al. (2010:651) state: “A symbolic memory strategy can be recognized in artifact production strategies; culturally transmitted innovative conventions for making artifacts imply symbolic behavior even if such artifacts did not function in symbolic ways in society. In other words, innovative and temporally constrained artifact traditions signal group identity”. While Wurz (2008) specifically addresses the Howiesons Poort complex of the southern African MSA, Villa et al. (2010) also include European Middle Paleolithic complexes such as the *Keilmesser* group and the Mousterian of Acheulian Tradition. This is of crucial importance since both European technocomplexes have been produced by Neanderthals and thus a certain degree of group identity would have to be admitted to that archaic hominin form as well (see also Uthmeier 2016).

Lyn Wadley (2003), viewing style as the repeated patterning that is geographically and chronologically restricted, argues that lithic assemblage variability can reflect active style, in the sense that stone tools can act as indices of social identity, but she emphasizes the tempo of technological change. We have to be aware, however, that variability in lithic assemblages not necessarily reflects style but may also be due to functional variability (see Bolus 2010). This notion was already made by Martin Wobst (1977), though meant in a general sense and not restricted to Paleolithic stone and organic tools, when he stated that style lacks meaning. For him, style either “is explicitly defined as a negative category (e.g., aspects of artifact variability which cannot be attributed to other agencies such as productive advantage, mechanical factors, or chance), or it is unmanageably multidimensional (e.g., aspects of artifact variability which are congruent with specific areas, time periods, or sets of personnel regardless of the cause for this congruence)” (Wobst 1977:317).

In a way similar to Wiessner in her analyses of (metal) San projectile points, many colleagues dealing with style in stone artifacts also concentrate on projectiles such as Solutrean leaf points or Paleoindian points. For the European Aurignacian this approach is not very promising since projectile technology is based on points made of organic materials there. It may be different for those early Upper Paleolithic assemblages which are generally labelled as Protoaurignacian or archaic Aurignacian (Laplace 1966; for a critique of the concept of the Protoaurignacian see Conard and Bolus 2015), or for assemblages of the Ahmarian where lithic points are present (Fig. 8.1). As Sackett (1982) points out, style may also be found in unmodified stone artifacts and strategies for core reduction, but systematic analyses are rare yet. An attempt has been made by Philip Nigst (2009) who interprets technological differences between lithic assemblages from the Middle to Upper Paleolithic transition and the early Upper Paleolithic in the Middle Danube region as seemingly related to the cultural identity stored in technological styles and passed on between generations (Nigst 2009:412). Another promising approach by Héloïse Koehler to infer identity (of Neanderthals) from technological behaviors, based on the detailed analysis of Middle Paleolithic lithic assemblages from the Paris Basin (Koehler 2009), should also be mentioned since it might be expanded to include Upper Paleolithic artifacts as well.

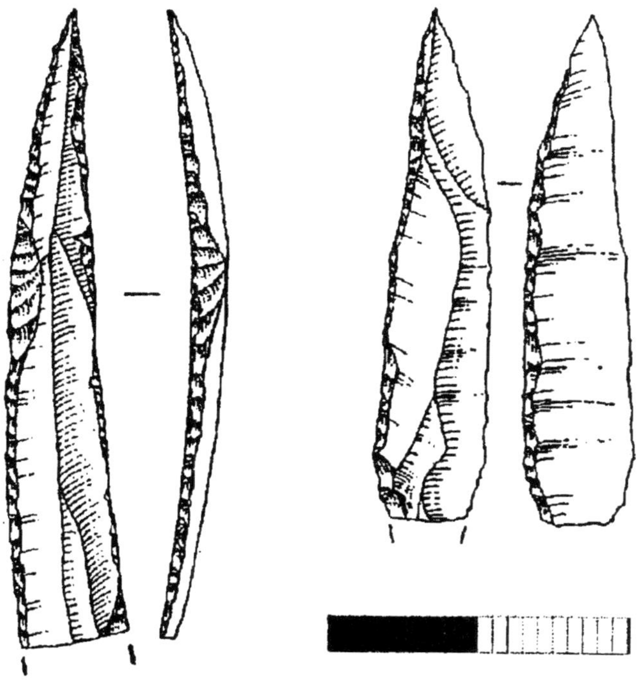

Fig. 8.1 Grotta di Fumane, Italy. Backed lithic points. Modified after Broglio et al. 2002

Another question to be asked is, if cultural identity might also become visible through the absence of specific tool types. The presence or absence of Dufour bladelets in Aurignacian assemblages might perhaps be interpreted in that way. The upper Aurignacian horizon (AH II) from Geißenklösterle Cave in the Swabian Jura yielded three Dufour bladelets of Roc-de-Combe sub-type, two of them fragments (see Moreau 2009:206–207), which in fact constitute the total number of Dufour bladelets from all Aurignacian assemblages in Swabia. Given that fact, it seems justified to speak of an absence of Dufour bladelets in the Swabian Aurignacian. In contrast to that, many Aurignacian sites in France, especially in the Périgord, yielded an abundance of Dufour bladelets. To give just one example: layer 8 from Abri Pataud in the French Périgord alone yielded 46 Dufour bladelets (Chiotti 2000) which means that this is the number of specimens from just one single Aurignacian layer of just one single site in the Périgord. What does it mean? Did the Aurignacians in the Périgord have an identity (or identities) different from that (those) of the Swabian Aurignacians? I think this was definitely the case. But can this been deduced from the presence or absence of Dufour bladets? Or does it in turn mean that all groups producing many Dufour bladelets and all groups not producing Dufour bladelets at all shared similar identities respectively? I guess not. My conclusion is that the presence or absence of specific lithic tool types provides little – if any – information about the group identity/identities of Upper Paleolithic groups. It rather reflects different performances within a common pool of cultural capacities.

What about the differences between the small, often twisted and very finely retouched bladelets of Dufour type and the larger, untwisted, mostly backed bladelets, sometimes named Fumanian points, found in assemblage belonging to the Protoaurignacian or archaic Aurignacian? Do they in a similar way reflect different performances within a common pool of cultural capacities rather than different identities on an extended group level? Perhaps different cultural identities on a supra-regional scale may become visible here (see discussion in the concluding paragraph).

This may also be true for another tool type to be mentioned here: the small segment-shaped backed points characteristic for the Uluzzian (Fig. 8.2), a so-called transitional industry which can be found in some parts of Italy and, sporadically, in other countries such as Greece, Romania, and perhaps Poland (see Bolus 2004). The points closely resemble the typical Howiesons Poort segments from the southern African MSA with an age of about 65 ka which have with some plausibility be viewed as parts of bow-and-arrow sets – the oldest evidence so far (Lombard and Phillipson 2010; Lombard 2011; Lombard and Haidle 2012). Though functional analyses have not been carried out for the Uluzzian segments yet, it is at least conceivable to view them as indicators for the use of bow-and-arrow in Europe at the transition from the Middle to the Upper Paleolithic. Usually the Uluzzian had been attributed to Neanderthals but a recent analysis of two human teeth from the Uluzzian of the eponymous site of Grotta del Cavallo proved the teeth to come from anatomically modern humans (Benazzi et al. 2011) which, with an age of about 43 ka calBP, would be the earliest modern humans in Europe so far. Similar to the Howiesons Poort technocomplex, some Uluzzian sites yielded personal ornaments and bone tools.

With regard to organic tools, some information is provided by special tool types such as slender pointed ivory objects, perhaps projectile points, found in the lowest Aurignacian layers from Hohle Fels and Geißenklösterle in the Swabian Jura (Bolus and Conard 2006) (Fig. 8.3). They differ strongly from the typical split-based points and seem to carry a fairly regional signature. As far as split-based points, themselves (Fig. 8.4), are concerned, their dimensions as well as their shape vary from site to site so that one may ask if these differences could be explained in terms of style or cultural identity.

In general, only few *bâtons percés* have been found in Aurignacian assemblages. While in central and eastern Europe they are exclusively made of ivory (Fig. 8.5), they are almost exclusively made of antler in Western Europe. In analogy to the case of the Dufour bladelets, one may again ask if group and/or cultural identity becomes visible through the presence (or absence) of specific tool types and/or

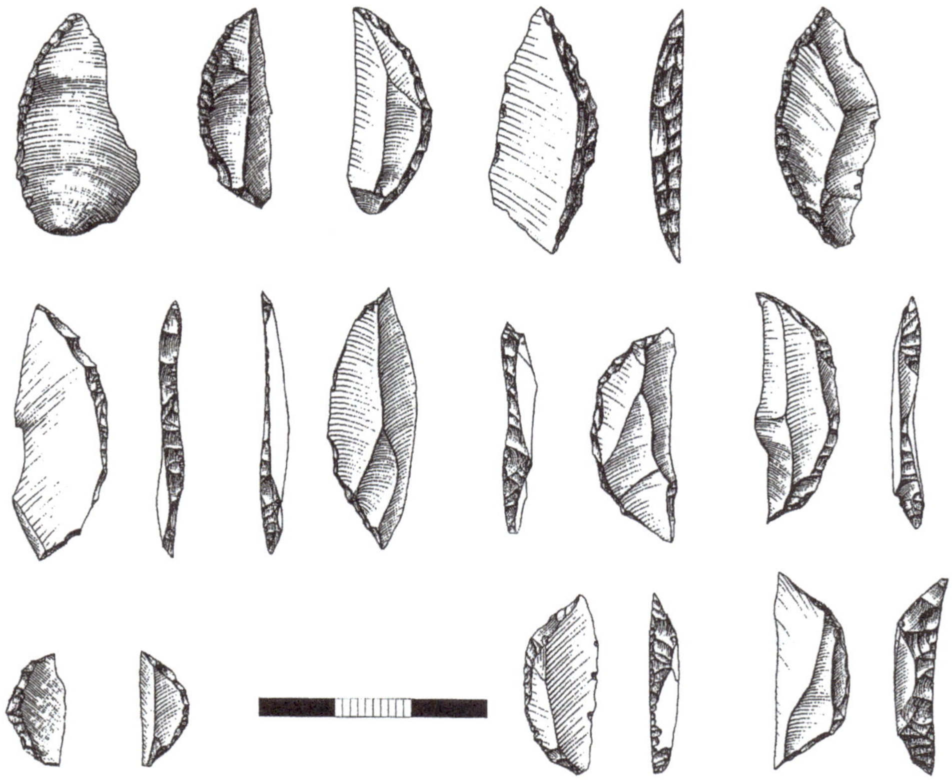

Fig. 8.2 Grotta del Cavallo, Italy. Uluzzian segment-shaped backed lithic points. Modified after Palma di Cesnola 1989

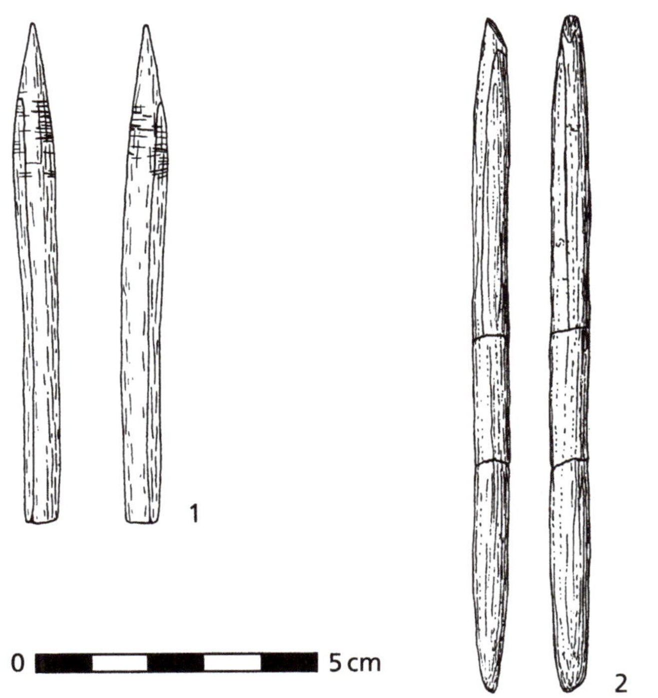

Fig. 8.3 Ivory points from the earliest Aurignacian assemblages of the Swabian Jura. (**1**) Hohle Fels archaeological horizon (AH) Va, (**2**) Geißenklösterle AH III. After Bolus and Conard 2006

Fig. 8.4 Vogelherd, Swabian Jura. Split-based points from Aurignacian layer V. University of Tübingen, photo: H. Jensen

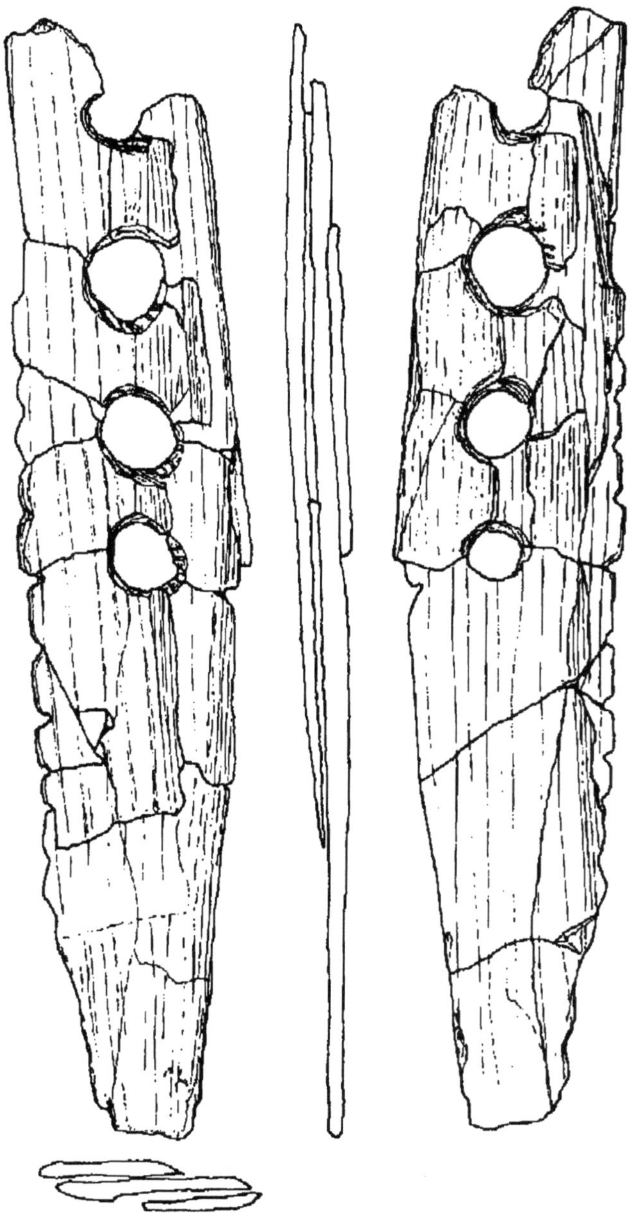

Fig. 8.5 Geißenklösterle, Swabian Jura. *Bâton percé* made of ivory from the upper Aurignacian horizon AH II. After Hahn 1988

specific raw materials. Different choices of raw materials in different regions might be another example of isochrestic style *sensu* Sackett (1982).

Discussion and Conclusions

It seems that aspects of style can help to trace cultural identity in the early Upper Paleolithic. Definitely, stone artifacts and organic tools deserve a stronger consideration in future discussions on style and identity. For instance, differences between retouched bladelets from Aurignacian and Protoaurignacian assemblages, respectively, and segment-shaped backed points from Uluzzian contexts are of relevance on a larger regional or supra-regional scale. An analogy may be seen in the different ways in which Neanderthals produced bifacial tools (see Uthmeier 2016). The presence of specific organic tool types (e.g., ivory points, specifically shaped split-based points, *bâtons percés* made of ivory) in some Aurignacian assemblages of Swabia seems to indicate a certain degree of group identity/ies (or even personal identities?) in the Swabian Aurignacian, which means on a somewhat more restricted regional scale.

A similar phenomenon can be observed with regard to personal ornaments from the Swabian Aurignacian. Several types of personal ornaments carved from mammoth ivory show a distinctive regional signal in that they are strictly limited to the Swabian Aurignacian and lacking in other Aurignacian sites, such as double perforated ivory beads. Other types, such as small disc-shaped ivory beads, are even limited to a single site (Hohle Fels) within the Swabian Aurignacian (Conard 2003; Wolf 2015). Identity on a group level (disc-shaped ivory beads limited to a single Swabian Aurignacian site) or on an extended group level (double perforated ivory beads limited to the Swabian Aurignacian) seems to be expressed through this, thus adding to the conclusions drawn with regard to some types of organic tools in the Swabian Aurignacian.

Seen from a regional point of view, a certain degree of variation and flexibility exists within the Swabian Aurignacian, even within the sequence of a single site. Identity on the group level may be reflected by this. On the other hand, there is a high degree of cultural continuity from the oldest to the youngest Aurignacian assemblages in the Swabian Jura, especially when art objects and ornaments are considered (Bolus 2010), and different assemblage types may be interpreted in terms of functional rather than cultural (and/or chronological) variability.

The French *Aurignacien ancien* shows close similarities and some (cultural or only stylistic?) differences with the Swabian Aurignacian, while assemblages of Protoaurignacian type, besides similarities, also show differences (Conard and Bolus 2015). This may reflect different cultural identities on a regional or even supra-regional scale.

Given these differences observed in various regions, which *sensu* Wiessner (1983) might be viewed as expressions of different emblemic styles shared by relatively large groups respectively, one may ask to what extent differences between more or less contemporary technocomplexes such as early Aurignacian (e.g., Swabian Aurignacian, French *Aurignacien ancien* etc.), Protoaurignacian, and Ahmarian may reflect different cultural identities. In addition to this more synchronous view regarding larger parts of Europe and the Near East, a stronger diachronic view for sites within a clear-defined region such as the Swabian Jura, the Périgord, or northern Italy seems to be promising.

Parameters which should be regarded in more detail in the future concern lithic raw materials and their transport distances. Can the use of specific raw materials for the production of specific tools indicate (cultural/group) identity? In her study on Gravettian organic tools from the Swabian Jura, Martina Barth (2007) could demonstrate that among the Swabian sites there are differences in the raw materials used for the production of specific tool types and that this picture in turn differs from the patterns observed in Western Europe and in eastern Central Europe. It will be a future task to find out if similar tendencies are visible with regard to the Aurignacian (lithic and organic) assemblages.

Having delivered these somehow optimistic considerations, one serious dilemma has to be addressed which is mostly neglected in the discussion about identity in the Paleolithic and which Jürgen Richter has called to my attention. When dealing with differences in stone and organic tools, is it always a question of different 'styles' or rather a question of different tool 'types'? Do, for instance, backed points of Fumane type and Uluzzian segments represent examples of isochrestic behavior? In the sense of Sackett (1982) one might agree. But is it possible that Sackett in fact uses 'style' instead of type? As Jürgen Richter (personal communication, June 2014) correctly stressed, an Uluzzian (or a Howiesons Poort) segment is in the same way a tool 'type' as is a backed point from the Protoaurignacian. This being said, is it really possible to deduce group identity, even on a supra-regional level, from it? This touches the problem of identifying group identity/identities as discussed for younger prehistoric periods (e.g., Bronze and Iron Ages, Middle Ages) since much earlier time than within Paleolithic archaeology (see Burmeister and Müller-Scheeßel 2006 and the papers therein, especially Müller-Scheeßel and Burmeister 2006). Especially within Medieval archaeology, different types of artifacts, for instance different kinds of early Medieval brooches, are often interpreted as mirroring different ethnic groups with different group identities (see Müller-Scheeßel and Burmeister 2006:26). While other researchers involved in the archaeology of the Metal Ages or the Middle Ages reject ethnic interpretations of that kind, they seem to be completely excluded from Paleolithic research today.

Fully aware of the problems and shortcomings resulting from more or less arbitrarily mapping single elements from the material culture and interpreting them in terms of displaying social, cultural or group identities, Müller-Scheeßel and Burmeister (2006:29–31) provide an interesting approach. They try to reconstruct intensities of communication between neighboring groups by means of spatial comparisons of archaeological entities. Instead of focusing on similarities, they argue that if the material culture of two groups shows distinct differences even if the groups live in close proximity, these differences probably result from deliberately avoiding close contacts, and they conclude that this deliberate separation presumably strengthened the identity of both respective groups. To elaborate on this point, this means that differences in the material culture of neighboring groups, whether visible in different 'types' of artifacts or in different 'styles', can be indicators of different identities, but following the argumentation of Müller-Scheeßel and Burmeister (2006) it must not necessarily be the case.

What does it mean for the question(s) raised in this paper? If we view differences between, for instance, backed (projectile) points of Fumane type, Uluzzian segments (presumably projectiles as well), and organic projectile points from the Swabian Aurignacian in terms of different 'styles' or isochrestic behavior and in terms of different 'types', respectively, what do distribution maps showing these artifacts tell us? Do they mirror different styles or even different ethnic groups? Do these distributions tell us anything about group identities? Following the approach by Müller-Scheeßel and Burmeister (2006) it seems possible to conclude that the makers of the Protoaurignacian, those of the 'classic' Aurignacian and those of the Uluzzian, which existed more or less contemporaneously, might have had contacts to a certain degree but watched to keep their respective group identities. One has to admit, though, that the spatial distribution of the technocomplexes mentioned does not overlap too much so that the groups did not live in real close proximities. Moreover, given the caveat with ethnic interpretations of material culture as discussed by Müller-Scheeßel and Burmeister (2006) in general, one ought to be more than cautious. What we definitely need are clear parameters to define 'style' and clearly differentiate 'style(s)' from 'type(s)'.

In conclusion, this paper presents questions rather than answers. It is meant to arouse discussion and to direct the focus on stone and organic tools which are often neglected when discussing style and identity in the Paleolithic. It shall also draw attention to the problems that arise when trying to interpret single elements of material culture. In any case, it seems obvious that stone and organic tools alone will only be weak indicators – if they are indicators at all – to trace group identity in early Upper Paleolithic assemblages. Adding data from other artifact categories such as personal ornaments, decorated objects and art objects may help to learn more about group organization, social differences, and, hopefully, identity in the early Upper Paleolithic. Major problems result from the fact that "identity is to be seen as an open process rather than a static entity "(Müller-Scheeßel and Burmeister 2006:13) and from the "difficulty to wrench unambiguous identity markers from material culture" (Müller-Scheeßel and Burmeister 2006:9; both translations from German by MB). This being said and with the discussion about 'style' versus 'types' in mind, I remain pessimistic about the possibility to unambiguously trace group identity in the early Upper Paleolithic.

Acknowledgements I thank Miriam Haidle, Jürgen Richter, and Thorsten Uthmeier for very helpful comments.

References

Barth, M. (2007). *Familienbande? Die gravettienzeitlichen Knochen- und Geweihgeräte des Achtals (Schwäbische Alb)*. Tübinger Arbeiten zur Urgeschichte 4. Rahden/Westf.: Verlag Marie Leidorf.

Barton, C. M. (1997). Stone tools, style, and social identity: An evolutionary perspective on the archaeological record. *Archaeological Papers of the American Anthropological Association, 7*, 141–156.

Bar-Yosef, O. (2002). The Upper Paleolithic revolution. *Annual Review of Anthropology, 31*, 363–391.

Benazzi, S., Douka, K., Fornai, C., Bauer, C. C., Kullmer, O., Svoboda, J., et al. (2011). Early dispersal of modern humans in Europe and implications for Neanderthal behaviour. *Nature, 479*, 525–529.

Bolus, M. (2004). Der Übergang vom Mittel- zum Jungpaläolithikum in Europa. Eine Bestandsaufnahme unter besonderer Berücksichtigung Mitteleuropas. *Germania, 82*, 1–54.

Bolus, M. (2010). Continuity or Hiatus? The Swabian Aurignacian and the transition to the Gravettian. In C. Neugebauer Maresch & L. R. Owen (Eds.), *New Aspects of the Central and Eastern European Upper Palaeolithic – methods, chronology, technology and subsistence* (pp. 139–150). Wien: Verlag der Österreichischen Akademie der Wissenschaften.

Bolus, M., & Conard, N. J. (2006). Zur Zeitstellung von Geschossspitzen aus organischen Materialien im späten Mittelpaläolithikum und Aurignacien. *Archäologisches Korrespondenzblatt, 36*, 1–15.

Bolus, M., & Conard, N. J. (2008). What can we say about the spatial-temporal distribution of early Aurignacian innovations? *Eurasian Prehistory, 5*(2), 19–29.

Broglio, A., Bertola, S., de Stefani, M., & Marini, D. (2002). L'Aurignaziano della Grotta di Fumane. In A. Aspes (Ed.), *Preistoria Veronese. Contributi e Aggiornamenti* (pp. 29–36). Memorie del Museo Civico di Storia Naturale di Verona – 2^{A} Serie. Sezione Scienze dell'Uomo 5. Verona: Museo Civico di Storia Naturale di Verona – Museo Tridentino di Scienze Naturali.

Burmeister, S., & Müller-Scheeßel, N. (Eds.) (2006). *Soziale Gruppen, Kulturelle Grenzen. Die Interpretation sozialer Identitäten in der Prähistorischen Archäologie*. Tübinger Archäologische Taschenbücher 5. Münster, New York, München & Berlin: Waxmann.

Chiotti, L. (2000). Lamelles Dufour et grattoirs aurignaciens (carénés et à museau) de la couche 8 de l'abri Pataud, Les Eyzies-de-Tayac, Dordogne. *L'Anthropologie, 104*, 239–263.

Close, A. E. (1978). The identification of style in lithic artefacts. *World Archaeology, 10*, 223–237.

Conard, N. J. (2003). Eiszeitlicher Schmuck auf der Schwäbischen Alb. In S. Kölbl & N. J. Conard (Eds.), *Eiszeitschmuck – Status und Schönheit* (pp. 15–49). Blaubeuren: Urgeschichtliches Museum.

Conard, N. J. (2008). A critical view of the evidence for a Southern African origin of behavioural modernity. *South African Archaeological Society Goodwin Series, 10*, 175–179.

Conard, N. J., & Bolus, M. (2015). Chronicling modern human's arrival in Europe. *Science, 348*, 754–756.

Hahn, J. (1988). *Die Geißenklösterle-Höhle im Achtal bei Blaubeuren I. Fundhorizontbildung und Besiedlung im Mittelpaläolithikum und im Aurignacien*. With contributions by H. Gollnisch, A. Scheer, N. Symens, R. Whallon, & J. Weißhaupt. Stuttgart: Theiss.

Haidle, M. N., Bolus, M., Collard, M., Conard, N. J., Garofoli, D., Lombard, M., et al. (2015). The nature of culture – An eight-grade model for the evolution and expansion of cultural capacities in hominins and other animals. *Journal of Anthropological Sciences, 93*, 43–70.

Koehler, H. (2009). *Comportements et identité techniques au Paléolithique moyen (Weichsélien ancien) dans le Bassin parisien: une question d'échelle d'analyse?* Ph.D. Dissertation, University of Paris-Nanterre. Published electronically. http://www.theses.fr/2009PA100185/document/.

Laplace, G. (1966). *Recherches sur l'origine et l'évolution des complexes leptolithiques*. Paris: Éditions de Boccard.

Lombard, M. (2011). Quartz-tipped arrows older than 60 ka: Further use-trace evidence from Sibudu, KwaZulu-Natal, South Africa. *Journal of Archaeological Science, 38*, 1918–1930.

Lombard, M., & Haidle, M. N. (2012). Thinking a bow-and-arrow set: Cognitive implications of Middle Stone Age bow and stone-tipped arrow technology. *Cambridge Archaeological Journal, 22*(2), 237–264.

Lombard, M., & Phillipson, L. (2010). Indications of bow and stone-tipped arrow use 64 000 years ago in KwaZulu-Natal, South Africa. *Antiquity, 84*, 635–648.

Moreau, L. (2009). *Geißenklösterle. Das Gravettien der Schwäbischen Alb im europäischen Kontext*. Tübingen: Kerns.

Müller-Scheeßel, N., & Burmeister, S. (2006). Einführung: Die Identifizierung sozialer Gruppen. Die Erkenntnismöglichkeiten der Prähistorischen Archäologie auf dem Prüfstand. In S. Burmeister & N. Müller-Scheeßel (Eds.), *Soziale Gruppen, Kulturelle Grenzen. Die Interpretation sozialer Identitäten in der Prähistorischen Archäologie* (pp. 9–38). Tübinger Archäologische Taschenbücher 5. Münster, New York, München & Berlin: Waxmann.

Nigst, P. R. (2009). *The Early Upper Palaeolithic in the Middle Danube region: A regional study using the evidence of Willendorf II, Stratzing 94, Vedrovice V, and Stránská skála IIIc*. Ph.D. thesis, University of Leipzig (published version 2012: *The Early Upper Palaeolithic in the Middle Danube Region*. Leiden: Leiden University Press).

Palma di Cesnola, A. (1989). L'Uluzzien: faciès italien du Leptolithique archaïque. *L'Anthropologie, 93*, 783–811.

Sackett, J. R. (1982). Approaches to style in lithic archaeology. *Journal of Anthropological Archaeology, 1*, 59–112.

Uthmeier, T. (2016). Neanderthal utilitarian equipment and group identity: The social context of bifacial tool manufacture and use. In M. N. Haidle, N. J. Conard, & M. Bolus (Eds.), *The nature of culture* (pp. 65–77). Dordrecht: Springer.

Vanhaeren, M., & d'Errico, F. (2006). Aurignacian ethno-linguistic geography of Europe revealed by personal ornaments. *Journal of Archaeological Science, 33*, 1105–1128.

Villa, P., Soriano, S., Teyssandier, N., & Wurz, S. (2010). The Howiesons Poort and MSA III at Klasies River main site, Cave 1A. *Journal of Archaeological Science, 37*, 630–655.

Wadley, L. (2003). How some archaeologists recognize culturally modern behavior. *South African Journal of Science, 99*, 247–250.

Wiessner, P. (1983). Style and social information in Kalahari San projectile points. *American Antiquity, 48*, 253–276.

Wobst, H. M. (1977). Stylistic behavior and information exchange. In C. E. Cleland (Ed.), *For the director: Research essays in honor of James B. Griffin* (pp. 317–342). Anthropological paper 61. Ann Arbor: Museum of Anthropology, University of Michigan.

Wolf, S. (2015). *Schmuckstücke. Die Elfenbeinbearbeitung im Schwäbischen Aurignacien*. Tübingen: Kerns Verlag.

Wurz, S. (2008). Modern behavior at Klasies River. *South African Archaeological Society Goodwin Series, 10*, 150–156.

Chapter 9
Childhood, Play and the Evolution of Cultural Capacity in Neanderthals and Modern Humans

April Nowell

Abstract The life history pattern of modern humans is characterized by the insertion of childhood and adolescent stages into the typical primate pattern. It is widely recognized that this slowing of the maturational process provides humans with additional years to learn, transmit, practice and modify cultural behaviors. In both human and non-human primates a significant amount of their respective dependency periods are spent in play. In contrast to modern humans, Neanderthals experienced shorter childhoods. This is significant as there is extensive psychological and neurobiological evidence that it is during infancy, childhood and adolescence that milestones in social and cognitive learning are reached and that play and play deprivation have a direct impact on this development. Faster maturation rates and thus shorter childhoods relative to modern humans lessen the impact of learning through play on the connectivity of the brain. In the context of play behavior, humans are unique in that adult humans play more than adults of any other species and they alone engage in fantasy play. Fantasy play is part of a package of symbol-based cognitive abilities that includes self-awareness, language, and theory of mind. Its benefits include creativity, behavioral plasticity, imagination, apprenticeship and planning. Differences in the nature of symbolic material culture of Neanderthals and modern humans suggest that Neanderthals were not capable of engaging in human-grade fantasy play.

Keywords Middle Paleolithic • Upper Paleolithic • Life history • Brain • Behavioral plasticity • Cognitive development • Fantasy • Imagination • Archaeology of children

A. Nowell (✉)
Department of Anthropology, University of Victoria, STN CSC, P.O. Box 3050, Victoria, BC V8W 3P5, Canada
e-mail: anowell@uvic.ca

Introduction

In prehistoric societies children likely comprised at least forty to sixty-five percent of the population (Baxter 2005, 2008), yet the archaeological literature, if not the archaeological record itself, has largely been silent about the lives they lived and the contributions they made (e.g., see discussions in Sofaer Derevenski 1997; Kamp 2001; Shea 2006). Children's play and their unconventional use of material culture were, until recently, believed to introduce a randomizing and distorting element into the archaeological record (Baxter 2005, 2008). Children were not only unknown, they were unknowable. This chapter, however, argues that children's play is in fact serious business and that an understanding of how experiences garnered in childhood shape the brain may be key to documenting the evolution of unique aspects of human cognition and behavior. While this chapter is not about 'finding' children and their material culture in the Paleolithic record per se it does focus on childhood as a crucial ontological stage for cognitive maturation and on the implications for behavior and cultural capacity in having a longer period of neural development.

This chapter begins by describing the slowing of the human maturation process and the emergence of childhood and adolescence stages in human life history. Then, maturation rates and lengths of childhood are compared between Neanderthals and modern humans based on game changing studies of dental, brain and somatic growth and development. Next, extensive psychological and neurobiological evidence is presented that suggests that it is during infancy, childhood and adolescence that milestones in social and cognitive learning are reached and that play and play deprivation have a direct impact on this development. By way of discussion and conclusion, it is argued that Neanderthals had shorter childhoods and this influenced the adults they became and the material culture they produced. It is further argued that the key to understanding hominin

Miriam N. Haidle, Nicholas J. Conard and Michael Bolus (eds.), *The Nature of Culture: Based on an Interdisciplinary Symposium 'The Nature of Culture', Tübingen, Germany*, Vertebrate Paleobiology and Paleoanthropology, DOI 10.1007/978-94-017-7426-0_9

cognitive abilities lies not in absolute brain size or encephalization quotients but rather in our ability to reconstruct the childhoods of our hominin ancestors and to use the biological and archaeological evidence to piece together how they spent that time.

Human Life History

Life history can be defined as "the allocation of an organism's energy for growth, maintenance and reproduction… [and is] a life strategy adopted by an organism to maximize fitness in a world of limited energy" (Dean and Smith 2009:115). Life history is sometimes referred to as reproductive turnover or 'the speed of life' (Stearns 1992; Nowell 2010). Following this metaphor, it is often said that primates have the slowest life histories of all the mammals (Harvey and Clutton-Brock 1985; Zimmermann and Radespiel 2007; Robson and Wood 2008) and, by extension, one might infer that humans experience the slowest life histories of all the primates but the emerging picture of human life history is at once more complex and more elegant (Nowell 2010).

As paleoanthropologists, we are accustomed to thinking in terms of k-strategists and r-strategists. Relative to r-strategists, k-strategists reach the age of reproduction more slowly, have fewer numbers of offspring, and put greater investment in each offspring. While k-strategists are thought to produce high quality offspring, the drawback to being a slow reproducer is that you risk dying before you reproduce (Robson and Wood 2008; Dean and Smith 2009:101). But humans are unique in that they have elements of an r-strategy within the larger life history of a k-strategist. Specifically, they have a long gestation period, a large brain, mature more slowly with females reaching the age of reproduction later, experience an extended dependency period, and enjoy increased longevity, but at the same time they have evolved shorter birth spacing, they wean sooner, and have more dependents than expected for an ape that matures at the age modern humans do (Robson and Wood 2008). Essentially, modern humans are characterized by a reproductive pattern that works twice as fast as that of the great apes (Dean and Smith 2009:115) – something that Wood (1994) has described as secondary r-selection. Added to this unique combination of features are a suite of derived elements – concealed ovulation, helpless young, rapid postnatal brain growth in infants, continued dependency after weaning, paternal care, and vigorous post-menopausal life in females (Bogin 1997; Kaplan 2002; Hawkes et al. 2003; Leigh 2004, 2012; Zimmermann and Radespiel 2007; Dean and Smith 2009:115; Robson and Wood 2008). Thus, modern humans have developed a strategy of producing high quality offspring while at the same time reducing the risk of dying before maturation by 'living fast' in some respects, within an overall pattern of 'living slow'.

This slowing of the human maturation process relative to other primates has led to the development of two unique life history stages in humans. Human biologists divide human life history into five stages – infancy (from birth to weaning), childhood (from weaning to the eruption of M1), juvenile (from M1 to puberty and the onset of the adolescent growth spurt), adolescence (from the onset of the adolescent growth spurt to the cessation of the growth and maturation) and adulthood (Mace 2000; Bogin 2003, 2009) with the stages of childhood and adolescence being unique to humans. It should be noted that the term childhood (and adolescence for that matter) is often thought of as a cultural construct and depending on the context it can be. We know that cross-culturally and over time what it means phenomenologically to 'be a child' varies greatly (e.g., Kamp 2001; Baxter 2008; Konner 2010) but the terms 'childhood' and 'adolescence' are used here as biologically defined stages of human development dependent on anatomical markers that permit comparisons between species.

The slowing of the maturation process in humans, as Bogin (2003:32) notes, is significant as the childhood stage "adds an additional four years of relatively slow growth and allows for behavioral experience that further enhances developmental plasticity" (see also Neubauer and Hublin 2012) while adolescence further extends this period of growth and development with additional years to learn, practice, transmit, modify and innovate upon aspects of their culture. Data from pediatric fMRI studies of prefrontal cortical activity suggest that attention, memory and inhibition – all key elements of working memory – continue to develop during childhood and adolescence (Casey et al. 2000) with the dorso-lateral prefrontal cortex, the region most closely associated with working memory, the last region of the brain to mature and develop (Casey et al. 2000; Vuontela et al. 2003; Paus 2005; Durston and Casey 2006; Neubauer and Hublin 2012).

Childhood in Neanderthals and Modern Humans

Current data suggest that while Neanderthals matured more slowly than earlier hominins such as *Homo erectus*, they matured more quickly than modern humans and experienced a shorter childhood as a result. For instance, Smith et al. (2010) concluded based on a synchrotron virtual histology study of a Middle Paleolithic dental sample from Neanderthal and early modern human juveniles that Neanderthals matured more quickly than their modern human counterparts do. They found that both have an "extended duration of dental development [but this] period of dental immaturity is particularly prolonged in modern humans" (Smith et al. 2010:20923). Similarly, in a detailed review of studies of Neanderthal somatic growth, Thompson and Nelson (2011) conclude that Neanderthals experienced a compressed adolescence and growth spurt compared to modern humans. Additional evidence for faster maturation rates and thus shorter childhoods in Neanderthals derives from studies of brain growth patterns. Neubauer and Hublin (2012); see also Coqueugniot and Hublin (2007, 2012), estimate that Neanderthal neonates had brain sizes that slightly exceeded those of modern human neonates. This is indicative of greater pre-natal brain growth than is found in humans. Post-natal brain growth in Neanderthals was more accelerated. While shorter in duration, this accelerated brain growth led to larger overall adult brain sizes in Neanderthals and to the attainment of adult brain size more quickly (Neubauer and Hublin 2012). These findings concerning growth rate and duration are particularly important because "morphological development of the brain occurs simultaneously with cognitive development while the young individual learns and interacts with his or her environment" (Neubauer and Hublin 2012:568).

Cognitive Development in Childhood

In the context of the above studies, the fact that Neanderthals experienced shorter childhoods (and adolescence) is salient because learning, and in particular learning through play, may be key to understanding cognitive differences between them and modern humans. Liu and colleagues (2012) studied post-mortem brain samples from humans, chimpanzees and macaques at all stages of life (pre-natal to senescent stages). They found that there were significant differences in the developmental trajectories of synaptogenesis between these three species (Cohen 2012). Specifically, in the pre-frontal cortex (PFC), a region of the brain implicated in social behavior, abstract thinking and reasoning, the genes responsible for synaptogenesis are turned on in humans slightly after birth, peaking at five years of age. By contrast, in non-human primates the expression of these genes peaks during the last few months of fetal development and are turned off just after birth. Furthermore, there are twelve times the number of genes involved in the development and functioning of synapses in the PFC in humans than in chimpanzees. Similarly, researchers counted more than 7000 synapses in the three species at different ages and found that in non-human primates they dramatically increase in number just after birth whereas in humans they peak around 4 years of age. Similar results were found when the researchers studied gene expression and synapse formation in the lateral cerebellar cortex (CBC), a region associated with language, attention and manual abilities (Liu et al. 2012). What this research suggests is that even when corrected for differences in lifespan, humans have much more time to form synapses and that synapse formation is particularly influenced by experiences garnered during the first five years of life (Cohen 2012; Liu et al. 2012).

It is likely that the extended dependency period in human children evolved in parallel with the extension of synaptic development in the PFC because of this region's involvement in higher order cognitive processing (Liu et al. 2012). Thus, it is significant that it is during early and late childhood that milestones in social and cognitive learning are reached (Casey et al. 2000; Neubauer and Hublin 2012). The results of Liu et al.'s (2012) study further suggest that the timing of synaptic differentiation and synaptic pruning is similarly extended. These results support the findings of other researchers who have noted that through late childhood and into adolescence there is a gradual decrease in synaptic density in the pre-frontal cortex with concomitant strengthening of the remaining synapses (Casey et al. 2000). This plateauing and pruning of synapses in the prefrontal cortex likely represents "the behavioral, and ultimately, the physiological suppression of competing, irrelevant behaviors" (Casey et al. 2000:246).

The Evolutionary Importance of Play

Humans, like most mammals, spend a great deal of their dependency period in play. Play increases steadily as "young become more mobile and begin to interact with litter mates and other young ones in a group or herd and then decreas[es] as sexual maturity approaches, stable dominance positions are being acquired and the animal is typically engaged in serious competition for mates and resources" (Smith 2010:54). Play forms part of a bio-behavioral package involving prolonged immaturity, opportunities for learning and parental investment in such learning and is characteristic of more encephalized species, particularly k-strategists (Konner 2010; Smith 2010). Among invertebrates, play has been documented only among captive octopuses who, after a period of object exploration, seem to enjoy object play such as squirting water bottles at objects (Smith 2010) and playing with Lego (Kuba et al. 2006). This is perhaps not surprising given that octopuses are the most highly encephalized invertebrates. There is very little, if any, credible evidence for play among fish with the exception of sharks, or among reptiles and amphibians (Smith 2010). The frequency and variety of play behaviors increase significantly among bird species who have been documented engaging in locomotor, object and social play (Smith 2010).

Among mammals, play behavior is truly widespread. In fact, play is an almost universal mammalian characteristic with seventeen of nineteen orders of placental mammals engaging in play (Burghardt 2005). These behaviors include locomotor, object, social, and sexual play. Of all the mammals, primates engage in play the most with the apes adding 'play mothering' to the mammalian repertoire of play behaviors. Play mothering mostly refers to juveniles who pick up and 'mother' infants (Smith 2010) but there may be instances where this behavior involves the transference of species-specific mothering behaviors to inanimate objects such as sticks or logs (see, for example, Kahlenberg and Wrangham 2010). This particular form of play will be discussed further below in relation to the significance of fantasy play.

Evidence suggests that experiences garnered through play have a dramatic impact on synaptic formation, differentiation and pruning. For example, in a study involving weanling rats, one group of young rats was housed and reared with one peer, another group had access to three peers, and a third group had only an adult female for company (Pellis and Pellis 2009). All three groups of weanlings were exposed to normal socialization except that the latter group did not engage in play because it is rare for adult rats to play with juveniles – even with their own offspring. Post-mortem studies conducted on the rats' brains revealed that the major difference between the three groups was in the development of the prefrontal cortex. Specifically, there was greater synaptic density in the medial prefrontal cortex of rats that were prevented from interacting with members of their peer group than in the groups of rats that were permitted such interaction (Pellis and Pellis 2009). In other words, that all important pruning did not take place. This study provides the first direct evidence that play deprivation actually altered the "anatomy of the neurons", leading Pellis and Pellis (2009:92, 94) to conclude that "the brain not only shapes play but that play shapes the brain." In fact, they observed that play deprivation reduced the ability of animals "to formulate and engage behavioral options dependent on the executive functions of the prefrontal cortex – the same kinds of problems that animals have when they are reared normally but are subjected to experimental damage of their prefrontal cortex" (ibid).

What Is Play?

While approximately 60 attributes of play have been identified (see Pellegrini et al. 2007), following Smith (2010:5), play is defined here as being comprised of familiar behaviors such as running, climbing, and manipulating objects but that are fragmented, repeated, exaggerated or reordered in some fashion. Behaviors are normally identified as play if they have positive affect (i.e., there is enjoyment as indexed by play signals such as laughter), are flexible in form and content; are intrinsically motivated (i.e., performed for their own sake); if they are a means to an end (i.e., children are often more interested in the performance of the behavior than its outcome) and non-literal (this is most relevant for fantasy play which is discussed below) (Smith 2010:6). The importance of play is supported by the fact that it will rebound in frequency and intensity in offspring deprived of play (Smith 2010) and the lack of play experience can seriously disadvantage an animal later in life (Bateson 2005).

Costs and Benefits of Play

There are both costs and benefits to engaging in play behavior. First, a great deal of time is spent in play that could be put to other uses such as resting, eating, watching or exploring (Smith 2010:64). Second, play is often quite vigorous and energetically demanding (Bateson 2005; Smith 2010). According to Konner (2010), play increases food requirements by 10–20 %. Third, in bouts of rough and tumble play, the likelihood of injury increases (Bateson 2005; Smith 2010). Finally, there is risk associated with neglect of predator danger. Individuals absorbed in play are often less vigilant while at the same time their behavior can make them more conspicuous (Bateson 2005; Smith 2010). For selection to take place, however, the benefits of play must outweigh these costs. The apparent benefits of play include the attainment of knowledge, skills and experience through engagement with the environment (Bateson 2005), physical fitness, development of technical and social skills, cognitive development, behavioral plasticity, enhanced problem solving abilities and increased ability to innovate.

During play, animals are able to practice behaviors that they will use once they reach adulthood and to learn from their mistakes safely (Bateson 2005:17). According to researchers, this has the greatest effect on adult behaviors characterized by the greatest risk such as fighting, mating (when there is serious competition), catching prey, avoiding being prey and moving efficiently in familiar environments (Bateson 2005:17). Play also increases physical fitness. High intensity bouts of play are good exercise for animals, while low intensity bouts build general physical capacity (Smith 2010). Furthermore, there is a positive relationship between the frequency of social play and cerebellum size across species. As Smith (2010:72) observes, the fact that "the cerebellum is strongly implicated in the coordination and control of motoric activities [...] suggests that some aspects of social play, much of which is locomotor, have been selected for aspects of motor development," the tuning of musculature and sculpting of the nervous system. One critical feature of the mammalian nervous system is its excess number of neuronal connections (Bateson 2005). As an animal develops, unused connections are lost, thus the "sculpting of the nervous system reflects the steadily improving efficiency of the body's classification command and control systems" (Bateson 2005:16). Furthermore, these changes are reflected in behavior. Bateson (2005:16) argues that movements practiced during play become more efficient and better coordinated.

Animals also develop technical and social skills through play. Object play is well documented among primates (Smith 2010) as they use a large number of objects in their natural environment for food extraction, prey catching, agnostic displays and for a variety of other tasks. Thus, during play these animals are able to practice and explore tool behavior (Smith 2010). Social play provides opportunities for animals to cement social relationships (Bateson 2005), evaluate competitors, develop behavioral flexibility and coping skills and improve motor skills necessary for fighting (Bateson 2005; see also Ragir and Savage-Rumbaugh 2009). Furthermore, during social play, human children socialize and enculturate each other and this may be at least as important if not more so than parental nurturing. In fact, researchers argue that the "vertical transmission of knowledge from older to younger children parallels, complements or undermines adult-child transmission" and thus children can be seen as significant "agents of cultural change" (Konner 2010:661). Furthermore, peaks in social play activity correlate with the development of neural networks that serve as the basis for shared systems of communication (Ragir and Savage-Rumbaugh 2009 and references therein). Taken together, this means that social play provides the evolutionary context within which meanings can be generated and shared by convention, hence the emergence of symbols and with them of creativity. This will be explored further below in relation to fantasy play.

Play can also have a significant effect on learning in general. Phenomenological and neurobiological evidence suggests that during play individuals feel positive and that play is pleasurable and rewarding (Konner 2010:511). Studies show that individuals learn better and in a greater variety of ways when they are feeling playful (Konner 2010:512). Furthermore, researchers have studied neurotransmitter systems involved in play and it appears that opioid systems are implicated. As Konner (2010:510) notes, "during social play activity increases in the nucleus accumbens and other reward mediating brain areas."

Finally, play, as we have seen is key to cognitive development, problem solving and innovation (Bateson 2005). For example, play affords opportunities for the generation of new and possibly adaptive responses to novel environments (Pellegrini et al. 2007; Smith 2010). It reproduces culture but can also change culture over time (Smith 2010). It gives individuals an opportunity to probe particular behaviors or explore potential solutions (Bateson 2005:18). Play rearranges previously unrelated thoughts and ideas into new combinations and while most are discovered to be fruitless, this is seen as a "powerful means of gaining insights and opening up possibilities that had not been previously recognized" (Bateson 2005:18–19). Thus, play allows for the breaking away from established patterns (Bateson 2005). Bateson (2005:22) argues that "aspects of play can increase the total sum of spontaneously developing behavioral structures that serve to solve complex problems" and it is these characteristics of play that many researchers argue were selected for evolutionarily. In fact, some researchers assert that this is the most important aspect of play. They argue that many tasks could be practiced and learned through direct observation without engaging in play but because play can lead to innovation and these innovations can be derived and transmitted within the social group, play is especially important evolutionarily (Pellegrni et al. 2007).

Why Stop Playing?

The overwhelming benefits of play beg the question of why the frequency of play behaviors declines later in life around sexual maturity. Adults of all species seem to engage in play far less often than younger individuals. Most bouts of play in adults are associated with infants and this interaction likely contributes to the overall healthy development of their offspring (Lewis 2010; Smith 2010). But there appears to be a significant shift in the benefit to cost ratio against play as animals reach maturity (Smith 2010). It is seems clear that costs of play might increase as play fighting becomes rougher, for example, but do the concomitant benefits of play decrease with age as well (Smith 2010)? I believe that there are two things happening. First, the ways in which play shapes the developing brain may be less effective or actually impossible in an adult brain. Second, there is something special about the kind of learning and information transfer that takes place during play. Some researchers have argued that the innovative outcomes of play discussed above could result in changes in gene frequencies through the process of organic selection (Bateson 2005). Organic selection is a "process that allows for environmentally acquired behaviors to influence the genotype" without having to invoke Lamarckian inheritance (Smith 2010:72). In other words, in each "generation a new response is socially learn[ed] or at least environmentally acquired but over a number of generations, Natural Selection will favor those genotypes that more readily acquire this new behavior" (Smith 2010:76). Thus, characteristics acquired extragenetically via play and learning processes may over time result in genotypic change (Smith 2010; see also Bateson 2005). In particular, "innovative behaviors associated with play during [the childhood] and juvenile periods should be especially prone to this process because of the protection and provisioning associated with play during" these periods (Smith 2010:76). By contrast, late adolescence and adulthood are not protected in the same way and in this case learning through direct observation is a better strategy (Smith 2010:76). As Bateson (2005:16) observes, play is like scaffolding, once the "job is done, it largely falls away."

Adult humans, however, are a striking exception to this general rule. They play more frequently than mature animals of any other species and this may be because in humans the learning, information transfer and overall shaping of the brain that takes place through play continues throughout an individual's lifetime even if at a lower intensity. In the study by Liu and colleagues (2012) discussed above, researchers discovered that while the expression and number of genes related to synapse formation and function declines during adulthood among all three species studied, the average expression levels of these genes were always higher in humans. According to Liu et al. (2012:619), "this implies that humans might sustain higher levels of synaptogenesis or synaptic activity throughout adult life." Thus, at least some of the benefits of play connected to learning and innovation appear to continue in adult humans.

Fantasy Play

When it comes to play, humans are different in another important way. Humans are the only species to engage in fantasy play, or imaginative/pretend play. Fantasy play is part of a package of symbol-based cognitive abilities that includes self-awareness, language, and theory of mind (Smith 2010:151). The definition of fantasy play depends on the intention and awareness of the participant(s) – you have to know you are pretending (Smith 2010) and this, of course, can make it difficult to identify in non-human species. Fantasy play differs from deception in that fantasy play occurs when a person or animal openly signals pretense whereas deception occurs when the person or animal tries to make others believe the pretense is reality (Smith 2010). Researchers argue that fantasy play has been sustained evolutionarily because of its contribution to human psychological development and growth. The benefits of fantasy play include the development of creativity (Carruthers 2002), behavioral plasticity, and imagination as well as apprenticeship and planning.

Fantasy play appears to be exclusive to humans as there are virtually no reliable examples of fantasy play among the apes with the possible exception of some cases where maternal behaviors are applied to inanimate objects (Gómez and Martín-Andrade 2005). While there is some limited evidence of these behaviors in the wild as noted above (e.g., when a juvenile chimpanzee 'mothers' an infant that is not her own or more rarely a log), the best albeit controversial examples (see Pellegrini et al. 2007) are provided by the so-called linguistic apes. Much like human children, these apes have been documented bathing, wrestling and otherwise playing with dolls (including stuffed baby gorillas). The reason for this, according to researchers such as Premack (1983), is that when apes are introduced to human language they learn a "new way of representing the world – a more abstract representational code – that allows them to tackle problems in a novel way" (Gómez and Martín-Andrade 2005:166). They have what Premack (1983) refers to as an 'upgraded mind'. Conversely, Tomasello (1999a) suggests that it is interacting with humans that is key. Specifically, humans "shape their attentional and action patterns in a characteristic way that somehow may result in higher cognitive processes" (Gómez and Martín-Andrade 2005:166). For Gómez and Martín-Andrade (2005:167), however, it is the combination of learning how to use symbols and human interaction that can explain the apparently superior imaginative play of these apes.

A third factor not directly addressed by the researchers is the enriched material world that these apes find themselves in and this may be especially important given that object play is already well developed in primates. Play often involves things, and things (at least partly) shape imagination, facilitate exploration and the novel combination of elements (see Iriki and Sakura 2008; papers in Malafouris and Renfrew 2010). In fact in humans, fantasy play begins just as children begin to verbalize and this is facilitated by encouragement from adults and by the presence of toys and other objects designed for symbolic play. This may explain why linguistic apes also exhibit this behavior at times (Gómez and Martín-Andrade 2005:167). In the wild, the apes have the "spontaneous ability to produce action schema out of context including possibly object substitution" (Gómez and Martín-Andrade 2005:167) but mothering a log instead of an infant in the wild does not have to involve any symbolic intent or imagined representations of the usual context of these objects (Gómez and Martín-Andrade 2005). Nonetheless, symbolic training working on these types of behavior could result in more complex play episodes (Gómez and Martín-Andrade 2005:167). Symbolic play in children emerges in "parallel with and as a reflection of the development of a number of specific adaptations for symbolic behavior such as imitation, nonverbal referential communication, labeling, and elaborate object manipulations and categorization." As Gómez and Martín-Andrade (2005: 167–168) note, "apes lack several of these adaptations [...] but when placed in an enriched environment that offers new pressures and new opportunities [...] the cognitive makeup of these apes seems to be able to generate some instances" of true fantasy play.

Evolutionary Implications of Fantasy Play

These observations among non-human primates have important evolutionary implications. It appears that when placed in a typical human environment, apes' minds demonstrate three of the components from which fantasy play may have emerged in human evolution (Gómez and Martín-Andrade 2005). First, they demonstrate some ability to separate action schema from the contexts in which they are executed. Second, they demonstrate some ability to recognize similarities between objects and physical representations of real objects such as dolls. Third, they demonstrate the ability and motivation to play with and explore action schema.

Drawing on Finke et al. (1992), Carruthers (2002:230) argues that fantasy play involves two distinct stages. First, there is the creation of a new idea or hypothesis that can then be explored and developed. Individuals work out the consequences of this idea before finally rejecting it or accepting it and putting into practice. While Carruthers (2002:241) argues that fantasy play and imaginative thinking require the "creation of a whole new type of propositional attitude", in my view fantasy play appears to draw on, practice and integrate aspects of working memory such as sequential memory, temporal-order memory, and the integration of information across space and time. These are all necessary for sequencing activities in their proper order to attain a goal (Casey et al. 2000; Coolidge and Wynn 2005, 2007). Furthermore, working memory is correlated with other abilities such as emotional reasoning, storytelling, general intelligence, and fluid intelligence (Coolidge and Wynn 2005, 2007) and these abilities would seem to be necessary for fantasy play.

Carruthers (2002:241) argues there are two forms of imagination underlying fantasy play – experiential imagination and propositional imagination. Experiential imagination is "the capacity to form and manipulate images relating to a given sense modality" (e.g., visual, auditory, tactile). He describes this in relation to the extensive neural feedback systems that come into play when an individual attempts to discern what he or she is looking at when the input is somehow unclear or degraded. Carruthers (2002) believes that this kind of imagination is a free by-product of the conceptualizing processes of these sensory input systems. Propositional imagination "is the capacity to form and consider a propositional representation without commitment to its truth or desirability" (Carruthers 2002:241). This too is a 'free' by-product but this time of language. Therefore, through fantasy play humans draw on these abilities and generate not only new but relevant and innovative ideas, images and sentences (Carruthers 2002). These two capacities allow humans the ability to frame and consider a possibility without yet endorsing it. In the context of all of this research then, how likely is it that Neanderthals with their faster maturation rates and shorter childhoods engaged in a human level of fantasy play with all of the concomitant benefits of innovation and imaginative problem solving?

Discussion and Conclusion – The Implications of a Shorter Childhood in Neanderthals

Recent genetic evidence for interbreeding between Neanderthals and modern humans (Green et al. 2010) raises the question of how great an impact a shorter childhood might have had on Neanderthal cognition and behavior if these two populations were members of the same species. There are two issues – (1) whether evidence of interbreeding precludes Neanderthals and modern humans from belonging to separate species; and (2) whether differences in maturation rates within a species or between subspecies can have tangible and significant cognitive and behavioral effects. First, the evidence of interbreeding between Neanderthals and modern humans does not tell us whether they formed one or two species. Given how closely related Neanderthals and modern humans were phylogenetically, it is possible that gene-level reproductive isolating mechanisms had not yet fully evolved between the two lineages. Species of the genus *Canis* are a well-known example of this phenomenon (Vila et al. 1997; Wayne et al. 1997; Vila and Wayne 1999) as all species of this genus can interbreed and produce fertile offspring. If Neanderthals and modern humans did belong to different species it is clear that a shorter childhood, as a species specific trait, would be significant in the context of the evidence discussed above for the relationships between decelerated maturation rates, synaptogenesis, synaptic differentiation and pruning, learning through play and behavioral plasticity and other social/cognitive skills.

Second, if on the other hand, Neanderthals represented one end of human variation or more likely were a sub-species of modern humans then differences in the rate and pattern of brain growth during ontogeny could still have had enormous impacts on cognition and behavior. Studies of variation within our own species illustrate this relationship. Autistic children are characterized by brain overgrowth (Courchesne and Pierce 2005). By three to four years of age, their brains are 5–10 % larger than typically developing children and "this process results in a reduced influence of the environment on the connectivity of the developing brain" (Neubauer and Hublin 2012). This, of course, includes the benefits garnered through play – particularly social play as this brain overgrowth does not affect the brain uniformly but affects the frontal and temporal lobes the most often resulting in specific social (e.g., inability to 'read' social cues, lack of empathy) and linguistic disabilities (Courchesne and Pierce 2005; Gunz et al. 2010; Neubauer and Hublin 2012). According to Neubauer and Hublin (2012), accelerated brain growth in the first few years of life may be related to over-connectivity locally and under-connectivity between brain regions that are further apart.

While I am not suggesting that Neanderthals were disabled in any way, as Neubauer and Hublin (2012:573) note, "the case of autistic patients shows that differences in early development can have large effects on cognitive and behavioral differences." Not only do Neanderthals experience accelerated rates of pre- and especially post-natal brain growth, their *pattern* of brain growth differed from modern humans as well (Neubauer and Hublin 2012; see also Ponce de León and Zollikofer 2001). The human neonate brain is not simply a smaller version of an adult brain. Rather, during early post-natal growth, a period important for establishing neuronal connections, the modern human brain experiences several stages of morphological change resulting in the globular shape that is characteristic of our species. Neanderthals lack this globularization phase (Gunz et al. 2010) and because this "difference occurs during a vulnerable period of brain development when neural connections are built, it is likely that the connectivity of the brain was different in Neanderthals and modern humans" (Neubauer and Hublin 2012:586) leading to differences in interaction with and perception of the environment. It is interesting to note that Green et al. (2010) found in their study of the Neanderthal genome that the few nucleotide substitutions in modern humans associated with cognitive development, when mutated in extant humans, can lead to schizophrenia, autism or Down syndrome. As Neubauer and Hublin (2012:579) observe, the serious consequences of mutations in these genes "suggest that [Neanderthal] brains developed differently and, therefore, that [they] may have had different cognitive and behavioral abilities than modern humans."

Fifteen years ago Tomasello (1999b:512) wrote that human children are unique in that they grow up in "the midst of the accumulated wisdom of their social group as embodied in its material artifacts, symbolic artifacts and conventional processes", something he referred to as their ontological niche. He further argued that children are specially adapted to acquire knowledge in these forms through social cognition skills such as joint attention and imitative learning. But how this came about was somewhat of a black box for Tomasello (1999b) who suggested that these skills might be the result of a 'simple' genetic change. Based on the evidence reviewed here, I would argue that life history, ontological development of the brain and the science of play provide a more robust mechanism for understanding the development of much is what is unique about human cognition.

The more limited ability of Neanderthals to engage in and benefit from the advantages of play relative to modern humans is apparent in Neanderthal material culture. While the archaeological records of Neanderthals and modern humans during the Middle Paleolithic and early Upper Paleolithic were quite similar in many ways (Nowell 2013), one of the major differences between them lies in the material expression of symbolic behavior. Late Neanderthals (i.e., those younger than 50 ka) do appear to create items of personal adornment to a greater extent than earlier members of this species possibly as the result of increased social pressures (Nowell and Chang 2012) but the beads, feathers, pigment and raptor talons (e.g., Zilhão 2007; Zilhão et al. 2010; Peresani et al. 2011; Morin and Laroulandie 2012) associated with these hominins are fundamentally different in nature from the contemporaneous symbolically mediated material culture associated with modern humans. In particular, I am referring to the creation of fantastical creatures such as the human-lion(ess) from Geißenklösterle, Hohle Fels, and Hohlenstein-Stadel in Germany (Conard 2011) (Fig. 9.1) and the bison-human from Chauvet in France (Clottes 2010) (Fig. 9.2). The creation of these types of artifacts requires the combination of two different elements to create something that does not exist in nature.

Fig. 9.1 Löwenmensch figure from Hohlenstein-Stadel (Germany), new reconstruction. *Photo credit* Dagmar Hollmann/Wikimedia Commons, *Licence* CC BY-SA 3.0

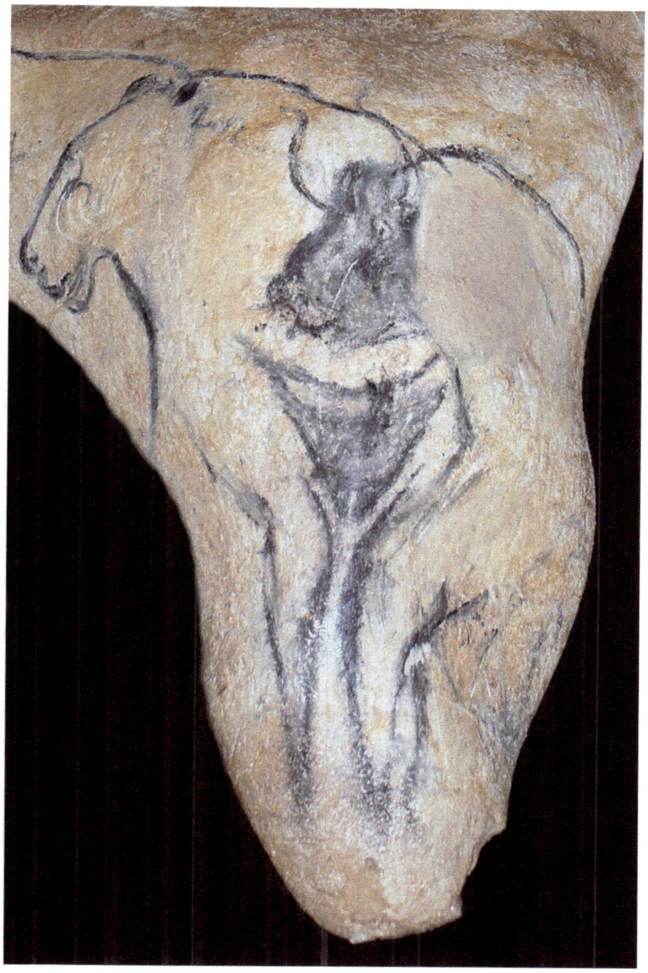

Fig. 9.2 Painting from Grotte Chauvet (France) showing the lower part of the body of a woman and, next to her, a fantastical creature – a bison with a human arm and hand. *Photo credit* Yanik Le Guillou/Ministère de la Culture (France); *Courtesy* Jean Clottes

To expand this to cave art more broadly, the ability to reproduce a three dimensional form on a two dimensional surface or to 'see' a figure in ivory necessitates a completely different way of imagining the world. These are more than 'cultural' differences – all of these endeavors associated with modern humans necessitate a human-grade imagination indicative of fantasy play and this speaks to broader issues of brain plasticity, the ontogenetic context of brain plasticity and general implications for learning through play as discussed in this chapter. In sum, whether it is large brained Neanderthals or small brained hobbits, the key to understanding their cognitive abilities lies not in absolute brain size or encephalization quotients but rather in our ability to reconstruct their childhoods and to use the biological and archaeological evidence to piece together how they spent that time.

Acknowledgements I thank Miriam Haidle, Nicholas Conard, and Michael Bolus for inviting me to participate in the ROCEEH (The Role of Culture in Early Expansions of Humans) symposium at the University of Tübingen and for their support and patience. I also thank Helen Kurki, Felix Riede, and Tanya Smith for helpful suggestions. Finally, I am grateful to Iain Davidson for his encouragement and insightful comments throughout the process of writing this chapter.

References

Baxter, J. E. (2005). *The archaeology of childhood: Children, gender and material culture*. Walnut Creek, CA: AltaMira Press.

Baxter, J. E. (2008). Archaeology of childhood. *Annual Review of Anthropology, 37*, 159–195.

Bateson, P. (2005). The role of play in the evolution of the great apes and humans. In A. D. Pellegrini & P. K. Smith (Eds.), *The nature of play* (pp. 13–26). New York: Guilford Press.

Bogin, B. (1997). Evolutionary hypotheses for human childhood. *Yearbook of Physical Anthropology, 40*, 63–89.

Bogin, B. (2003). The human pattern of growth and development in paleontological perspective. In J. L. Thompson, G. E. Krovitz, & A. J. Nelson (Eds.), *Patterns of growth and development in the genus Homo* (pp. 15–44). Cambridge: Cambridge University Press.

Bogin, B. (2009). Childhood, adolescence, and longevity: A multilevel model of the evolution of reserve capacity in human life history. *American Journal of Human Biology, 21*, 567–577.

Burghardt, G. (2005). *The genesis of animal play: Testing the limits*. Cambridge, MA: MIT Press.

Carruthers, P. (2002). Human creativity: Its cognitive basis, its evolution and its connections with childhood pretense. *British Journal for the Philosophy of Science, 53*, 225–249.

Casey, B. J., Giedd, J. N., & Thomas, K. M. (2000). Structural and functional brain development and its relation to cognitive development. *Biological Psychology, 54*, 241–257.

Cohen, J. (2012). Human Brains wire up slowly but surely. February 1, 2012. *Science Now*. http://news.sciencemag.org/2012/02/human-brains-wire-slowly-surely.

Clottes, J. (2010). Les mythes. In M. Otte (Ed.), *Les aurignaciens* (pp. 237–251). Paris: Éditions Errance.

Conard, N. J. (2011). The demise of the Neanderthal cultural niche and the beginning of the Upper Paleolithic in southwestern Germany. In N. J. Conard & J. Richter (Eds.), *Neanderthal lifeways, subsistence and technology* (pp. 223–240). New York: Springer.

Coolidge, F. L., & Wynn, T. (2005). Working memory, its executive functions, and the emergence of modern thinking. *Cambridge Archaeological Journal, 15*, 5–26.

Coolidge, F. L., & Wynn, T. (2007). The working memory account of Neandertal cognition: How phonological storage capacity may be related to recursion and the pragmatics of modern speech. *Journal of Human Evolution, 52*, 707–710.

Coqueugniot, H., & Hublin, J.-J. (2007). Endocranial volume and brain growth in immature Neandertals. *Periodicum Biologorum, 109*, 379–385.

Coqueugniot, H., & Hublin, J.-J. (2012). Age-Related changes of digital endocranial volume during human ontogeny: Results from an osteological reference collection. *American Journal of Physical Anthropology, 147*, 312–318.

Courchesne, E., & Pierce, K. (2005). Brain overgrowth in autism during a critical time in development: Implications for frontal pyramidal neuron and interneuron development and connectivity. *International Journal of Developmental Neuroscience, 23*, 153–170.

Dean, C. M., & Smith, H. (2009). Growth and development of the Nariokotome youth, KNM-WT 15000. In F. E. Grine, J. G. Fleagle, & R. E. Leakey (Eds.), *The first humans: Origin and early evolution of the genus* Homo (pp. 101–120). New York: Springer.

Durston, S., & Casey, B. J. (2006). What have we learned about cognitive development from neuroimaging? *Neuropsychologia, 44*, 2149–2157.

Finke, R., Ward, T., & Smith, S. (1992). *Creative cognition*. Boston, MA: MIT Press.

Gómez, J.-C., & Martín-Andrade, B. (2005). Fantasy play in apes. In A. D. Pellegrini & P. K. Smith (Eds.), *The nature of play: Great apes and humans* (pp. 139–172). New York: Guilford Press.

Gunz, P., Neubauer, S., Maureille, B., & Hublin, J.-J. (2010). Brain development after birth differs between Neanderthals and modern humans. *Current Biology, 20*(21), R921–R922.

Green, R. E., Krause, J., Briggs, A. W., Maricic, T., Stenzel, U., Kircher, M., et al. (2010). A draft sequence of the Neandertal genome. *Science, 328*, 710–722.

Harvey, P. H., & Clutton-Brock, T. H. (1985). Life history variation in primates. *Evolution, 39*, 559–581.

Hawkes, K., O'Connell, J. F., & Blurton-Jones, N. J. (2003). Human life histories: Primate trade-offs, grandmothering socioecology, and the fossil record. In P. M. Kappeler & M. E. Pereira (Eds.), *Primate life histories and socioecology* (pp. 204–231). Chicago: University of Chicago Press.

Iriki, A., & Sakura, O. (2008). The neuroscience of primate intellectual evolution: Natural selection and passive and intentional niche construction. *Philosophical Transactions of the Royal Society B, Biological Sciences, 363*, 2229–2241.

Kaplan, H. S. (2002). Human life histories. In M. Pagel (Ed.), *Encyclopedia of evolution* (Vol. 2, pp. 627–631). Oxford: Oxford University Press.

Kahlenberg, S. M., & Wrangham, R. (2010). Sex differences in chimpanzees' use of sticks as play objects resemble those of children. *Current Biology, 20*(24), R1067–R1068.

Kamp, K. A. (2001). Where have all the children gone? The archaeology of childhood. *Journal of Archaeological Method and Theory, 8*, 1–34.

Konner, M. (2010). *The evolution of childhood*. Cambridge: The Belknap Press of Harvard University Press.

Kuba, M. J., Byrne, R., Meisel, D. V., & Mather, J. A. (2006). When do octopuses play? Effects of repeated testing, object type, age, and food deprivation on object play in *Octopus vulgaris*. *Journal of Comparative Psychology, 120*, 184–190.

Leigh, S. R. (2004). Brain growth, life history, and cognition in primate and human evolution. *American Journal of Primatology, 62*, 139–164.

Leigh, S. R. (2012). Brain size growth and life history in human evolution. *Evolutionary Biology, 39*, 587–599.

Lewis, K. P. (2010). From landscapes to playscapes: The evolution of play in humans and other animals. In R. R. Sands & L. R. Sands (Eds.), *The anthropology of sport and human movement: A biocultural perspective* (pp. 61–89). Lanham, MD: Lexington Books.

Liu, X., Somel, M., Tang, L., Yan, Z., Jiang, X., Guo, S., et al. (2012). Extension of cortical synaptic development distinguishes humans from chimpanzees and macaques. *Genome Research, 22*, 611–622.

Mace, R. (2000). Evolutionary ecology of human life history. *Animal Behaviour, 59*, 1–10.

Malafouris, L., & Renfrew, C. (2010). *The cognitive life of things. Recasting the boundaries of the mind*. Cambridge: McDonald Institute for Archaeology.

Morin, E., & Laroulandie, V. (2012). Presumed symbolic use of diurnal raptors by Neanderthals. *PLoS ONE, 7*(3), e32856.

Neubauer, S., & Hublin, J.-J. (2012). The Evolution of Human Brain Development. *Evolutionary Biology, 39*, 568–586.

Nowell, A. (2010). Working memory and the speed of life. *Current Anthropology, 51*, S121–S133.

Nowell, A. (2013). Cognition, behavioral modernity, and the archaeological record of the Middle and Early Upper Paleolithic. In G. Hatfield & H. Pittman (Eds.), *The evolution of mind, brain, and culture* (pp. 235–262). Philadelphia: University of Pennsylvania Museum of Archaeology and Anthropology Press.

Nowell, A., & Chang, M. (2012). *Symbolism in Late European Neanderthals: Detection and evolutionary context*. Paper presented at the American Association for Physical Anthropologists meeting, Portland, OR.

Paus, T. (2005). Mapping brain maturation and cognitive development during adolescence. *Trends in Cognitive Sciences, 9*(2), 60–68.

Pellegrini, A. D., Dupuis, D., & Smith, P. K. (2007). Play in evolution and development. *Developmental Review, 27*, 261–276.

Pellis, S., & Pellis, V. (2009). *The playful brain: Venturing to the limits of neuroscience*. Oxford: Oneworld Publications.

Ponce de León, M. S., & Zollikofer, C. P. (2001). Neanderthal cranial ontogeny and its implications for late hominid diversity. *Nature, 412*, 534–538.

Peresani, A., Fiore, I., Gala, M., Romandinni, M., & Taggliacozzo, A. (2011). Late Neandertals and the intentional removal of feathers as evidenced from bird bone taphonomy at Fumane Cave 44 ky B.P., Italy. *Proceedings of the National Academy of Sciences of the USA, 108*, 3888–3893.

Premack, D. (1983). The codes of man and beasts. *The Behavioral and Brain Sciences, 6*, 125–167.

Ragir, S., & Savage-Rumbaugh, S. (2009). Playing with meaning: Normative function and structure in play. In R. Botha & C. Knight (Eds.), *The Prehistory of language* (pp. 122–141). Oxford: Oxford University Press.

Robson, S. L., & Wood, B. (2008). Hominin life history: Reconstruction and evolution. *Journal of Anatomy, 212*, 94–425.

Shea, J. J. (2006). Child's play: Reflections on the invisibility of children in the Paleolithic record. *Evolutionary Anthropology, 15*, 212–216.

Smith, P. K. (2010). *Children and play*. Malden, MA: Wiley-Blackwell.

Smith, T. M., Tafforeau, P., Reid, D. J., Pouech, J., Lazzari, V., Zermeno, J. P., et al. (2010). Dental evidence for ontogenetic differences between modern humans and Neanderthals. *Proceedings of the National Academy of Sciences of the USA, 107*, 20923–20928.

Sofaer Derevenski, J. (1997). Engendering children, engendering archaeology. In J. Moore & E. Scott (Eds.), *Invisible people and processes. Writing gender and childhood into European Archaeology* (pp. 192–202). London: Leicester University Press.

Stearns, S. C. (1992). *The evolution of life histories*. Oxford: Oxford University Press.

Thompson, J. L., & Nelson, A. J. (2011). Middle childhood and modern human origins. *Human Nature, 22*, 249–280.

Tomasello, M. (1999a). *The cultural origins of human cognition*. Cambridge, MA: Harvard University Press.

Tomasello, M. (1999b). The human adaptation for culture. *Annual Review of Anthropology, 28*, 509–529.

Vila, C., & Wayne, R. K. (1999). Hybridization between wolves and dogs. *Conservation Biology, 13*, 195–198.

Vila, C., Savolainen, P., Maldonado, J. E., Amorim, I. R., Rice, J. E., Honeycutt, R. L., et al. (1997). Multiple and ancient origins of the domestic dog. *Science, 276*, 1687–1689.

Vuontela, V., Steenari, M.-R., Carlson, S., Koivisto, J., Fjällberg, M., & Aronen, E. T. (2003). Audiospatial and visuospatial working memory in 6–13 year old school children. *Learning and Memory, 10*, 74–81.

Wayne, R. K., Geffen, E., Girman, D. J., Koepfli, K. P., Lu, L. M., & Marshall, C. R. (1997). Molecular systematics of the Canidae. *Systematic Biology, 46*, 622–653.

Wood, B. A. (1994). *Dynamics of human reproduction: Biology, biometry and reproduction*. New York: Aldine de Gruyter.

Zilhão, J. (2007). The emergence of ornaments and art: An archaeological perspective on the origins of "Behavioral Modernity". *Journal of Archaeological Research, 15*, 1–54.

Zilhão, J., Angelucci, D. E., Badal-García, E., d'Errico, F., Daniel, F., Dayet, L., et al. (2010). Symbolic use of marine shells and mineral pigments by Iberian Neandertals. *Proceedings of the National Academy of Sciences of the USA, 107*, 1023–1028.

Zimmermann, E., & Radespiel, U. (2007). Primate life histories. In W. Henke & I. Tattersall (Eds.), *Handbook of paleoanthropology, vol 1: Principles, methods and approaches* (pp. 1163–1205). Berlin: Springer.

Chapter 10
Stone Tools: Evidence of Something in Between Culture and Cumulative Culture?

Iain Davidson

Abstract This paper goes back to some first principles about what culture might be and how it can be investigated in order to ask questions about the Last Common Ancestor and the role of stone tools in changing the nature of culture. In doing so it considers the relations between learned behavior, tradition, culture, cumulative culture, and cultures: I juxtapose models used by ROCEEH with an alternative model that shows how creatures which can be argued to have such behaviors, and thus the behaviors are related to each other through time and across the animal world.

Keywords Tradition • Artifacts • Primates • Early hominins • Learning • Oldowan • Acheulean • Levallois • Mousterian

Introduction

For the symposium that was the source of the papers in the volume, I was asked to write about stone tools in the evolution of culture. I was specifically asked to put my presentation into the context of some hierarchical models of culture produced by the ROCEEH project. In this paper, I will confine myself to issues relevant to that question, but will propose a variation on one of the ROCEEH models.

What Is Culture?

The plausible claim that non-human animals have culture (Laland 2008) presents a challenge and an opportunity to anthropologists (sensu lato). On the one hand there seems to be a challenge to the uniqueness of humans as culture-bearing organisms, and the frisson of excitement among those who have found culture in other animals that they have, rather cheekily, disturbed one of the central tenets of another discipline. On the other hand there is an opportunity to see how the uniquely human aspects of culture emerged through evolutionary processes to produce the situation familiar to any first year student of anthropology (Herrmann et al. 2007).

The proposition of this paper is that the Last Common Ancestor (LCA) of humans and chimpanzees had something more like those behaviors that are called chimpanzee culture than those that are like human culture. The addition of stone tools did not initially make much difference to that situation, and the period over which this was true was probably longer than is generally recognized. But the acts involved in making and using stone tools were one of the selective contexts in which cognitive change took place and made a difference to survival prospects such that expanded populations with the new cognition were capable of sustaining innovations and expanding into new environments. This paper concentrates on using the methodology applied to the identification of ape culture to the early record of stone tools. Consideration of the cognitive implications will be presented elsewhere.

The fundamental issues are ones of definition. We cannot avoid them. On one hand are social and cultural anthropologists who emphasize that people living in societies produce values and pass them on in ways which have fundamental

Dedication In memory of Lewis Binford, whose many contributions about the nature of culture informed us all and stimulated us to think more carefully about the way we do archaeology. I particularly remember his phrase "tool-assisted animal behavior" as a way of thinking about early hominins. It inspires many of the thoughts in this paper.

I. Davidson (✉)
School of Humanities, University of New England, 10 Cluny Rd, Armidale, NSW 2350, Australia
e-mail: iain.davidson@live.com.au

and

Department of Archaeology, Flinders University of South Australia, GPO Box 2100, Adelaide, SA 5001, Australia

Miriam N. Haidle, Nicholas J. Conard and Michael Bolus (eds.), *The Nature of Culture: Based on an Interdisciplinary Symposium 'The Nature of Culture', Tübingen, Germany,* Vertebrate Paleobiology and Paleoanthropology, DOI 10.1007/978-94-017-7426-0_10 influence on all aspects of society, whether they appear to be part of that value system or not (e.g., Kuper 1999). On the other are biologists who consider that such values are just a special case of the more general set of conditions for learning and constraining behavior (Bonner 1980; Enquist and Ghirlanda 2007). Some have advocated studying living animal species by concentrating on the relative importance of social and asocial learning (Laland and Janik 2006), removing the issue from the question of values. It is this conflict that needs to be addressed by archaeologists looking at the evolution of culture. We might do this in two ways: one, by assuming that apes have culture, and so the common ancestor (whenever that was) would also have had culture and in consequence, investigating the different ways in which culture changed during the process by which hominins and apes evolved in separate directions; the second, by taking a strict anthropological view that culture involves values (socially defined such that, for example, good values are determined by comparison with bad ones; see for example Kuper 1999:57–58) and consider how such values could have emerged from the "almost-culture" of the Last Common Ancestor. I prefer the second course of action (Noble and Davidson 1996).

Original Definitions

The concept of culture has a long history in many disciplines including anthropology (Bennett 2005) and the variation in meaning ranges from its early use in association with agriculture (Goddard 2005) to that associated with "High" culture: "culture being a pursuit of our total perfection by means of getting to know, on all the matters which most concern us, the best which has been thought and said in the world," (Arnold 1869) – what I call "the Sydney Opera House (SOH)" sense of the word culture. The definition may have been produced a long time ago, but this sense underlies most modern vernacular uses of the word. The original anthropological definition was given a year after Arnold by Tylor (1870): "Culture or civilization, taken in its wide ethnographic sense, is that complex whole which includes knowledge, belief, art, morals, law, custom and other capabilities and habits acquired by man as a member of society" such that the much more ethnocentric SOH definition can be considered a special case of Tylor's, which was more comprehensive (assuming he included woman with man).

For both of these nineteenth century definitions, many of the criteria of recognition cannot be identified either for archaeology or for ethology of non-humans since both studies have no access to the thoughts, sayings, beliefs and intentions of the actors concerned. There is no sign that either chimpanzees or orangutans evaluate their behavior in order to pursue the "best" or to distinguish individual or small groups within the larger group. The exception to this generalization can be found in some incidents in chimpanzee social interaction at Gombe that involved dominance displays (Goodall 1986:426–427). The best documented is the use by Mike of kerosene tins to startle other chimpanzees while grooming,[1] and his switch to other objects when the tins were removed. Most remarkably for this discussion, the adolescent Figan was twice seen appearing to practise using tins in such a display, but he never actually used them (Goodall 1986:426). It may be that what this example demonstrates is that the "pursuit of the best" definition really needs to be qualified by saying that the best can be recognized (or evaluated) because it is achieved by successive members of the society, who learned to behave in the appropriate way, with minor variations on the excellence of performance. The chimpanzee examples might not be included in that definition – it is moot, for example, whether they could evaluate the merits of one grooming hand-clasp (McGrew and Tutin 1978) against another. Of course, the capacity of humans to represent particular examples of behavior as "the best" is a product of symbolic representation of value, since that judgment depends on convention and is to an extent arbitrary. Whether Mike's successful use of kerosene tins can be considered peak performance, or Figan's unconsummated practice could be regarded as a step towards "the best", it remains difficult to use the Arnold definition (or modern, vernacular variants of it) in seeking to identify the evolutionary emergence of culture.

Both archaeologists and ethologists, therefore, have turned to another set of definitions, particularly those collected by Kroeber and Kluckhohn (1963). This collection formed the basis for an argument to the effect that some behaviors of chimpanzees differed so much between the populations in Gombe and Mahale that they involved social custom (McGrew and Tutin 1978) (Table 10.1). Then and later, McGrew stopped just short of claiming that chimpanzees had culture (e.g., McGrew 1992), but became more certain (McGrew 2004) when the debate among ethologists shifted (Whiten et al. 1999, 2001). When Davidson and McGrew (2005) extended these criteria to the consideration of early hominin stone tools, it seemed possible that by the same criteria, culture was not readily identified for Oldowan industries.

Various alternative discussions have been offered, including, for example, on the one hand those that recommend abandoning the static concept of culture in favor of understanding how infant apes acquire shared meanings

[1] I am very grateful to both Bill McGrew and Barbara King who independently drew my attention to the example of Mike clattering kerosene cans. I am sure they would not want me to bind them to the use I have made of the example.

Table 10.1 Criteria used by McGrew and Tutin (1978) and by Davidson and McGrew (2005) to decide whether chimpanzees or early hominin stone-knappers could be said to have met criteria for recognizing culture derived from Kroeber and Kluckhohn (1963)

Criterion	Chimpanzees	Stone industries	Questions for later archaeology
Innovation	Evidence slight, short observation time	Yes – but then long stasis	What evidence is sufficient to recognize innovation among stone tools?
Dissemination	Circumstantial (but getting better)	Not certain – no studies of learning	What precision of chronology would be needed, and what type of pattern in the stone tools?
Standardization	Slight, but maybe	Probable but not certain – problem of equifinality	Need to understand the constraints on knapping (see Kuhn in Nowell and Davidson 2010)
Durability	Quite good	Not certain – tools survive, but does tradition?	Do similarities in stone artifact form arise from tradition or equifinality?
Diffusion	Not at the time of McGrew and Tutin (1978)	Independent discovery possible	How do you account for the Movius line and the "impoverishment" of early Australian tools?
Tradition	Yes	Probable, but archaeologists have only assumed it	Was the Acheulean invented more than three times? (Africa, Europe and Australia)
Non-subsistence	Grooming hand clasp	Probably	Woodworking at Koobi Fora 1.8 Ma
Natural adaptiveness	Yes (but behaviors in laboratory and captivity are much more remarkable)	Uniquely hominin – experimental evidence with apes irrelevant	At what point does variation in stone artifact form stop being a simple product of contingencies of manufacture, on one hand, and use on the other?

through patterned interactions with adults (King 2002), or on the other those that want to retain the concept as a descriptor for the processes by which children come to have certain sorts of ideas and not others (Brown 2002). Unlike most of the traits used for identifying the "cultural" nature of ape or early hominin behavior, the emphasis here is on the socially-mediated inter-generational context of learning, which, again, is likely to be difficult for the study of archaeology and of apes in the wild. For reasons of this sort, emphasis shifted to socially-learned behaviors and their products and away from an emphasis on "knowledge, belief, art, morals, law, custom" and especially on the "best" among such traits.

Learned Behavior

How can such social learning aspects of a definition of culture be applied to non-humans? Such an approach might emphasize three requirements: consistency within groups, variation between groups and tradition over a number of generations. I should begin this discussion with a caveat about the word "social" which some purists might restrict to language-mediated interactions among humans. Clearly using such a definition would automatically exclude animals that do not use language from entry into the classification "social learning", so in this context the word is taken to have a more general meaning which encompasses groups of animals consistently interacting with each other. Under this definition, social learning occurs where it is not individual learning alone, nor learning by vertical transmission from parent to offspring alone, but includes observational learning which may be vertical, horizontal among peers in the "social" group so-defined, as well as oblique from adults to members of the next generation who are not their offspring (Box and Gibson 1999).

All learning tends to produce similarity between the behavior of the learner and the model – it is one of the ways in which we recognize that learning has happened. In this sense there is always likely to be consistency between behaviors that result from learning. But this is also true of behaviors that result from genetic determination, the functional requirements of dealing with specific environmental circumstances, or the equifinality involved in the limited means of producing certain types of artifacts (Davidson 2002; Moore 2011). The problem is not simple, as indicated by the example of Japanese macaques washing potatoes that is often cited as an example of social learning (Nishida 1986), but seems more likely to have resulted from individual discovery or learning (Visalberghi and Fragaszy 1990). The problem is how to tell whether there is social learning.

Teaching, the mirror image of learning, is commonplace among humans. Yet it is rare to find examples in which there seem to be elements of teaching among chimpanzees. The best documented case involves chimpanzee adults apparently scaffolding the actions of infants trying to crack nuts (Boesch 1991; Boesch and Boesch 1993; Boesch et al. 1994), but the length of time over which the process operates to achieve success is so long that it may be that chimpanzees learned by observational learning among the "social" group rather than as a result of the teaching. Such scaffolding actions may be no more than a rare behavior on which selection could operate, and that what is important about it is the observation that the adult chimpanzee was behaving as if it recognized the infant's lack of understanding.

One of the ways in which social learning has been demonstrated for chimpanzees (Whiten and Boesch 2001) and orangutans (van Schaik et al. 2003) is in identifying the differences between groups, while holding other variables equal. There is still a question of the scale at which genetic constraints might be operating (assuming genetic variation between the studied groups is enough to separate them whatever the variation within groups; see Gagneaux et al. 2001), and there seems to be a role for some ecological constraints. The problem seems to be very difficult (Laland and Janik 2006), but there is at least an argument that for chimpanzees many of the supposedly cultural differences reflect genetic differences between populations (Langergraber et al. 2011) or geographic ones (Kamilar and Marshack 2012). One implication of these studies would be an expectation that as culture-bearing organisms moved apart the genetics of populations differentiated by drift and as their geographic circumstances became distinctive their socially learned behavior might also have shown differences.

Among humans, it may be problematic whether there is good evidence for social learning in some domains (Whiten et al. 2003), but there are other situations where the extent of social learning is undeniable and very strong. Two important cases of such social learning are language and aspects of ritual behavior, and for these there is no doubt that learning is social and there is absolutely no parallel in non-humans. To suggest otherwise is to misunderstand the nature of social learning in humans, and failure to be clear about the nature of such social learning is a major contributor to the confusion about whether it is present in non-humans. In all cases, language is the usual means of communication among humans and only present among other free-living animal species by stretching the definitions of language (see, for example, extensive discussions in Davidson 1999). As for ritual, in many cases recognition of membership of societies derives from performance of rituals such as *rites de passage* by which the relationships of individuals to other members of society are determined (see, for example, Rappaport 1999). This is not to suggest that social learning can only be identified through ritual, but that if ritual can be identified then social learning must have taken place. There seems to be no evidence of ritual or other social practices among non-human primates. The practice can be identified in archaeological evidence (Ross and Davidson 2006). In other words, social learning is manifest in human societies in these and other ways which fundamentally affect the human-based criteria for recognition of culture, but which will always be distinct from learning in other animals. Some primatologists might say we just do not know about such things, but in our heart of hearts, we know non-human primates do not have such learning.

Among humans, the SOH doctrine that Arnold was expressing was about the application of the word "culture" only to what he could see of his own culture. This usage has a long history among people who have not interacted with other cultures, and the definition of SOH culture is about a special, privileged section of "our" own society (people who think of themselves as the best). What is generally missing from such definitions is the different perception of privilege in different sections of society – the language-based symbolic recognition of differences in value. From these different sections, what is seen as "best" can vary even within what would otherwise be considered one society (Bach versus the Beatles; Botticelli versus Banksy). Clearly there is ethnocentrism and other sorts of special pleading about such definitions. What is important here is that these vernacular understandings of the word "culture" are part of the reason why so many scientists have, historically, preferred not to use the word or its related concepts at all.

This variable meaning of culture can be expressed in a simple diagram (Fig. 10.1). Here, within the field of all learned behavior (which I will call Culture-L, C-L), there are much smaller areas which represent the learned behavior in a particular society or community (Culture-C, C-C). Within that, a small portion represents Arnold's "best" (Culture-SOH, C-SOH). For another society, there will be differences in what is learned, but for some societies, there will also be overlap. Differences at the level of what is learned may be a result of no more than different practice with no particular assessment on the part of the learners that one thing is better than another. However, giving importance to the "best" portion of what is learned is often a way in which one society distinguishes itself from another (Opera versus Grand Ole Opry). As this process of evaluation of cultural knowledge emerged during the evolution of human society, there arose the potential for rapid diversification of cultures. The presence of such diversification, therefore, could be said to be a result of the evaluation of the quality of what is learned – making the multi-millennial monotony of the Acheulean an unlikely candidate for such evaluation. It is a question that cannot be dealt with here, whether such evaluations could only have taken place in societies that had language.

Culture and Material Culture

Faced with the invisibility of the social and ontogenetic processes that make culture possible, and despite the certainty of the evolution of ontogeny due to changes in life history (Locke and Bogin 2006; Nowell 2010), archaeologists and primatologists have both appealed to characteristics of material culture to identify aspects of culture, as McGrew did (McGrew and Tutin 1978; McGrew 1992). The method

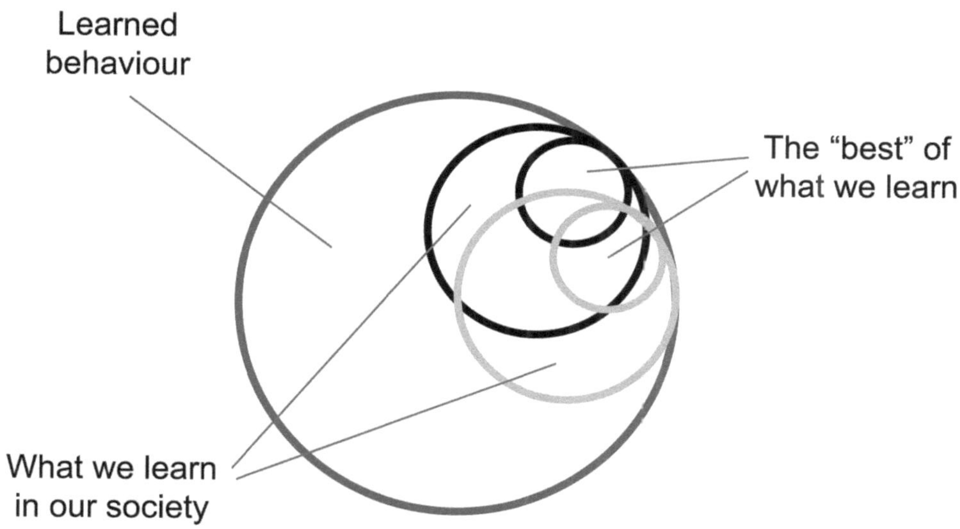

Fig. 10.1 Three dimensions of culture: *1* Learned behaviour; *2* What we learn in our society; *3* The "best" of what we learn

is full of pitfalls, but does return the emphasis to what archaeology is good at: material things. Here I will attempt to concentrate the discussion not only on the thing itself but on the social and learning processes that must have operated to achieve the qualities of such things.

One of the features of the list of behaviors deemed to be cultural (C-L) among chimpanzees is the large number that seem to involve materials removed from their fixed position (e.g., sticks broken off trees) on which actions are performed: 29 of the 39 (74.4 %) of the behaviors in the authors' group D (Whiten et al. 1999), as well as 12 out of 19 (63.2 %) among orangutans (van Schaik and Pradhan 2003). One measure of the cultural difference between humans and other apes is that, among humans, it is impossible to quantify the numbers or proportions of material objects that are removed from their fixed positions because there are just so many of them. Whatever the qualitative differences between the learned behaviors of humans and other animals there are huge quantitative differences, too. Clearly one of the things we can look towards is trying to understand the ways in which hominins modified their environments through cultural constructions in this physical way.

Production of stone tools first required procurement of tool stone raw material (which became a core) and of hammer stone. Some aspects of procurement are similar to those for other primates (especially chimpanzees; see Wynn and McGrew 1989; Wynn et al. 2011). Among chimpanzees at most times, most raw materials are at hand to secure for their elementary technology. Literally, they arrive at a termite mound and reach out and pluck vegetation to make into a probe (Goodall 1986). Sometimes they pluck lots of pieces of raw material (thus indicating foresight?) when at least tens of metres away from the use site (thus indicating planning?) (Bill McGrew personal communication, December 2003). It is well-documented that chimpanzees carry stones for nutcracking from the last used nutting tree to a new nutting tree as it becomes useful, as shown by keeping track of where stones circulate (Boesch and Boesch 1984a, b). This can amount to another order of magnitude greater in terms of transport distance. The knapped stones from the 2.34 Ma site of Lokalalei were removed from sources said to be "available close to the site", although some cores seem to have been carried to the site already prepared (Delagnes and Roche 2005). This selection of raw materials close to the site seems to have been a consistent pattern for early sites (Goldman-Neuman and Hovers 2012) but shortly after 2 Ma, at Oldupai[2] Gorge, raw materials were deposited several kilometers from their source (Hay 1976).

Tool stone raw materials have the advantage that they are connected to their source by petrology. A large amount of data has now accumulated showing the distances over which raw materials travelled at different times. Often this is difficult to interpret, but some general principles can be established. One study collected various estimates and sought to relate it to the network sizes and hence the communicative abilities at particular time periods (Marwick 2003, using data from Féblot-Augustins 1997; Roebroeks et al. 1988, and others). For most regions and periods, there was a commonest distance which was very short – raw materials were acquired relatively locally; there was a group of raw materials from distances not too far from the most abundant source and this group often had an upper limit of distance that was separated from more distant sources; and

[2]The archaeologically well-known name Olduvai has recently been shown to have been a corruption of the original Maasai name Oldupai.

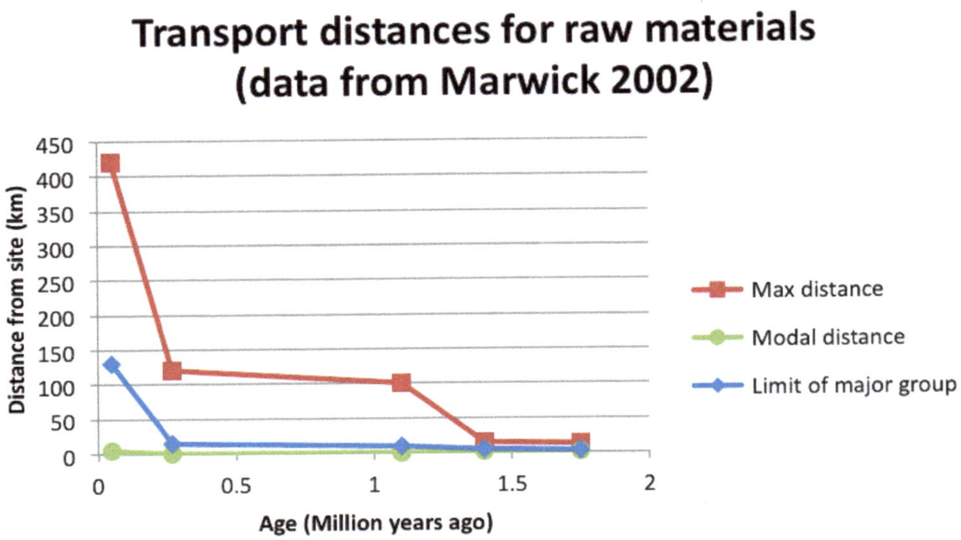

Fig. 10.2 Distances between sources of tool stone and sites where the stones were discarded at different periods (data from Marwick 2003)

there were small amounts of raw materials that were transferred relatively larger distances. Figure 10.2 plots these distances against time (using the mean of the time intervals in Marwick's data).

The important point here is that, in Marwick's study, the distances from which the bulk of the tool stone was acquired remained small until late in prehistory (the data Marwick plotted were described as "Late Middle Palaeolithic and early Upper Palaeolithic", say about 50 thousand years ago).

Cultural artifacts acquire meaning first as indexical signs – they are consistently associated with particular situations or conditions. Thus, the debris left from flaking episodes could be seen by the knapper or by other hominins as a sign that knapping had taken place there previously (Davidson and McGrew 2005). At some indeterminate time later, such signs came to be seen as symbols of identity through comparison of people in those situations or conditions (the case for this is often made by analogy with the situation among modern people of the Kalahari described by Wiessner 1983, but the analogy cannot tell us anything about the timing of that appearance). Such a possibility has been explored for Acheulean sites (Pope et al. 2006). There is nothing like this process of transformation of index into symbol in the non-human world, and it is arguable that it has not even been demonstrated that non-humans comprehend created indexes (but see one such claim by Savage-Rumbaugh et al. 1996). Such cultural artifacts that can function, to third parties, as markers of group identity, do not seem to be in the same category as the common practices in food-getting and tool use among non-human primates. The most well-known apparent exception to this, the chimpanzee hand-clasp (McGrew and Tutin 1978), is not an artifact and there is little evidence that it functions as a sign to third parties (other than primatologists). One of the central challenges in applying the concept of culture to any group other than modern humans is how the use of symbols first occurred, and then transformed culture (for a related argument about the origins of pictures, see Davidson 2012).

Culture in Non-human Animals and Evolution

When McGrew (1992) argued for culture in the material products of chimpanzees, apes and early hominins were much less well-known. It was straightforward to argue that this level of culture would be a feature of the last common ancestor of chimpanzees and humans. A similar argument had been put forward previously (Isaac 1978) and many times before and since (e.g., Noble and Davidson 1996). But when similar claims were made for orangutans (van Schaik et al. 2003) problems emerged: was there culture (C-L) in the last common ancestor of chimpanzees and orangutans? Is there, for example, any evidence for it among Ramapithecines (because orangutans, unlike chimpanzees, seem to have fossil ancestors) (Lipson and Pilbeam 1982; Prasad 1982). If we are looking at an LCA with orangutans, what about the question of culture in bonobos and gorillas (see the phylogenetic tree in Fig. 10.3)? The bonobo question is now settled (Hohmann and Fruth 2003; Savage-Rumbaugh et al. 2004), but something needs to be said about gorillas[3]. If they do not

[3]Since this paper went to the publisher, tool use for food acquisition has been reported (Kinani and Zimmerman 2014) among free-living mountain gorillas.

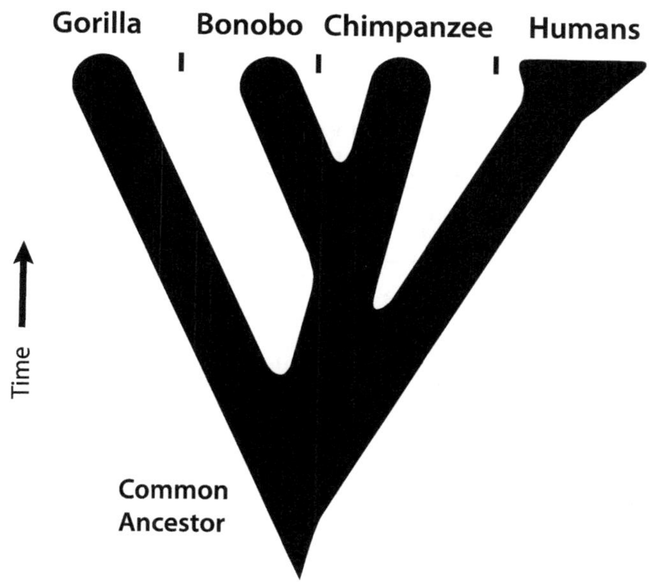

Fig. 10.3 Phylogenetic relationships of African apes

have anything that can be called culture this says something very interesting about culture or about gorillas, or both.

But the story is more complicated, and therefore more interesting. If South American capuchins do culture-like things (Perry 2011), does that mean that the whole common ancestor story goes back to their common ancestor with African apes and monkeys? And what happened to the other African monkeys and to gibbons and gorillas? If dolphins and whales have something like culture (Rendell and Whitehead 2001), is the whole cladistic story wrong? Are we going to argue that culture (C-L) is plesiomorphic – primitive – back to the common ancestor between dolphins and humans, or are these homoplastic – convergent – traits?

Again, the resolution of this question tells us more about what we are dealing with. If it is plesiomorphic, then it may be that the level at which the argument about non-human culture (C-L) is pitched is trivial. This is just a statement about what happens when social learning takes place and all species back to the last common ancestor of humans and guppies (which some have argued have learning that could be called cultural, for example Laland and Hoppitt 2003) might have it or not, probably depending principally on the social nature of the interactions among conspecifics. If it is homoplastic, then arguably it is irrelevant to find "Culture-L" in guppies, because it can emerge in any creature capable of learning to do what other members of the species do. It may be that culture as social learning will be found in any reasonably social species once it is studied appropriately. Under those circumstances, finding "culture" in early hominins will be telling us principally about the nature of the social interaction of hominins at that stage of evolution.

All of this is the essential background to, and changes the nature of, the question: how did the elements of culture (C-L), which may be presumed to have existed since some common ancestor, get transformed into the meaning-laden sort of culture (C-C) that is characteristic of humans? It would, of course, be possible to ignore the claims for culture among non-humans completely and concentrate only on this question, but in doing so, we would lose some of the subtlety. In all likelihood, there has been a continuous pattern of social learning which is plesiomorphic among non-human primates since at least the LCA of humans and chimpanzees. But since social learning can be found among animals which do not share a common ancestry only with other "culture bearing organisms", there are, at the same time, probably many convergent elements of any case of social learning, making it difficult to unravel what is relevant about comparison with chimpanzees, and what is not. In other words, the cultural behavior of early hominins contains some elements that are plesiomorphic and others that are homoplastic with other instances of cultural behavior.

A landmark comparison between chimpanzee behavior and inferences about the Oldowan showed strong similarities between the two (Wynn and McGrew 1989). Since then, more and more similarities have been found in the improved data on cases, and over the same period, there is an enhanced understanding of the elements of cognition for chimpanzees and hominins (Wynn et al. 2011). As compared with a comparison in 1989, chimpanzees seem more similar to a human ancestor, but at the same time those human ancestors seem more like apes. This similarity can be seen in the use of tools to access and process food (but, for example, there is an absence of cutting among chimpanzees, and it is pre-eminent among hominins using stone flakes as argued by Davidson and McGrew 2005). Among chimpanzees, tool use is crucial in the quality of nutritional intake; in the selection of raw materials for tools and of tools prior to use; in carrying tools and food; in re-use of activity areas; in flexibility of procedures to solve immediate problems; and in hierarchical organisation of procedures, as analysed in the operational sequences of Haidle's cognigrams (Haidle 2010). The important point about this analysis is its specific details about behavior in the comparisons, as well as the explicit formulation of chains of related actions in Haidle's cognigrams.

Models of Culture in Hominin Evolution

The conference began with some models (e.g., Fig. 10.4) and was presented with others (e.g., Fig. 10.5). If the inquiry into culture is to progress, we need to understand what such models can achieve.

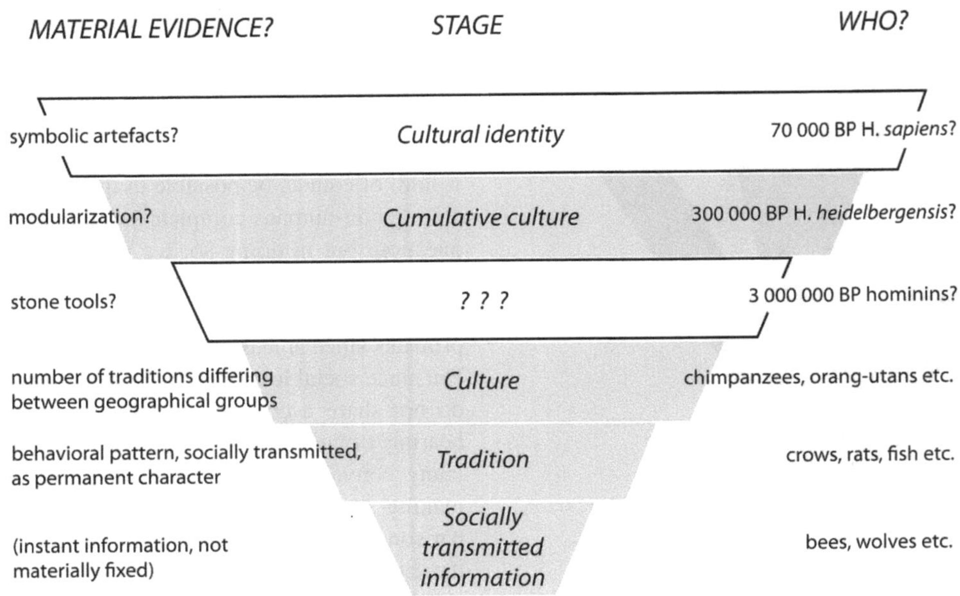

Fig. 10.4 The ROCEEH 1 model of cultural change and associated material evidence precirculated before the workshop

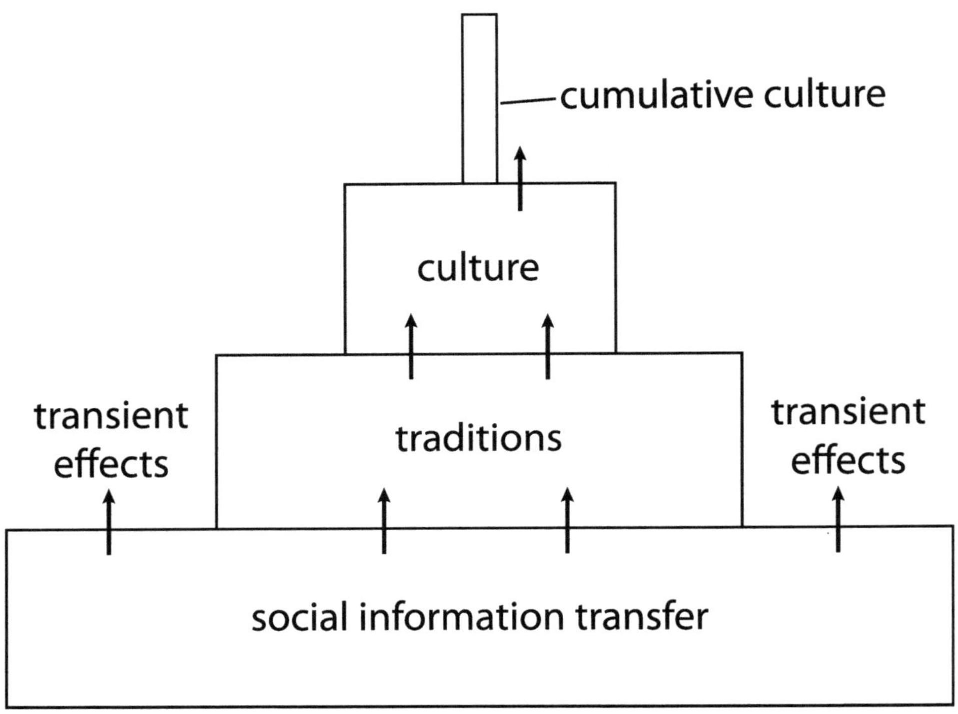

Fig. 10.5 The Whiten and van Schaik (2007) model of culture

The ROCEEH 1 model seemed to suggest that there was a sequence of inevitable stages through which non-human animals and then hominins passed. Eventually they moved beyond what apes were doing (C-L). The attached indications of date and material evidence for each stage suggested that stone tools were crucial for hominins moving beyond what apes could achieve. I have argued this in previous publications (Davidson and McGrew 2005; Davidson

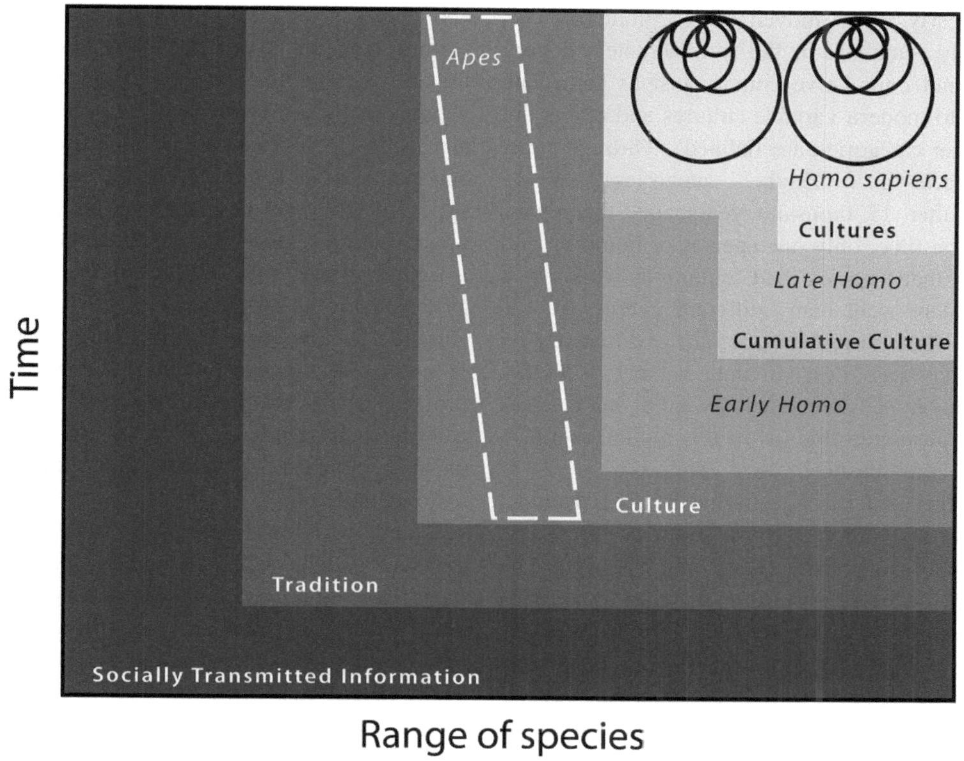

Fig. 10.6 Iain Davidson's modification of ROCEEH 1 model for this paper. See text for explanation

2010a, b; Nowell and Davidson 2010), but it is salutary to remember that there was nothing inevitable about it. The recent revisiting of the comparison between what apes and early hominins could achieve demonstrated close similarities in lots of ways (Wynn et al. 2011), such that, while it may be true that knapping created a new environment of opportunity for hominins (Davidson and McGrew 2005), there is considerable uncertainty about when they benefitted from that new opportunity. That is the nature of evolution: that which seemed unimportant at one time can be selected to be of fitness-changing importance later. By contrast, the model gives the impression that as all creatures progressed onwards and upwards towards the fully symbolic cultural identity of modern humans, the world became dominated by their success – a caricature of the misrepresentations of evolution by its opponents for the last 150 years. The point here is that we should avoid a stage model of this sort because it glosses over the very variation that made some hominin behaviors successful *and others not*. In other words, it is counter-productive to understanding the role of the things that make up culture in the evolution of human behavior.

By the same token, the alternative which reverses the direction of expansion (Fig. 10.5 taken from Haidle and Conard 2011 from an original in Whiten and Van Schaik 2007) has the virtue that it seems to suggest that there are fewer species that have cumulative culture than those that have culture, but it still suggests "progress" and that culture will always trump tradition. Extinction shows that progress is fleeting, and the survival of different species descended from a common ancestor shows that the ideology of progress depends on the point of view of the observer – it is always solipsistic. The continued existence of animals with social learning but little else in the way of culture, or with traditions – the very basis of the observation that culture is a concept that should be extended beyond humans – is an indication that the stages are wrongly conceptualized. While it may be objected that these are only indicative models, what they indicate is wrong in many ways (stage thinking, directionality, lack of emphasis on variation, ignoring selection etc.).

As an alternative, I propose the model in Fig. 10.6 which while still partly in stages, suggests that when traditions arise from social learning (etc.), social learning continued and can still be found among several species (although some of the examples are said to be culture (C-L), it would be better to think of them as social learning in Laland and Hoppitt 2003; Madden 2008), such as guppies. Likewise, tradition was not completely replaced by culture (see e.g., Janik and Slater

2003). As some early *Homo* moved on to cumulative culture,[4] others likely remained in the intermediate category between culture and cumulative culture, just as some late *Homo* moved on to modern variable cultures and others did not (however these categories are defined). Through time, the number of species in the "higher" categories got fewer and fewer until after 17 thousand years ago, there was, perhaps for the first time, only one species of hominin, and all members of that species had culture-as-we-know-it (C-C). By then there were many different cultures (C-C), some of which (but probably not all) evaluated and acknowledged what was "best" in their society (C-SOH). A refinement of this model should acknowledge that C-L has elements that are plesiomorphic (as in this model) and also elements that are homoplastic.

At the end of the conference, a further ROCEEH model was discussed which suffered from some of the same problems I have outlined here (in Haidle and Conard 2011). This model went successively from socially transmitted information through tradition, basic culture, modular culture, composite culture to collective culture. But an interesting addition was a set of achievements that supposedly accompanied those stages. These included, for example, expansion of population size, expansion of childhood, expansion of planning depth and expansion of "ecospace".[5] All of these "expansions" were represented to be continuous from an origin in socially transmitted information and there was no suggestion that any of them had a particular importance for transition from one stage to another. Such a model is counter-productive for generating understanding of the narrative of culture in hominin evolution. It can only ever appear to be like an egregious caricature produced by the opponents of understanding.

Finally, there are problems about the classification as cultural (in any sense of the word) of the behavior of species or in stages. At its worst, culture (in whichever sense is relevant) is used as an explanatory variable for particular aspects of behavior. It would be a mistake to say, for example, "Early *Homo* moved out of Africa because it had culture." Rather, the aggregate of the behavioral evidence about early *Homo* might be evaluated to see which aspects of cultural behavior it was consistent with. As McGrew and I found, it may be that the behavior of early *Homo* fell short of unequivocally satisfying all of the criteria by which McGrew and Tutin (1978) had assessed the grooming hand-clasp of chimpanzees. It would then be a separate issue whether the appearance beyond the modern geographers' limits of Africa of archaeological sites that might be attributed to early *Homo* is a product only of that near-cultural behavior and not just an expansion of range similar to that of other non-cultural animals. The danger is that in trying to refine the sub-types of culture and assigning them to stages of an inevitable progress, the subtleties of the argument about the role of cultural behavior can be lost. Whatever the particular meaning attached to the concept, culture is not an explanatory variable.

Finally, it is essential that, in studying cultural behavior archaeologically we understand the fallacy built into one of the most famous archaeological definitions: "Culture is patterned… [Therefore] the patterning which the archaeologist perceives in his material is a reflection of the patterning of the culture which produced it" (Deetz 1967:7). Deetz, of course, was a historical archaeologist and this may be true for historical material culture (though I suspect it is not). But it is most certainly not true for most of the archaeological record. In particular, a large part of the patterning of flaked stone tools is a product of the constraints of knapping stone combined with the effects of use (e.g., Moore 2010, 2011). Some of the patterning occurred not because of any cultural habits of the knappers, but because there were not many options. In addition, the practice of archaeologists is to identify and name patterns (which they sometimes call "cultures") and then to suggest that the patterns have some ancient objective reality.

Stone Tools and the Variables of the Evolutionary Transformation of Culture

A couple of characteristics of culture should be stressed. Characteristic 1: the essence of social learning is about the transmission, by whatever means, of information among members of a single population. Characteristic 2 follows from this means of transmission: in almost all cases, the information acquired is not exactly the same as the information available for acquisition as Henrich (2004) has argued. To the extent that culture (in any of the three senses I have defined here) is a result of social transmission, variation, albeit clustered variation, is expected to be the norm.

The story of the Paleolithic has been dominated by discussion of types of industry given a restricted number of

[4]This term was used in the original ROCEEH model and derives from Tomasello (1999) to refer to the characteristic of human culture (C-C) that cultural knowledge does not always need to be re-invented but can build up from one generation to another.

[5]This word is an unfortunate neologism and could promote a conflation of the well-established concepts of habitat and niche. It is unfortunate because it is arguable that early hominins were enabled to expand their habitat while retaining the same niche, but that aspects of culture enabled them to change their niche and thus to move into habitats that were otherwise unavailable. Failure to distinguish these separate processes by referring instead to expansion of ecospace seriously threatens a project aimed at understanding the roles of culture in hominin adaptation.

names (Oldowan, Acheulean, Levalloisian, Mousterian, Upper Paleolithic – OALMUP). The simplification of classification into the OALMUP sequence glosses over variability that created stability or led to change. This glossing occurred for the convenience of archaeological analysis (Kuhn 2010) derived from a time when different questions were being asked (Davidson 2002, 2014), but there are important points about the nature of culture involved in this classification.

First, in light of Characteristic 1 of cultural transmission, it is something of a surprise in the record associated with early stone industries that all of them seem to have been made by at least three different species of hominin each (Davidson 2003). There are good reasons for being sceptical of the usefulness of the classification of hominin skeletal remains (Collard and Wood 2000; Gibbs et al. 2000; Davidson 2014), but nevertheless, any revision of the way they are grouped is not likely to suggest that hominins that now show so much variation that some people think they were different species were actually sufficiently closely related as to constitute the social groups among which information was transmitted. At very least, however, it is difficult to argue that material culture at this resolution is closely tied to genetics (assuming the hominin species really do reflect genetics and not just the habits of physical anthropologists).

Characteristic 2 is also problematic for most stone industries of the Paleolithic. It is a prediction of a theory of culture based on social transmission that within a group of communicating creatures, and over time among them, there should be some similarities of form of materials where learning is required (in the same way that heritability is a feature of biological reproduction). In addition, there should be numerous errors and variations arising from those errors which may give rise to new forms (with some similarities to mutation in biological transmission). The proposition here is that the similarity of form of Acheulean bifacial handaxes seems unlikely to result from the sharing of cultural values about shape over distances and times over which hominins or people were unlikely to have been able to communicate at all. Similarity of form is an unreliable guide to cultural similarity, particularly when there may have been some common constraints imposed by the physics of flaking, yet it is the mainstay of archaeological classification into those groups that end up being called "cultures". The real issue, therefore, is: "what was it that was learned within the group, such that the process of knapping produced a similarity of outcome?"

Isaac (1972:175–178) discussed the various forces that affected the outcome of knapping, emphasizing the physical properties of the raw material, the "tradition" within which knapping was learned and the functional requirements, but the lessons have not always been heeded. The apparent clustering of variation in stone industries seems not to meet the two principal characteristics of culture I have outlined here, and, I suggest (for example in Davidson 2010b), the clustering of forms that has been the mainstay of typological characterisation for more than a century seems more likely to be an outcome of equifinality of knapping processes than a convergence on an intended tool type that represents any form of culture (similar suggestions have been made by Clark and Riel-Salvatore 2006; Moore 2011).

As a way ahead for future research, examination of workings of the concepts of culture will need to be as empirically detailed as that by Wynn et al. (2011) if progress is to be made in understanding the role of culture in hominin behavior. In order to achieve this, we need a clear understanding of the variables that are relevant to culture and many of these were discussed or alluded to during the course of the conference (although my views are at variance with much of the way they are used, these variables are discussed in Haidle and Conard 2011). At this stage, only preliminary observations are possible.

A discussion of stone tools could include three independent sets of variables:

(1) Those concerned with the structure of stone tool actions

   a. the various aspects of learning (Hovers 2009a)
   b. patterning (Lycett and von Cramon-Taubadel 2008)
   c. innovation (Henrich 2004)
   d. structure of motor actions (Moore 2010)

(2) Those concerned with the process of stone tool production

   a. the procurement of tool stone (Goldman-Neuman and Hovers 2009)
   b. production of tools (Delagnes and Roche 2005)
   c. cognitive capabilities (Barnard 2010) including combinations (modularity) that abbreviated the learning patterns (Wynn and Coolidge 2010)
   d. use and function (Dominguez-Rodrigo et al. 2001)
   e. the reasons for abandonment of flaked stone

(3) The role of these variables in the context of

   a. the impact on population size (Stiner 2006; Mellars and French 2011)
   b. evolution of hominin life history (Kaplan et al. 2000; Locke and Bogin 2006), including the extension of the period of childhood (Nowell and White 2010; and see Nowell 2016)
   c. changes to habitat and niche of hominins (Dennell et al. 2011; Reynolds et al. 2011)
   d. intentionality of these impacts and the extent of planning (Moore 2013).

There are 13 variables, as well as others I have not mentioned, but as they are in three more or less independent sets, they actually constitute 80 variable combinations (e.g., learning tool stone sources and its impact on the ability to colonise new habitats; patterning of the function of stone tools and its impact on population size) to be considered over the more than 2.5 Ma history of stone tools. Even if we limited the discussion to the Oldowan, Acheulean, Levalloisian, Mousterian and Upper Paleolithic (lumping acknowledged regional variants, such as the Middle Stone Age of Africa or the early industries of Sahul, for the sake of argument) there would be 400 conditions to be considered.

Behavioral Issues in the OALMUP Sequence

Just to make things more complicated, although there are certainly consistencies and competencies of the Oldowan in Africa (de la Torre 2010), it is generally acknowledged that Oldowan artifacts, which might be properly called that if found in Africa in sediments dated to the early Pleistocene, can be found in other places and at other times. Thus, for example, when Australian Aboriginal people were asked to cut trees as an exercise in experimental archaeology (Hayden 1979), and supplied with, effectively, an unlimited supply of raw material, they did not fashion a polished axe (as Hayden hoped) but used expedient flaking to produce a chopper of a form reminiscent of the Oldowan (Toth 1985) and similar forms have been described from the archaeological record (Moore 2003a). It would seem that Oldowan industries are not so much an indication of a particular set of cultural circumstances as a default condition, an expedient one that arises whenever flakes are produced by simple flaking. There is evidence that the final form of many of the artifacts in the Oldowan was dependent on the initial form of the tool stone (Toth 1985), and that the product that was used was the flakes (Keeley and Toth 1981). It is clear that knapping can be interpreted as skilled and structured, particularly when there are large conjoin sets (Delagnes and Roche 2005), but it is less clear in what ways those skills were necessarily a product of the social learning by which we recognize Culture-L.

In terms of the requirements we defined for identifying culture (C-L), there have, so far, been relatively few studies that show the consistency of patterning within groups in a limited area and time period, and consistent variation between such groups that would enable us to define traditions in this earliest stage of stone tool making. This may not mean that such conditions will not eventually be found, but the difficulties of finding intact sites and refining chronologies for such remote periods suggest that it may be very difficult. At the same time, that taphonomic problem should not be used to imply that, therefore, the necessary patterning did exist and we just have not found evidence of it.

Acheulean

Many sites around the world have produced often quite large numbers of artifacts, generally called handaxes, which are symmetrical about a plane that forms the platform from which flakes were removed, and symmetrical about an axis on that plane but on the orthogonal axis. They are often rounded and broad at one end and pointed at the other. They also have in common some of their metric characteristics but scholars differ about the significance of this (Gowlett 1984; McPherron 2000; Davidson 2002). Such artifacts are typical of those industries called Acheulean.

It is generally claimed (Klein 2005) that the Acheulean was a tradition that spread widely (but not universally e.g., Lycett and Norton 2010) around the world, and lasted for about 1.5 Ma based on the remarkable uniformity of geometric characteristics wherever handaxes are found (Marshall et al. 2002; Lycett and von Cramon-Taubadel 2008). Some of the assumptions of this characterisation are certainly exaggerated (Dibble 1989; Davidson 2002) but the one that has not tended to be considered is the question of whether the similarities depend on a continuous tradition (see preliminary discussion in Davidson 2014; and see Clark and Riel-Salvatore 2006:36–40).

Handaxes were made by at least two different techniques. One involved flaking the margins of large flakes, such that if the distal end of the flake was not modified, an artifact called a "cleaver" was produced (Sharon 2010); the other technique involved the production of a bifacial core from a nodule of flint by the removal of cortex and subsequent reduction (for conjoining of archaeological discards and modern experimental knapping, see Bradley and Sampson 1986). On other occasions, similar forms were produced by removing flakes from the margins of cobbles. There are several ways to interpret these different pathways to the handaxe: (1) they may all have converged on the similar final product by chance outcome of the removal of flakes for use, implying no connection in culture or tradition; (2) there may have been a mental template of what a well-formed handaxe should look like and knappers passed on this cultural knowledge from generation to generation for a period of about 1.5 Ma; (3) making "handaxes" on large flakes was an incidental outcome of flaking sharp edged cores and this is unconnected to the knapping habits that led to "handaxes" made from nodules or cobbles. If we are asking about the nature of cultural knowledge at different times in prehistory, these are fundamental issues that need to be decided.

In some remarkable cases, such as Boxgrove, there seems to be a full sequence from acquisition of nodular raw material, cortex removal, core preparation and flake removal, and with some "finished" handaxes (Roberts et al. 1997). The source of raw material was within a few tens of metres of the knapping floor. In one area, GTP17, there were eight scatters for which conjoining was possible. These all involved steps in bifacial flaking, but there were many knapping episodes which cannot be said to be handaxe-related, some represented the early stages of production of bifaces, others were late in the flaking of artifacts that were brought to the site as bifaces (Pope 2002, Chapter 4). In addition, there were some less diagnostic artifacts used for cutting and butchering a horse. Unfortunately, very many analyses elsewhere have concentrated on the characteristics of the handaxes at the expense of understanding in this way the behavioral contexts of the full set of artifacts from sites where they were found (but Isaac 1977 gave a complete description of whole assemblages at Olorgesailie, including the cores that were not handaxes).

This belief that large numbers of bifacial handaxes can be attributed to a single tradition, and that analysis of them can show the similarity within the category, is responsible for much of the narrative about stone industries for over a million years over much of the world. But there are some awkward facts that keep being ignored. First, bifacially flaked handaxes have been found in Australia (Sahul[6]), particularly in western Queensland (see analysis in Moore2003a) and the Northern Territory (Rainey 1991). As no such artifacts have been found in Indonesia (but see the claim – unsupported with evidence in the paper – that there are Acheulean large flake industries early at Sangiran in Java in Mishra et al. 2010) or the islands west of Sahul (for discussion of the deficiencies in analysis of artifact assemblages in this region, see Moore and Brumm 2007), and as Sahul was colonized long after the end of the so-called Acheulean, there cannot be any relationship to an Acheulean "tradition" spatially and chronologically (Davidson 2002). Handaxes were produced both in the Acheulean and the chronologically substantially later Mousterian of "Acheulean Tradition" (Mellars 1969) although the strategies of repeated removal of flakes from the handaxes seem to be slightly different between the two (Iovita and McPherron 2011).

This chronological and spatial separation suggests that the particular outcome – a bifacially flaked handaxe – may not be the result of a tradition at all, but one end product of particular knapping strategies. Clark and Riel-Salvatore (2006) even agree with the suggestion that many of them were principally produced as cores. Others were the result of retouching strategies which concentrated on the repeated production of useful edges on the margins of large flakes (Brumm and McLaren 2011). The Australian evidence suggests that, rather than a result of a continuous tradition, bifacial flaking, which produced things that archaeologists call "handaxes," must have been invented at least three times and, I would suggest that if we add the differences in production methods between Africa (made on flakes) and Europe (made on nodules of flint) and the sporadic occurrence in China (Zhang et al. 2010) that number should be at least five. Given the long periods when handaxes were absent from particular regions, it seems more likely that (and the null hypothesis should be), over 1.5 Ma, the outcome occurred independently thousands of times without an intention to produce handaxes. This does not preclude the possibility that in one or other of these examples there *was* an intention to produce artifacts that we would call handaxes, but it seems highly unlikely that that was always true. The appearance of a tradition from these disparate occurrences has as much to do with the traditional practices of archaeologists as it has with the behavior of hominins – but it tells us almost nothing about the cultural behavior of hominins.

Something accounts for the similarity of form but it may not (always) have been a desire to produce the particular shape (although that seems to be the suggestion for the handaxes of the Mousterian of "Acheulean Tradition") but at least as much as anything a product of repeated removal of flakes from a core with acute edges symmetrically-located around the plane of the plan. The fact that there are specimens in Acheulean assemblages which show the pattern of removal of flakes from a symmetrically-located acute edge for only a single edge (Davidson 2010a) (Fig. 10.7) suggests that the concentration on the shape of the plan of the core may be a distraction to analysts. The final form which has fascinated archaeologists since the end of the eighteenth century may be a product of what Tennie and others (see, for example, Tennie et al. 2009, 2016) would call the Zone of Latent Solutions (ZLS) for flaking from an acute edged core. The argument would be, then, that some of the similarity of forms of the distinctive artifacts of the Acheulean is a product of this ZLS rather than the pursuit of the best form of artifact, or the product of a cultural tradition. If that is an acceptable argument, it may be that what was learned socially was not the making of a handaxe, but the habit of flaking from sharp edges of flakes and appropriate rotations of the core in and about the plane to maximise the production of flakes.

Finally, one of the reasons it is important to understand the constraints on the form of bifacial handaxesis that there have been claims that production of artifacts with "imposed form" was only possible with modern cognition (Mellars 1989, 1996), or by hominins who had language (Davidson and Noble 1993). If it could be demonstrated that Acheulean handaxes were made to some intentional plan, then we would probably have to concede that modern cognition first

[6]Sahul is the name that has been given to the continent which includes the islands of New Guinea, Tasmania and the mainland Australia. Since these islands are principally separated during the brief interglacial periods (as now) Sahul is their normal condition.

Fig. 10.7 Bifacially flaked core from Slindon, UK, illustrating that bifacial flaking was important even when the "finished artifact" did not become a handaxe. (*Photo* I. Davidson)

appeared several hundred thousand years earlier than is generally thought (see, for example, the positions about modern human behavior and cognition by Noble and Davidson 1996, Henshilwood and Marean 2003, and even Coolidge and Wynn 2009). That is why it has been so important to understand the non-cultural constraints that make the form appear to be standardized (Davidson 2002). This has been understood by some (Sharon 2008) and ignored by others (Henshilwood and Marean 2003). I suggest that no form was imposed, rather the similarity of form was the almost incidental result of the outcome of simple decisions in knapping together with high selectivity by archaeologists. It may be that we should consider those decisions about knapping as cultural, but not the planning for a final handaxe form.

In terms of the three requirements, it certainly seems to be the case that there was some consistency within groups, though it is less clear that this was matched by differences from one contemporary group to another. As for tradition, in this the most famous of stone age traditions, the argument is that much of the appearance of tradition is an illusion resulting from a little learning, and a lot of equifinality.

Levallois

The literature which produces scepticism about some of the claims about the Levallois technique has been summarized on previous occasions (Davidson and Noble 1993; Davidson 2010a, b).

In the Levallois technique, in contrast to the Acheulean, the plane between the two surfaces from which flakes were removed was not symmetrically placed. There was instead, a relatively flat surface on the thin side of the plane, and a thick side of the plane opposite this surface. Such plano-convex cores (sometimes classified as discoids) have been found from very early assemblages at Oldupai Gorge DK (Leakey 1971), and at Peninj (de la Torre et al. 2003). Flakes seem to have been removed from the platforms present at the asymmetric plane. Eventually this knapping technique produces angles at the plane that are too obtuse for more flake removal. At Dmanisi, cores with a similar morphology were present, but not all of the flakes were removed from the platform at the plane, but rather some were removed in the direction of the platform (Baena et al. 2010).

The standard story about in dustries using this technique is that the asymmetrically-located plane was set up by the knappers by removing a long series of anticipatory flakes (which, in the standard story, were not intended for use). It is suggested that the objective was that a large, thin flake (generally a "Levallois flake") could be removed from the flat surface on the thin side of the plane (e.g., Boëda 1988). Such flakes were not removed from the Levallois-like cores at Oldupai and Peninj. In industries said to have the Levallois technique there were several patterns of removal of flakes from the cores, some of which seemed to involve the removal of successive flakes from the thin side of the core (Boëda 1988). Apart from this, the cores were produced in a routine which was known many hundreds of thousands of years earlier. This suggests that one of the issues about the removal of the Levallois flake was the renewal of the core so that more flaking could continue. What was missing at Oldupai and Peninj was this sense of renewal of the core when the angles became too obtuse. That cores could be renewed may indeed have been one of the technological innovations that made a difference to tool production during the course of hominin evolution, but it seems unlikely that the whole process of production up to the removal of the "core rejuvenation flake" was conceptualized by early hominins as a path to rejuvenation of the core!

There is substantial evidence to suggest that the standard story is not plausible (Davidson 2003): in some cases, the "final flake" is preserved beside the core (Schäfer 1990), notably when the flake does not follow through to the other edge of the core, so that no renewal of the platform angles has been achieved; in some sites cores were preserved which can be fitted back together to show that the supposedly anticipatory flakes were removed, presumably for use (Van Peer 1992); studies of the wear on edges resulting from use of artifacts shows that non-Levallois flakes were more often used than Levallois flakes (Beyries 1987). Analysis of some refitted artifacts, particularly from one core known as "Marjorie's core" (Schlanger 1996) has demonstrated that the knapping activities involved in the preparation of Levallois cores could be considered as "chunked" into subroutines as happens in expert learning in living humans (Wynn and Coolidge 2010). Nevertheless, all of the other caveats about the standard story make it unlikely that the chain of thought processes involved planning all of the stages towards the production of the "final" flake.

As with the story about Acheulean bifacial handaxes, another feature of the Levallois technique has been discussed as indicating an aspect of modern human behavior: that learning all of the stages of production of Levallois flakes was so complex that it must have required teaching by means of language (e.g., Bar-Yosef and Van Peer 2009). The Wynn and Coolidge modular learning argument goes some way to counter this suggestion, but in addition, the argument is difficult because there is so little evidence about what people can be taught without language (for a survey see Shennan and Steele 1999), and because there is little theorizing about how this aspect of cognition evolved from an ape-like common ancestor to humans. However, Davidson (2009), following the model of cognitive evolution of Barnard et al. (2007), suggested that learning could have been guided by the use of vocal utterances without the reflective component of language.

A technique that has been called Levallois by experienced analysts (de Sonneville-Bordes 1986) was used in Australia (Moore2003a, b, c), but some of the artifacts that might be identified as Levallois points elsewhere were made by a blade-making technique on single-platform cores known as horsehoof cores (Binford and O'Connell 1984). The equifinality involved weakens the case that the Levallois and related techniques were the fundamental technologies carried by modern humans from Africa to the rest of the world (see e.g., Foley and Lahr 1997). The fact that, as observed by Binford and O'Connell, these artifacts were made in 1974 on horsehoof cores otherwise said to characterise the earliest stone industries of Australia (Bowler et al. 1970), further undermines the use of both "Levallois" points and horsehoof cores as type fossils of early industries. It seems more likely, given the persistence of the core form across almost the whole period of stone tool production, that the form results from common practices of flaking cores with an asymmetrically-located plane. The ability to do that had been present from the earliest levels at Oldupai, so that the real discovery was about a way of continuing to flake the core when the angles at the plane had become too obtuse.

There is widespread agreement that the Levallois technique required learning, although the production of similar cores in the early Oldowan at Oldupai and Peninj suggests that this is exaggerated. Because the technique is not confined to one of the major taxa (it occurs with both Acheulean industries and Mousterian ones), it is relatively difficult to say that consistency within groups has been demonstrated. For example, Levallois knapping is widely distributed in Northern France and Southern France, but not continuously (Delagnes and Meignen 2006). By the same token it is difficult to define variation between groups and a tradition. It would be ironic, indeed, if the technique which has the most elaborate narrative about preconceptions of the finished product turned out to be difficult to analyze as a cultural product by the criteria we are using here. I suspect that if we were to shift the attention away from the standard story and on to techniques of core rejuvenation, it would be easier to identify the extent of social learning, variation between groups and the traditions in the Levallois technique.

The Mousterian

In its emphasis on flakes, the Levallois technique is commonly associated with the industries known as Mousterian, although the perspective presented here suggests that there is substantial evidence that the flakes were as important in the earliest assemblages as well as in assemblages that have bifacially flaked cores. I have previously recounted a contrast between the knapping of François Bordes when making a handaxe (who taught himself to reproduce the forms in which he classified artifacts and ignored the flakes he produced as he made a handaxe) and that of Irari Hipuya, a New Guinean brought up to use stone flakes (who made many flakes and ignored the cores) (Davidson and Noble 1993:365–366). Whether you emphasize the cores or flakes may well depend on cultural decisions.

The association of Mousterian industries with Neandertals was established in the nineteenth century (Trinkaus and Shipman 1993:121–123)[7] but later it was asserted that: "The only workable definition of Neanderthal man and period seems to be, for the time being, the man and period of the Mousterian culture" (Hrdlicka 1927:251). The equation of Mousterian industries with Neandertals (M = N) has been remarkably persistent, surviving even the demonstration that the skeletal remains at Skhul and Qafzeh about 100 ka were closer to modern humans than to Neandertals (Vandermeersch 1981), yet the stone industries were Mousterian (Hovers 2009b). Little attention has been paid to the significance of breaking the nexus between the stone industries and the hominin forms (M ≠ N). Potentially, any Mousterian site in Europe later than 100 ka, without Neandertal remains to suggest otherwise, could have been made by the descendants of Qafzeh and Skhul.

In dealing with the information in the Mousterian flaked stone, the history has been recounted many times (e.g., Binford 1972; Clark and Riel-Salvatore 2006) of the appearance of a variety of artifact types (Bordes 1961); the interpretation in terms of cultures of the assemblage types with different relative frequencies of those artifact types (C-C) (e.g., Bordes and de Sonneville-Bordes 1970); the challenge by the Binfords to this "culture-historical" approach to those assemblage types in favor of variation resulting from intuitively determined functions (Binford and Binford 1966); the undermining by Dibble of the whole typology on which both studies were based because some of the typological variation was due to the differing degrees of resharpening (Dibble 1984, 1987) – well summed up in the question "[H]ow can implements be designed for, and be efficient in, a specific use if their morphology is continuously changing?" (Hiscock and Attenbrow 2005); the failure to demonstrate common functional use-wear for particular "types" (Beyries 1987); and the demonstration that much of the patterning in the degree of resharpening was due to variations in the availability of tool stone in different environmental conditions (Rolland and Dibble 1990). Although there appeared to be the first two requirements for culture in the consistency within groups and the variation between groups in the Bordes analysis, subsequent work has suggested that that may have been illusory. As with the previous three taxonomic entities, it may be that the standard story is quite unsuited to identifying the cultural component of the Mousterian.

Many of the assumptions about the regularities of forms of stone artifacts and the patterning of the assemblages in which they were found were challenged by these studies – and the challenge applies to more than just the European Mousterian (see, for example, Moore 2010, 2011). One problem with this critique is that, even if the operational sequences by which tools were made (and used) are identified, there may be no option than to suggest the stone tools allow "understanding of the simple facts of prehistoric life" (Bar-Yosef and Van Peer 2009) but that this might allow no more insight than that our predecessors made stone tools and used them (to cut meat or plants, or to scrape hide or wood) (Davidson 2009). The answer to this dilemma, as Binford (1972) argued many times, is in considering the systemic organisation of artifact use in the landscape at a particular time, and changes in that organization through time in the context of environmental change, as suggested by Rolland and Dibble for Southwestern France.

Some indication of the advantages of such an approach is given by the assessment of the contrast between Archaic and Modern Humans in the East Mediterranean (Lieberman and Shea 1994) in terms quite similar to the extremes of the spectrum Binford defined between foraging and logistic organization (Binford 1980). The analysis included the typology of artifacts, but also the functions of them; not only the species of animals obtained but the seasonality of the hunt. Through this analysis, the authors were able to show that the modern humans at Qafzeh had much more flexible mobility than the Neandertals at Kebara which occupied the site at several seasons. Here is a clear example of two species of hominin (even though there may have been some inbreeding between them according to the Neandertal genome sequencing of Green et al. 2010) occupying substantially similar habitats, but occupying different niches while using rather similar stone artifact assemblages. Subsistence choices of this sort, when consistently applied, as appears to be the case here, would have had ecological consequences in terms of the depletion of resources for the hominins at Kebara, and a more benign impact at Qafzeh, with superficially the same technology (Lieberman and Shea 1994).

A further example comes from Central Italy (Stiner 1994), where, again, the stone artifact assemblage, considered at the level of artifact types and their relative abundance seemed to have little relationship with the economy of meat acquisition. Instead the important interaction was between the "life histories of artifacts" and foraging strategies, and as with the East Mediterranean example, one of the key constraints was the amount of movement around the landscape and the opportunities that presented for obtaining tool stone

[7]Pat Shipman kindly drew my attention to the importance of the discovery of the Šipka mandible in stratigraphic context which established the association of Neandertals and the Mousterian.

(Stiner and Kuhn 1992). Examination of the technology suggests that different local solutions were found to foraging options within what generally would be classified as Mousterian industries (Kuhn 2006). Although often the same prey were killed, often at prime age, in Middle and in Upper Paleolithic contexts, the methods by which they were taken seem likely to have differed, with greater reliance on technology to increase reliable resource production in the Upper Paleolithic, probably increasing rates of child survivorship and hence population density (Stiner 2006).

In brief, the Late Pleistocene record allows analysis of the archaeological evidence freed from the worst problems of taphonomy and dating. As a result, the resolution of the record allows faunal remains and technology to be interpreted in terms of the plausible foraging strategies and consequences of hominins at different times and allows comparisons between hominins. The patterning is not well represented by the coarse taxonomic units derived from the nineteenth century classifications. In each region populations adapted to local circumstances and did not necessarily reach solutions that look the same within a simplistic analytical framework: they had what Kuhn called rugged fitness landscapes with many local and regional fitness peaks (Kuhn 2006). It may be that such fine-grained analysis is not possible for the earlier record, but these case studies show that, in addition to the other problems of interpretation outlined here, the analysis of stone artifact typology through time at one site, or by assigning assemblages to one or other class which is then interpreted in terms of the behavior of "cultures" defined by the existence of those classes, can miss the whole point that the tools were made for use and were part of the behavioral repertoire of hominins or people.

Binford, following White (1959:8), always claimed that culture was an extrasomatic means of adaptation (Binford 1962), but he did not mean that once the regularities in the archaeological remains have been given a name that is mistakenly called a "culture," the adaptation can be understood. Rather, the adaptation can be seen in the way in which patterning in the artifacts was reproduced from time to time and place to place, and used in consistent ways at one time and different ways at another time, all of which influenced the success of particular foraging strategies and impacted on the demographic success of the population. The culture is found in the way in which artifacts were used, not only in their form. The technology of stone tools was in the knowledge associated with (a) procurement of tool stone, (b) flaking the stone, (c) using the tool to make another tool, perhaps of wood or bone or skin, (d) using both in the acquisition of resources, (e) the knowledge of the resources that were made accessible through of the use of those tools, and (f) in preparing and consuming the resources once acquired, not just in the form of the product. It was technology that was socially learned, not just the form of the tool. And the cultural knowledge involved both the technology so-defined as well as the social configurations of the hominins or people who used it.

Conclusion

Primatologists and archaeologists share the problem that their subjects cannot speak to them about what they are doing. By adopting some of the methodology by which primatologists have determined whether social learning occurs among them, it has been possible to show up some of the deficiencies in the existing arguments about the "cultural" nature of early stone industries. Most importantly, it is not clear that archaeologists necessarily even understand what it was that was learned when knapping skills were passed from one generation to another. I tentatively suggest that for the Oldowan it may have been no more than the ability to remove a succession of flakes from a platform: the knappers "had the cognitive abilities to exploit angles when encountered but not to create new ones" (Delagnes and Roche 2005). For the Acheulean it may have been about the maximisation through core rotations of the number of flakes struck from the acute platform created symmetrically at the margins of a large flake. The Levallois technique enabled the repeated production of large numbers of flakes from asymmetric cores by the discovery that as core angles tended towards obtuseness those angles could be renewed by the removal of a large flake. The Mousterian represented the recognition that the use-life of flakes could be prolonged by repeated retouching and reshaping.

Using the criteria about the consistency of behavior within a group, the variation between groups and the persistence of a tradition through time, it is possible to highlight the weaknesses of arguments about the cultural nature of knapping behavior and set up a number of research questions that need to be investigated. For each major grouping, is it possible to determine the extent of consistency, variation and persistence of tradition in this way? What is the incidence of knapping errors at particular times and places in relation to different raw material availability that might indicate what was being learned and how?

Several studies suggest that, in addition, the way ahead requires an assessment of the organisation of behavior in relation to stone tool production and use, especially in the context of other behavior, such as food acquisition. The outcomes may well not produce simple unilinear progress towards modernity or the Upper Paleolithic, and, indeed, may produce different adaptive solutions which made such progress unlikely. Such subtleties will produce a way of looking at adaptation very different from the view derived from nineteenth century archaeology. At its best, such an approach may permit archaeologists to look at the impact of technology (tools as well as their role in society) on population size, and planning depth. One of the ways in which technology had such impacts was through the ability of hominins to expand either by finding a niche that was common to several habitats, or by finding new niches within a single habitat, or ultimately finding new niches that enabled them to move into new habitats.

The question of what labels should be given to such sorts of culture is complicated by the argument in this paper that much of the standard description of variation, in terms of artifact types, is not recognizing the appropriate variations. In Tomasello's (1999:5) words, cumulative culture, the distinctive property of modern humans that is absent in apes, requires not only creative invention but also, and just as importantly, faithful social transmission that can work as a ratchet to prevent slippage backward – so that the newly invented artifact or practice preserves its new and improved form at least somewhat faithfully until a further modification or improvement comes along.

If it is true that conventional accounts of stone artifact variation do not capture what are the creative inventions, it is difficult to see how archaeology can study whether cumulative culture was present or not. The simplified set of changes outlined in this paper may offer some chance of asking the right questions. One other possibility is represented by the possibility of modular learning. I indicated, above, the suggestion that "chunking" of sequences of removals permitted "expert learning" of the Levallois technique (Wynn and Coolidge 2010). There is an extent to which this conclusion is dependent on the belief in the standard interpretation of the Levallois technique, which seems implausible. I have previously suggested that a better candidate for such modular learning is the use of crested blades as a preparation for the production of blade cores in the Middle Stone Age of southern Africa and the European Upper Paleolithic (Davidson 2010b:197–198). In those cases, the cognitive significance came from the capacity to engage in a physical activity that appears unrelated to the final purpose of it (see discussion, also, in Haidle 2011:460–461). This suggestion could also lead to future research to establish whether there were similar modular knapping sequences in other industries.

In conclusion, the standard way of looking at stone artifacts has not yielded insights that are helpful in a study of social learning in hominin evolution but there are ways of looking at knapping that probably allow a way forward.

Acknowledgments I thank Miriam Haidle and Nick Conard for the invitation to the conference and all the conference participants for their friendly discussion and disagreements. Among those from whom I have received important guidance on issues related to this paper are Phil Barnard, Miriam Belmaker, Rob Gargett, Barbara King, Bill McGrew, Mark Moore, Kimberley Newman, April Nowell, and Pat Shipman. Kimberley Newman prepared the final version of the figures. I thank them all.

References

Arnold, M. (1869). *Culture and anarchy: An essay in social and political criticism.* Retrieved April 4, 2012 from http://www.gutenberg.org/catalog/world/readfile?fk_files=1724531&pageno=1

Baena, J., Lordkipanidze, D., Cuartero, F., Ferring, R., Zhvania, D., Martín, D., et al. (2010). Technical and technological complexity in the beginning: The study of Dmanisi lithic assemblage. *Quaternary International, 223–224*, 45–53.

Bar-Yosef, O., & Van Peer, P. (2009). The *Chaîne Opératoire* approach in Middle Paleolithic archaeology. *Current Anthropology, 50*, 103–131.

Barnard, P. J. (2010). Current developments in inferring cognitive capabilities from the archaeological traces left by stone tools: Caught between a rock and a hard inference. In A. Nowell & I. Davidson (Eds.), *Stone tools and the evolution of human cognition* (pp. 207–226). Boulder, CO: University Press of Colorado.

Barnard, P. J., Duke, D. J., Byrne, R. W., & Davidson, I. (2007). Differentiation in cognitive and emotional meanings: An evolutionary analysis. *Cognition and Emotion, 21*, 1155–1183.

Bennett, T. (2005). Culture. In T. Bennett, L. Grossberg, & M. Morris (Eds.), *New keywords. A revised vocabulary of culture and society* (pp. 63–69). Oxford: Blackwell Publishing.

Beyries, S. (1987). *Variabilité de l'industrie lithique au Moustérien.* Oxford: British Archaeological Reports International Series 328.

Binford, L. R. (1962). Archaeology as anthropology. *American Antiquity, 28*, 217–225.

Binford, L. R. (1972). Contemporary model building: Paradigms and the current state of Palaeolithic research. In D. L. Clarke (Ed.), *Models in Archaeology* (pp. 109–166). London: Methuen.

Binford, L. R. (1980). Willow smoke and dogs' tails: Hunter-gatherer settlement systems and archaeological site formation. *American Antiquity, 45*, 4–20.

Binford, L. R., & Binford, S. R. (1966). A preliminary analysis of functional variability in the Mousterian of Levallois facies. *American Anthropologist, 68*, 238–295.

Binford, L. R., & O'Connell, J. F. (1984). An Alyawara day: The stone quarry. *Journal of Anthropological Research, 40*, 406–432.

Boëda, E. (1988). Le concept Levallois et evaluation de son champ d'application. In M. Otte (Ed.), *L'Homme de Néandertal, Vol. 4: La Technique* (pp. 13–26). ERAUL 31, Liège: Univertsité de Liège.

Boesch, C. (1991). Teaching among wild chimpanzees. *Animal Behaviour, 41*, 530–532.

Boesch, C., & Boesch, H. (1984a). Mental map in wild chimpanzees: An analysis of hammer transports for nut cracking. *Primates, 25*, 160–170.

Boesch, C., & Boesch, H. (1984b). Possible causes of sex differences in the use of natural hammers by wild chimpanzees. *Journal of Human Evolution, 13*, 415–440.

Boesch, C., & Boesch, H. (1993). Aspects of transmission of tool-use in wild chimpanzees. In K. Gibson & T. Ingold (Eds.), *Tools, language and cognition in human evolution* (pp. 171–183). Cambridge: Cambridge University Press.

Boesch, C., Marchesi, P., Marchesi, N., Fruth, B., & Joulian, F. (1994). Is nut cracking in wild chimpanzees a cultural behaviour? *Journal of Human Evolution, 26*, 325–338.

Bonner, J. (1980). *The evolution of culture in animals.* Princeton: Princeton University Press.

Bordes, F. (1961). *Typologie du Paléolithique Ancien et Moyen.* Bordeaux: Delmas.

Bordes, F., & de Sonneville-Bordes, D. (1970). The significance of varibility in Palaeolithic assemblages. *World Archaeology, 2*, 61–73.

Bowler, J. M., Jones, R., Allen, H., & Thorne, A. G. (1970). Pleistocene human remains from Australia: A living site and human cremation from Lake Mungo, Western New South Wales. *World Archaeology, 2*, 39–60.

Box, H. O. & Gibson, K. R. (1999). Preface. In H. O. Box & K. R. Gibson (Eds.), *Mammalian social learning. Comparative and ecological perspectives* (pp. x–xiv). Cambridge: Cambridge University Press.

Bradley, B., & Sampson, C. G. (1986). Analysis by replication of two Acheulian artefact assemblages. In G. N. Bailey & P. Callow (Eds.), *Stone age prehistory* (pp. 29–45). Cambridge: Cambridge University Press.

Brown, P. (2002). Language as a model for culture: Lessons from the cognitive sciences. In R. G. Fox & B. J. King (Eds.), *Anthropology beyond culture* (pp. 169–192). Oxford: Berg.

Brumm, A., & McLaren, A. (2011). Scraper reduction and "imposed form" at the Lower Palaeolithic site of High Lodge, England. *Journal of Human Evolution, 60*, 185–204.

Clark, G. A. & Riel-Salvatore, J. (2006). Observations on systematics in Paleolithic archaeology. In E. Hovers & S. Kuhn (Eds.), *Transitions before the transition. Evolution and stability in the Middle Paleolithic and the Middle Stone Age* (pp. 29–56). New York: Springer.

Collard, M., & Wood, B. (2000). How reliable are human phylogenetic hypotheses? *Proceedings of the National Academy of the Sciences, 97*, 5003–5006.

Coolidge, F. L., & Wynn, T. (2009). *The rise of Homo sapiens. The evolution of modern thinking*. Oxford: Blackwell.

Davidson, I. (1999). The game of the name: Continuity and discontinuity in language origins. In B. J. King (Ed.), *The origins of language. What nonhuman primates can tell us* (pp. 229–268). Santa Fe, NM: School of American Research.

Davidson, I. (2002). The "Finished Artefact Fallacy": Acheulean handaxes and language origins. In A. Wray (Ed.), *Transitions to language* (pp. 180–203). Oxford: Oxford University Press.

Davidson, I. (2003). The archaeological evidence of language origins: States of art. In M. H. Christiansen & S. Kirby (Eds.), *Language evolution* (pp. 140–157). Oxford: Oxford University Press.

Davidson, I. (2009). Comment on Bar-Yosef and Van Peer. *Current Anthropology, 50*, 119–120.

Davidson, I. (2010a). The archaeology of cognitive evolution. *Wiley Interdisciplinary Reviews: Cognitive Science, 1*, 214–229.

Davidson, I. (2010b). Stone tools and the evolution of hominin and human cognition. In A. Nowell & I. Davidson (Eds.), *Stone tools and the evolution of human cognition* (pp. 185–205). Boulder, CO: University Press of Colorado.

Davidson, I. (2012). Origins of pictures: An argument for transformation of signs. In K. Sachs-Hombach & J. R. J. Schirra (Eds.), *Origins of Pictures. Anthropological Discourses in Image Science*. Cologne: Halem.

Davidson, I. (2014). It's the thought that counts: Unpacking the package of behaviour of the first people of Australia and its adjacent Islands. In R. Dennell & M. Porr (Eds.), *Southern Asia, Australia and the search for human origins* (pp. 243–256). Cambridge: Cambridge University Press.

Davidson, I., & McGrew, W. C. (2005). Stone tools and the uniqueness of human culture. *Journal of the Royal Anthropological Institute, 11*, 793–817.

Davidson, I., & Noble, W. (1993). Tools and language in human evolution. In K. Gibson & T. Ingold (Eds.), *Tools, language and cognition in human evolution* (pp. 363–388). Cambridge: Cambridge University Press.

de Sonneville-Bordes, D. (1986). Technique Levallois dans l'Australie du Sud-Est. *Arqueologia (Porto), 13*, 71–75.

de la Torre, I. (2010). Insights on the technical competence of the early Oldowan. In A. Nowell & I. Davidson (Eds.), *Stone tools and the evolution of human cognition* (pp. 45–65). Boulder, CO: University Press of Colorado.

de la Torre, I., Mora, R., Domínguez-Rodrigo, M., de Luque, L., & Alcalá, L. (2003). The Oldowan industry of Peninj and its bearing on the reconstruction of the technological skills of Lower Pleistocene hominids. *Journal of Human Evolution, 44*, 203–224.

Deetz, J. (1967). *Invitation to Archaeology*. New York: Natural History Press.

Delagnes, A. & Meignen, L. (2006). Diversity of lithic production systems during the Middle Paleolithic in France. Are there any chronological trends? In E. Hovers & S. Kuhn (Eds.), *Transitions before the transition. Evolution and stability in the Middle Paleolithic and the Middle Stone Age* (pp. 85–107). New York: Springer.

Delagnes, A., & Roche, H. (2005). Late Pliocene hominid knapping skills: The case of Lokalalei 2C, West Turkana, Kenya. *Journal of Human Evolution, 48*, 435–472.

Dennell, R. W., Martinón-Torres, M., & Bermúdez de Castro, J. M. (2011). Hominin variability, climatic instability and population demography in Middle Pleistocene Europe. *Quaternary Science Reviews, 30*, 1511–1524.

Dibble, H. L. (1984). Interpreting typological variation of Middle Paleolithic scrapers: Function, style, or sequence of reduction? *Journal of Field Archaeology, 11*, 431–436.

Dibble, H. L. (1987). Reduction sequences in the manufacture of Mousterian implements of France. In O. Soffer (Ed.), *The Pleistocene old world: Regional perspectives* (pp. 33–44). New York: Plenum Press.

Dibble, H. L. (1989). The implications of stone tool types for the presence of language during the Lower and Middle Paleolithic. In P. A. Mellars & C. B. Stringer (Eds.), *The human revolution: Behavioural and biological perspectives on the origins of modern humans* (pp. 415–431). Princeton, NJ: Princeton University Press.

Dominguez-Rodrigo, M., Serrallonga, J., Juan-Tresserras, J., Alcala, L., & Luque, L. (2001). Woodworking activities by early humans: A plant residue analysis on Acheulian stone tools from Peninj (Tanzania). *Journal of Human Evolution, 40*, 289–299.

Enquist, M., & Ghirlanda, S. (2007). Evolution of social learning does not explain the origin of human cumulative culture. *Journal of Theoretical Biology, 246*, 129–135.

Féblot-Augustins, J. (1997). *La circulation des matières premières au Paléolithique* (Vol. 1 et 2. ERAUL 75). Liège: Université de Liège.

Foley, R., & Lahr, M. M. (1997). Mode 3 technologies and the evolution of modern humans. *Cambridge Archaeological Journal, 7*(1), 3–36.

Gagneux, P., Gonder, M. K., Goldberg, T. L., & Morin, P. A. (2001). Gene flow in wild chimpanzee populations: What genetic data tell us about chimpanzee movement over space and time. *Philosophical Transactions of the Royal Society B: Biological Sciences, 356*, 889–897.

Gibbs, S., Collard, M., & Wood, B. (2000). Soft-tissue characters in higher primate phylogenetics. *Proceedings of the National Academy of Sciences of the USA, 97*, 11130–11132.

Goddard, C. (2005). The lexical semantics of 'culture'. *Language Sciences, 27*, 51–73.

Goldman-Neuman, T. & Hovers, E. (2009). Methodological considerations in the study of Oldowan raw material selectivity: Insights from A.L. 894 (Hadar, Ethiopia). In E. Hovers & D. R. Braun (Eds.), *Interdisciplinary approaches to the Oldowan* (pp. 71–84). New York: Springer.

Goldman-Neuman, T., & Hovers, E. (2012). Raw material selectivity in Late Pliocene Oldowan sites in the Makaamitalu Basin, Hadar, Ethiopia. *Journal of Human Evolution, 62*, 353–366.

Goodall, J. (1986). *The Chimpanzees of Gombe. Patterns of behavior*. Cambridge, MA: The Belknap Press of Harvard University Press.

Gowlett, J. A. J. (1984). Mental abilities of early Man: A look at some hard evidence. In R. A. Foley (Ed.), *Hominid Evolution and Community Ecology* (pp. 167–192). London: Academic Press.

Green, R. E., Krause, J., Briggs, A. W., Maricic, T., Stenzel, U., Kircher, M., et al. (2010). A draft sequence of the Neandertal genome. *Science, 328*, 710–722.

Haidle, M. N. (2010). Working memory capacity and the evolution of modern cognitive potential: Implications from animal and early human tool use. *Current Anthropology, 51*, S149–S166.

Haidle, M. N. (2011). Archaeological approaches to cognitive evolution. In D. B. Kronenfeld, G. Bennardo, V. C. de Munck, & M. D. Fischer (Eds.), *A companion to cognitive Anthropology* (pp. 450–467). Oxford: Wiley-Blackwell.

Haidle, M. N., & Conard, N. J. (2011). The nature of culture. Synthesis of an interdisciplinary symposium held in Tübingen, Germany, 15–18 June 2011. *Mitteilungen der Gesellschaft für Urgeschichte, 20*, 65–78.

Hay, R. L. (1976). *Geology of Olduvai Gorge*. Berkeley: University of California Press.

Hayden, B. (1979). *Palaeolithic reflections: Lithic technology and ethnographic excavations among Australian Aborigines*. Canberra: Australian Institute of Aboriginal Studies.

Henrich, J. (2004). Demography and cultural evolution: How adaptive cultural processes can produce maladaptive losses – the Tasmanian case. *American Antiquity, 69*, 197–214.

Henshilwood, C. S., & Marean, C. W. (2003). The origin of modern human behavior: Critique of the models and their test implications. *Current Anthropology, 44*, 627–651.

Herrmann, E., Call, J., Hernández-Lloreda, M. V., Hare, B., & Tomasello, M. (2007). Humans have evolved specialized skills of social cognition: The cultural intelligence hypothesis. *Science, 317*, 1360–1366.

Hiscock, P., & Attenbrow, V. (2005). Reduction continuums and tool use. In C. Clarkson & L. Lamb (Eds.), *Lithics 'Down Under': Australian perspectives on lithic reduction, use and classification* (pp. 43–55). Oxford: British Archaeological Reports, Archaeopress.

Hohmann, G., & Fruth, B. (2003). Culture in bonobos? Between-species and within species variation in behavior. *Current Anthropology, 44*, 563–571.

Hovers, E. (2009a). Learning from mistakes: Flaking accidents and knapping skills in the assemblage of A.L. 894, (Hadar, Ethiopia). In K. Schick & N. Toth (Eds.), *The cutting edge: New approaches to the archaeology of human origins* (pp. 137–150). Gosport, Indiana: Stone Age Institute.

Hovers, E. (2009b). *The lithic assemblages of Qafzeh Cave*. New York: Oxford University Press.

Hrdlicka, A. (1927). The Neanderthal phase of man. *The Journal of the Royal Anthropological Institute of Great Britain and Ireland, 57*, 249–274.

Iovita, R., & McPherron, S. P. (2011). The handaxe reloaded: A morphometric reassessment of Acheulian and Middle Paleolithic handaxes. *Journal of Human Evolution, 61*, 61–74.

Isaac, G. L. (1972). Early phases of human behaviour: Models in Lower Palaeolithic archaeology. In D. L. Clarke (Ed.), *Models in Archaeology* (pp. 167–199). London: Methuen.

Isaac, G. L. (1977). *Olorgesailie. Archaeological studies of a Middle Pleistocene lake basin in Kenya.* Chicago: The University of Chicago Press.

Isaac, G. L. (1978). The food-sharing behavior of protohuman hominids. *Scientific American, 238*, 90–108.

Janik, V. M., & Slater, P. J. B. (2003). Traditions in mammalian and avian vocal communication. In S. Perry & D. Fragaszy (Eds.), *The biology of tradition: Models and evidence* (pp. 213–235). Cambridge: Cambridge University Press.

Kamilar, J. M., & Marshack, J. L. (2012). Does geography or ecology best explain 'cultural' variation among chimpanzee communities? *Journal of Human Evolution, 62*, 256–260.

Kaplan, H., Hill, K., Lancaster, J., & Hurtado, A. M. (2000). A theory of human life history evolution: Diet, intelligence, and longevity. *Evolutionary Anthropology, 9*, 156–185.

Keeley, L. H., & Toth, N. (1981). Microwear polishes on early stone tools from Koobi Fora, Kenya. *Nature, 293*, 464–465.

Kinani, J. -F., & Zimmerman, D. (2014). Tool use for food acquisition in a wild mountain gorilla (Gorilla beringei beringei). *American Journal of Primatology*, doi:10.1002/ajp.22351.

King, B. J. (2002). Patterned interactions and culture in great apes. In R. G. Fox & B. J. King (Eds.), *Anthropology beyond culture* (pp. 83–104). Oxford: Berg.

Klein, R. (2005). Hominin dispersals in the Old World. In C. Scarre (Ed.), *The human past. World prehistory and the development of human societies* (pp. 84–123). London: Thames and Hudson.

Kroeber, A. L., & Kluckhohn, C. (1963). *Culture: A critical review of concepts and definitions*. New York: Vintage Books.

Kuhn, S. L. (2006). Trajectories of change in the Middle Paleolithic. In E. Hovers & S. Kuhn (Eds.), *Transitions before the transition. Evolution and stability in the Middle Paleolithic and the Middle Stone Age* (pp. 109–120). New York: Springer.

Kuhn, S. L. (2010). On standardization in the Paleolithic: Measures, causes, and interpretations of metric similarity in stone tools. In A. Nowell & I. Davidson (Eds.), *Stone tools and the evolution of human cognition* (pp. 105–134). Boulder, CO: University Press of Colorado.

Kuper, A. (1999). *Culture. The anthropologists' account*. Cambridge, MA: Harvard University Press.

Laland, K. N. (2008). Animal cultures. *Current Biology, 18*, R366–R370.

Laland, K. N., & Hoppitt, W. (2003). Do animals have culture? *Evolutionary Anthropology, 12*, 150–159.

Laland, K. N., & Janik, V. M. (2006). The animal cultures debate. *Trends in Ecology and Evolution, 21*, 542–547.

Langergraber, K. E., Boesch, C., Inoue, E., Inoue-Murayama, M., Mitani, J. C., Nishida, T., et al. (2011). Genetic and 'cultural' similarity in wild chimpanzees. *Proceedings of the Royal Society B: Biological Sciences, 278*, 408–416.

Leakey, M. D. (1971). *Olduvai Gorge. Volume 3. Excavations in Beds I and II, 1960–1963*. Cambridge: Cambridge University Press.

Lieberman, D. E., & Shea, J. J. (1994). Behavioral differences between archaic and modern humans in the Levantine Mousterian. *American Anthropologist, 96*, 300–332.

Lipson, S., & Pilbeam, D. (1982). *Ramapithecus* and Hominoid evolution. *Journal of Human Evolution, 11*, 545–548.

Locke, J. L., & Bogin, B. (2006). Language and life history. *Behavioral and Brain Sciences, 29*, 259–325.

Lycett, S. J., & Norton, C. J. (2010). A demographic model for Palaeolithic technological evolution: The case of East Asia and the Movius Line. *Quaternary International, 211*, 55–65.

Lycett, S. J., & von Cramon-Taubadel, N. (2008). Acheulean variability and hominin dispersals: A model bound approach. *Journal of Archaeological Science, 35*, 553–562.

Madden, J. R. (2008). Do bowerbirds exhibit cultures? *Animal Cognition, 11*, 1–12.

Marshall, G. D., Gamble, C. G., Roe, D. A., & Dupplaw, D. (2002). Acheulian biface database. from ADS, York. http://ads.ahds.ac.uk/catalogue/specColl/bifaces/bf_query.cfm

Marwick, B. (2003). Pleistocene exchange networks as evidence for the evolution of language. *Cambridge Archaeological Journal, 13*, 67–81.

McGrew, W. C. (1992). *Chimpanzee material culture. Implications for human evolution*. Cambridge: Cambridge University Press.

McGrew, W. C. (2004). *The cultured chimpanzee: Reflections on cultural primatology*. Cambridge: Cambridge University Press.

McGrew, W. C., & Tutin, C. E. G. (1978). Evidence for social custom in wild chimpanzees? *Man, 13*, 234–251.

McPherron, S. P. (2000). Handaxes as a measure of the mental capabilities of early hominids. *Journal of Archaeological Science, 27*, 655–663.

Mellars, P. A. (1969). The chronology of Mousterian industries in the Périgord region of South-West France. *Proceedings of the Prehistoric Society, 35*, 134–171.

Mellars, P. A. (1989). Technological changes across the Middle-Upper Palaeolithic transition: Economic, social and cognitive perspectives. In P. A. Mellars & C. B. Stringer (Eds.), *The human revolution: Behavioural and biological perspectives on the origin of modern humans* (pp. 338–365). Princeton, NJ: Princeton University Press.

Mellars, P. A. (1996). Symbolism, language, and the Neanderthal mind. In P. Mellars & K. R. Gibson (Eds.), *Modelling the early human mind* (pp. 15–32). Cambridge: McDonald Institute for Archaeological Research.

Mellars, P. A., & French, J. C. (2011). Tenfold population increase in Western Europe at the Neandertal-to-modern human transition. *Science, 333*, 623–627.

Mishra, S., Gaillard, C., Hertler, C., Moigne, A.-M., & Simanjuntak, T. (2010). India and Java: Contrasting records, intimate connections. *Quaternary International, 223–224*, 265–270.

Moore, M. W. (2003a). Australian Aboriginal biface reduction technique on the Georgina Rive, Camooweal, Queensland. *Australian Archaeology, 56*, 22–34.

Moore, M. W. (2003b). Australian Aboriginal blade production methods on the Georgina River, Camooweal, Queensland. *Lithic Technology, 28*, 27–54.

Moore, M. W. (2003c). Flexibility of stone tool manufacturing methods on the Georgina River, Camooweal, Queensland. *Archaeology in Oceania, 38*, 23–36.

Moore, M. W. (2010). 'Grammars of action' and stone flaking design space. In A. Nowell & I. Davidson (Eds.), *Stone tools and the evolution of human cognition* (pp. 13–43). Boulder, CO: University Press of Colorado.

Moore, M. W. (2011). The design space of stone flaking: Implications for cognitive evolution. *World Archaeology, 43*, 702–715.

Moore, M. W. (2013). Simple stone flaking in Australasia: Patterns and implications. *Quaternary International, 285*, 140–149.

Moore, M. W., & Brumm, A. (2007). Stone artifacts and hominins in Island Southeast Asia: New insights from Flores, Eastern Indonesia. [Review]. *Journal of Human Evolution, 52*, 85–102.

Nishida, T. (1986). Local traditions and cultural transmission. In B. B. Smuts, D. L. Cheney, R. M. Seyfarth, R. W. Wrangham, & T. T. Struhsaker (Eds.), *Primate societies* (pp. 462–474). Chicago: The University of Chicago Press.

Noble, W., & Davidson, I. (1996). *Human evolution, language and mind*. Cambridge: Cambridge University Press.

Nowell, A. (2010). Working memory and the speed of life. *Current Anthropology, 51*, S121–S133.

Nowell, A. (2016). Childhood, play and the evolution of cultural capacity in Neanderthals and Modern Humans. In M. N. Haidle, N. J. Conard, & M. Bolus (Eds.), *The nature of culture* (pp. 87–97). Dordrecht: Springer.

Nowell, A., & Davidson, I. (Eds.). (2010). *Stone tools and the evolution of human cognition*. Boulder, CO: University Press of Colorado.

Nowell, A., & White, M. J. (2010). Middle Pleistocene life history strategies and their relationship to Acheulean industries. In A. Nowell & I. Davidson (Eds.), *Stone tools and the evolution of human cognition*. Boulder, CO: University Press of Colorado.

Perry, S. (2011). Social traditions and social learning in capuchin monkeys (Cebus). *Philosophical Transactions of the Royal Society B: Biological Sciences, 366*, 988–996.

Pope, M. I. (2002). *The significance of biface-rich assemblages: An examination of the behavioural controls on lithic assemblage formation in the Lower Palaeolithic*. University of Southampton, Unpublished Ph.D. Thesis.

Pope, M. I., Russel, K., & Watson, K. (2006). Biface form and structured behaviour in the Acheulean. *Lithics, 27*, 44–57.

Prasad, K. N. (1982). Was *Ramapithecus* a tool-user? *Journal of Human Evolution, 11*, 101–104.

Rainey, A. (1991). Some Australian bifaces. *Lithics, 12*, 33–36.

Rappaport, R. A. (1999). *Ritual and religion in the making of humanity*. Cambridge, UK: Cambridge University Press.

Rendell, L., & Whitehead, H. (2001). Cultures in whales and dolphins. *Behavioral and Brain Sciences, 24*, 309–382.

Reynolds, S. C., Bailey, G. N., & King, G. C. P. (2011). Landscapes and their relation to hominin habitats: Case studies from *Australopithecus* sites in Eastern and Southern Africa. *Journal of Human Evolution, 60*, 281–298.

Roberts, M. B., Parfitt, S. A., Pope, M. I., & Wenban-Smith, F. F. (1997). Boxgrove, West Sussex: Rescue excavations of a lower Palaeolitic landsurface. *Proceedings of the Prehistoric Society, 63*, 303–358.

Roebroeks, W., Kolen, J., & Rensink, E. (1988). Planning depth, anticipation and the organization of Middle Palaeolithic technology: The "archaic native" meet Eve's descendants. *Helinium, 28*, 17–34.

Rolland, N., & Dibble, H. L. (1990). A new synthesis of Middle Paleolithic variability. *American Antiquity, 55*, 480–499.

Ross, J., & Davidson, I. (2006). Rock art and ritual: An archaeological analysis of rock art in arid Central Australia. *Journal of Archaeological Method and Theory, 13*, 305–341.

Savage-Rumbaugh, E. S., Williams, S. L., Furuichi, T., & Kano, T. (1996). Language perceived: *Paniscus* branches out. In W. C. McGrew, L. F. Marchant, & T. Nishida (Eds.), *Great ape societies* (pp. 173–184). Cambridge: Cambridge University Press.

Savage-Rumbaugh, E. S., Fields, W. M., & Spircu, T. (2004). The emergence of knapping and vocal expression embedded in a Pan/Homo culture. *Biology and Philosophy, 19*, 541–575.

Schäfer, J. (1990). Conjoining of artefacts and consideration of raw material. Their application at the Middle Palaeolithic site of Schweinskopf-Karmelenberg. In E. Cziesla, S. Eickhoff, N. Arts, & D. Winter (Eds.), *The Big Puzzle. Proceedings of the international symposium on refitting stone artefacts* (pp. 83–100). Bonn: Holos.

Schlanger, N. (1996). Understanding Levallois: Lithic technology and cognitive Archaeology. *Cambridge Archaeological Journal, 6*, 231–254.

Sharon, G. (2008). The impact of raw material on Acheulian large flake production. *Journal of Archaeological Science, 35*, 1329–1344.

Sharon, G. (2010). Large flake Acheulian. *Quaternary International, 223–224*, 226–233.

Shennan, S. J., & Steele, J. (1999). Cultural learning in hominids: A behavioural ecological approach. In H. O. Box & K. R. Gibson (Eds.), *Mammalian social learning. Comparative and ecological perspectives* (pp. 367–388). Cambridge: Cambridge University Press.

Stiner, M. C. (1994). *Honor among thieves. A zooarchaeological study of Neandertal Ecology*. Princeton, NJ: Princeton University Press.

Stiner, M. C. (2006). Middle Paleolithic subsistence ecology of the Mediterranean region. In E. Hovers & S. Kuhn (Eds.), *Transitions before the transition. Evolution and stability in the Middle Paleolithic and the Middle Stone Age* (pp. 213–231). New York: Springer.

Stiner, M. C., & Kuhn, S. L. (1992). Subsistence, technology, and adaptive variation in Middle Palaeolithic Italy. *American Anthropologist, 94*, 306–339.

Tennie, C., Call, J., & Tomasello, M. (2009). Ratcheting up the ratchet: on the evolution of cumulative culture. *Philosophical Transactions of the Royal Society B: Biological Sciences, 364*, 2405–2415.

Tennie, C., Braun, D. R., Premo, L. S., & McPherron, S. P. (2016). The Island test for cumulative culture in the Paleolithic. In M. N. Haidle, N. J. Conard, & M. Bolus (Eds.), *The nature of culture* (pp. 121–131). Dordrecht: Springer.

Tomasello, M. (1999). *The cultural origins of human cognition*. Cambridge, MA: Harvard University Press.

Toth, N. (1985). The Oldowan reassessed: A close look at early stone artifacts. *Journal of Archaeological Science, 12*, 101–120.

Trinkaus, E., & Shipman, P. (1993). *The Neandertals. Changing the image of mankind*. New York: Alfred A. Knopf.

Tylor, E. B. (1870). *Primitive culture. Researches into the development of mythology, philosophy, religion, language, art, and custom*. London: John Murray.

Van Peer, P. (1992). *The Levallois reduction strategy*. Madison, Wisconsin: Prehistory Press.

van Schaik, C. P., & Pradhan, G. R. (2003). A model for tool-use traditions in primates: Implications for the coevolution of culture and cognition. *Journal of Human Evolution, 44*, 645–664.

van Schaik, C. P., Ancrenaz, M., Borgen, G., Galdikas, B., Knott, C. D., Singleton, I., et al. (2003). Orangutan cultures and the evolution of material culture. *Science, 299*, 102–105.

Vandermeersch, B. (1981). *Les hommes fossiles de Qafzeh (Israël)*. Paris: CNRS.

Visalberghi, E., & Fragaszy, D. (1990). Do monkeys ape? In S. Taylor Parker & K. R. Gibson (Eds.), *"Language" and intelligence in monkeys and apes* (pp. 247–273). Cambridge: Cambridge University Press.

White, L. A. (1959). *The evolution of culture*. New York: McGraw-Hill.

Whiten, A., & Boesch, C. (2001). The cultures of chimpanzees. *Scientific American, 284*(1), 48–55.

Whiten, A., & van Schaik, C. (2007). The evolution of animal 'cultures' and social intelligence. *Philosophical Transactions of the Royal Society B: Biological Sciences, 362*, 603–620.

Whiten, A., Goodall, J., McGrew, W. C., Nishida, T., Reynolds, V., Sugiyama, Y., et al. (1999). Cultures in chimpanzees. *Nature, 399*, 682–685.

Whiten, A., Goodall, J., McGrew, W. C., Nishida, T., Reynolds, V., Sugiyama, Y., et al. (2001). Charting cultural variation in chimpanzees. *Behaviour, 138*, 1481–1516.

Whiten, A., Horner, V., & Marshall-Pescini, S. (2003). Cultural panthropology. *Evolutionary Anthropology, 12*, 92–105.

Wiessner, P. (1983). Style and social information in Kalahari San projectile points. *American Antiquity, 48*, 253–276.

Wynn, T., & Coolidge, F. (2010). How Levallois reduction is similar to, and not similar to, playing chess. In A. Nowell & I. Davidson (Eds.), *Stone tools and the evolution of human cognition*. Boulder, CO: University Press of Colorado.

Wynn, T., & McGrew, W. C. (1989). An ape's eye view of the Oldowan. *Man, 24*, 383–398.

Wynn, T., Hernandez-Aguilar, R. A., Marchant, L. F., & McGrew, W. C. (2011). "An ape's view of the Oldowan" revisited. *Evolutionary Anthropology, 20*, 181–197.

Zhang, P., Huang, W., & Wang, W. (2010). Acheulean handaxes from Fengshudao, Bose sites of South China. *Quaternary International, 223–224*, 440–443.

Chapter 11
The Island Test for Cumulative Culture in the Paleolithic

Claudio Tennie, David R. Braun, L.S. Premo, and Shannon P. McPherron

Abstract Early Stone Age artifacts have long been assumed to reflect the material record of communities whose members possessed the ability to transmit ideas, behaviors, and technologies from individual to individual through high-fidelity transmission (i.e., involving teaching and/or imitation), much like humans do today. Recent experimental work has highlighted marked differences between great apes and modern humans in the capacity and/or motivation for some forms of cultural transmission. In particular, high-fidelity mechanisms of social learning, which are thought to underlie the capacity for cumulative culture, appear to be enhanced in – if not unique to – humans. Taken as a group, these experiments suggest it is plausible that a combination of genetic, environmental, and social factors that do not include high-fidelity social learning mediate the "cultures" described for great ape populations to date. It may be that, while the distribution of great ape behavioral variation in time and space is likely affected by low-fidelity social learning (which is widespread in the animal kingdom), the observed variants were invented (i.e., learned) independently by each individual rather than copied from other individuals. Behaviors that do not require high-fidelity transmission between individuals in order to increase in frequency in a population lie within the so-called "zone of latent solutions." Here, we begin to grapple with the hypothesis that much of the Early Stone Age archaeological record may reflect deeply "canalized" behaviors of hominin toolmakers – those that reside in each individual's zone of latent solutions – rather than behaviors that necessarily require high-fidelity transmission between individuals. We explore this possibility while eschewing the simplistic notion that variation in stone tool shape, for example, is entirely determined by the genetic variation found in the toolmakers. Instead, we suggest that the variation observed in Early Stone Age artifacts may simply reflect a heavier reliance on behaviors that reside within the zone of latent solutions than on behaviors that make use of high-fidelity social learning. We discuss a thought experiment, called the Island Test, which may be useful for distinguishing hominin behaviors that require high-fidelity transmission from behaviors that do not. We conclude that the Early Stone Age archaeological record is consistent with the possibility that latent solutions explain the behavioral variation inferred from available material culture. Furthermore, we explore reasons why the assumption of high-fidelity transmission associated with Paleolithic industries is difficult to support.

Keywords Behavior • Apes • Humans • Early Stone Age • Oldowan • Acheulean • Tools • Social learning • High-fidelity transmission

C. Tennie (✉)
School of Psychology, University of Birmingham, Birmingham, B15 2TT, UK
e-mail: c.tennie@bham.ac.uk

C. Tennie · L.S. Premo · S.P. McPherron
Max Planck Institute for Evolutionary Anthropology, Deutscher Platz 6, 04103 Leipzig, Germany

S.P. McPherron
e-mail: mcpherron@eva.mpg.de

D.R. Braun
Center for the Advanced Study of Human Paleobiology, Department of Anthropology, George Washington University, 2210 G Street NW, Washington, DC 20052, USA
e-mail: david_braun@gwu.edu

and

Department of Archaeology, University of Cape Town, Cape Town, South Africa

L.S. Premo
Department of Anthropology, Washington State University, Pullman, WA 99164-4910, USA
e-mail: luke.premo@wsu.edu

Miriam N. Haidle, Nicholas J. Conard and Michael Bolus (eds.), *The Nature of Culture: Based on an Interdisciplinary Symposium 'The Nature of Culture', Tübingen, Germany*, Vertebrate Paleobiology and Paleoanthropology, DOI 10.1007/978-94-017-7426-0_11

> "Most scholars assume that the skills necessary to manufacture Acheulean tools were transmitted culturally in the same way that stone tool traditions are transmitted among living foragers. However, this assumption is hard to reconcile with either theory or data. [...] perhaps we need to entertain the hypothesis that Acheulean bifaces were innately constrained rather than wholly cultural and that their temporal stability stemmed from some component of genetically transmitted psychology." (Richerson and Boyd 2005:142)

Introduction

In recent years it has become apparent that trying to understand modern human evolution without understanding some features of cultural evolution is untenable (Brown et al. 2011), for cultural evolution is inextricably linked to biological evolution in humans (Laland et al. 2010). These points further justify the use of gene-culture co-evolution, or dual inheritance, theory (Boyd and Richerson 1985; Richerson and Boyd 2005; Richerson et al. 2010) and human niche construction, or triple inheritance (genes, culture, and ecology), theory (Odling-Smee et al. 2003; Rendell et al. 2011). While these perspectives have undoubtedly opened new avenues for understanding human evolution, applications to earlier hominins tend to paint with a broad brush – as Richerson and Boyd acknowledge in the quote above – treating their subjects as more or less equivalent to modern humans with regard to the capacity for cultural transmission.

While numerous studies show that "tool use cultures" may be present in some great ape species (McGrew 1998; Whiten et al. 1999; Byrne 2002; Carvalho et al. 2008; Marshall-Pescini and Whiten 2008; Schöning et al. 2008; Haslam et al. 2009), suggesting that humans are not the only ones with culture writ large, it remains that modern human culture may be unique in important ways (Tomasello 1999; Hill et al. 2009). In particular, human cultural change is associated with a ratcheting effect, referred to as cumulative cultural evolution (Tomasello 1999), that may prove to be a unique feature of the hominin lineage (Tennie et al. 2009; but see Pradhan et al. 2012 for an alternative view) even if we do not yet know when it evolved. Although the list of cognitive ingredients thought to be sufficient for cumulative culture remains a source of debate, many agree that cumulative culture (at least the human version) is not possible without high-fidelity social learning mechanisms, such as teaching and imitation (Tomasello 1999). This is one reason why those searching for cumulative culture in non-humans have been interested in identifying examples of imitation and teaching in ape societies. While some argue for the presence of imitation in chimpanzees (De Waal 2001; Whiten et al. 2009a), others remain skeptical (Tennie et al. 2012) or insist that if chimpanzees do indeed imitate, they do so less regularly and with less fidelity than humans (Tennie et al. 2009; Whiten et al. 2009a). Currently there is no credible evidence to support the claim of active teaching in chimpanzees (Tennie et al. 2009).

Although there is still some debate over the presence of high-fidelity transmission in apes, it appears that chimpanzees do have some low-fidelity social learning abilities, as shown convincingly in laboratory settings (Whiten et al. 2009a). In addition, some behavioral patterns are different among populations of closely related chimpanzees who live in similar environmental contexts, suggesting that the behavioral variation is not well explained by a purely genetic or environmentally deterministic explanation (Gruber et al. 2009; Luncz et al. 2012). To paleoanthropologists on the outside looking in, the mounting evidence gathered from ape societies provides a unique view of what cultural variation looks like in the absence of the type of pervasive high-fidelity cultural transmission that characterizes humans today.

Research conducted by one of us (CT) on the mechanisms that drive cultural variation in ape societies (Tennie and Hedwig 2009; Tennie et al. 2009, 2010) identifies two important factors. First, many primate behaviors that are considered "cultural" in naturalistic settings can be (re)invented by a single naïve individual that did not have access to an experienced model, or teacher (Visalberghi 1987; Tennie et al. 2008; Menzel et al. 2013). The particular *form* that any such behavior takes thus appears to result from a complex and admittedly opaque interaction between the environmental conditions and genetic predispositions of the individual rather than from high-fidelity transmission between individuals. Second, low-fidelity mechanisms of social learning, including stimulus enhancement (Whiten and Ham 1992), can influence how often, by whom, where, to what and when such a behavior is expressed. Thus, low-fidelity social learning plays a role mainly in influencing the temporal and spatial distribution of variation in individually learned behaviors (Tennie et al. 2009).

Together these two factors constitute the "zone of latent solutions" hypothesis (hereafter ZLS).[1] Behaviors or variants that do not require high-fidelity social learning in order to increase in frequency in a population lie within a species' ZLS. Thus, behaviors that reside in the ZLS are not indicative of cumulative culture. A ZLS-based explanation provides a valid alternative for interpreting the currently available data concerning chimpanzee culture (both from field observations and lab experiments; Tennie et al. 2009).

[1]The term "solutions" is used because social learning experiments often focus on physical problems to solve. But the hypothesis may also be called more broadly "zone of latent behavior," as indeed the theory also entails social behaviors such as "hand-clasp grooming" (Tennie et al. 2009) that may not be best described as "solutions." Nevertheless, here we stick to the original term so as not to introduce ever more terms.

One of the reasons for introducing the ZLS here is that we think the concept might also provide a useful framework for studying and ultimately interpreting the variation in Early Stone Age archaeological assemblages.

Galef (2001) has claimed that unless imitation has actually been shown to underlie the spread of a behavior, that behavior should be considered a tradition rather than culture. We recognize that there should be some distinction between animal traditions and modern human cultures. However, here we follow Whiten and van Schaik (2007) in making a distinction instead between culture that does not require imitation (i.e., "a distinctive behavior pattern shared by two or more individuals in a social unit, which persists over time and that new practitioners acquire in part through socially aided learning…" [Fragaszy and Perry 2003] to which we would add: "…usually with the exception of imitation") and cumulative culture that accumulates through imitation or teaching (Tennie et al. 2009). This distinction seems heuristically pragmatic because culture is the term most often used in descriptions of great ape behavior (e.g., Whiten 2000). Following this definition, culture can be sustained by several social learning mechanisms and it does not necessarily require imitation. Cumulative culture is distinctive in that it requires the underlying mechanism of high fidelity transmission (imitation or teaching; Tomasello 1999).

To date, primatologists have spent far more time studying cultures that do not exhibit pervasive high-fidelity social learning (e.g., McGrew 1998; Whiten et al. 1999; Matsuzawa et al. 2001; Biro et al. 2003; Horner and Whiten 2005; Lycett et al. 2007; Schöning et al. 2008; Whiten et al. 2009a) than have Paleolithic archaeologists, but maybe this should change. We might begin by asking whether the Early Stone Age archaeological record is consistent with the predictions of an explanation that places the behaviors responsible for the manufacture, use, and discard of stone tools within the ZLS of the hominin toolmakers. Considering the many similarities between humans and their closest living relatives today, we can expect that some basic – though not yet cumulative – cultural patterns existed deep in the past (and maybe that variation looked similar to what we see within and between ape societies today). However, at the point in hominin evolution when high-fidelity social learning became the rule rather than the exception, culture became cumulative. Identifying when this occurred and explaining why it happened when it did, strike us as extremely important yet neglected paleoanthropological research questions. Given that technology and associated selective advantages are one of the most pervasive features of hominin adaptation for at least the last 60 ka, it seems imperative that paleoanthropologists address the mechanisms of how it arises and spreads.

The assumption that early stone tools represent cumulative culture is commonly made (Gamble and Porr 2005; Shipton 2010), if not explicitly stated (although see the critique in Richerson and Boyd 2005; Acerbi et al. 2011). Models that treat early stone tool technology as representative of the toolmakers' ability to transfer the information needed to manufacture implements through high- (or at least, medium-) fidelity copying also view high-fidelity cultural transmission as having been a feature of hominins for at least 2.5 Myr (Whiten et al. 2011). In a bit of a departure, McNabb et al. (2004) suggest that while patterns of variation in large stone cutting tools from southern Africa may imply that Acheulean tools were mimetic constructs (and that their main "idea" needed to be transmitted rather than learned individually), they argue against the notion that what was being culturally transmitted was a detailed mental template.

Here, we entertain an alternative hypothesis that suggests that much of the Early Stone Age cultural material record may be explained by latent solutions rather than by behaviors that require high-fidelity forms of social learning, such as imitation or teaching. This perspective holds that the flint-knapping techniques for making Oldowan and even Acheulean tools fit squarely within the ZLS of Early Stone Age hominins. This explanation also assumes that, although the technological know-how did not require high-fidelity transmission, low-fidelity social learning such as stimulus enhancement and product emulation (sensu Tennie et al. 2009) could have played a role in the spatial and temporal distributions of the otherwise independently invented (i.e., individually learned) toolmaking behaviors as well as the archaeological assemblages those behaviors left behind.

If the appearance of simple flaked stone implements at 2.5 Ma is not sufficient to signal the presence of cumulative culture and the high-fidelity social learning mechanisms that underlie it, then what is? When using the Paleolithic archaeological record to address this question we must be careful to consider whether other mechanisms might have been responsible for the spatial and temporal variation in cultural material before settling on cultural transmission as the best working hypothesis. A more conservative approach, given the apparent rarity (or quite possibly, absence) of cumulative culture in apes, would be to treat a ZLS explanation as the default for Early Stone Age hominins until it can be shown that the archaeological data demonstrate otherwise.

The Archaeological Record of Oldowan and Acheulean

The Oldowan is well dated in sites that are at least as old as 2.5 Ma in the Gona region of Ethiopia (Semaw et al. 1997; Semaw 2000). However, this is the only locality where sites of this antiquity are known. By 2.3 Ma localities elsewhere in Ethiopia and in various parts of Kenya are known (Kimbel et al. 1996; Roche et al. 1999). By 1.8 Ma much of the

African continent appears to be inhabited by tool using hominins and shortly after this we begin to see their appearance throughout the Old World (Kuman and Clarke 2000; de Lumley et al. 2005; Sahnouni et al. 2009). Although there are a few that would suggest linear patterns of increasing complexity throughout the Oldowan (Carbonell et al. 2006), the dominant and constant feature of the Oldowan is pebble tools (or "cores") that have been knapped to produce flakes with sharp edges. Experimental studies have shown that most of these cores are not useful for many of the activities we assume early hominins were engaged in (e.g., butchery, woodworking (Toth 1987)). Importantly, the most prominent feature of the Oldowan is that there does not appear to be variation in stone tool technologies at a given time that cannot be explained by environmental factors (but see Whiten et al. (2009b) for possible explanatory scenarios), and time transgressive variation does not display patterning that differs from variation among assemblages that have been dated to similar time periods. By some measures, the degree of Oldowan behavioral variation revealed by stone tools appears even less patterned than that described for chimpanzees (Lycett et al. 2007).

Hominins likely invested substantial energy in the development and maintenance of stone tool kits (Stout et al. 2005; Braun et al. 2008, 2009), and the production of these artifacts was likely bound by certain rules of manufacture (Delagnes and Roche 2005). However, whether or not any of these rules benefitted from or, more to the point at hand, *required* high-fidelity transmission processes remains unknown. We find it difficult to identify characteristics of the Oldowan record that preclude the possibility that multiple bouts of individual learning (i.e. naïve individuals applying largely biologically based skills) were responsible for these artifacts. Oldowan tool manufacture and use may have been deeply canalized behaviors that resided within the ZLS of Early Stone Age hominins. Here, just like in the case of living chimpanzees, it is also possible that some low-fidelity social learning mechanisms facilitated the distribution and frequency of behaviors. Such social learning most likely increased the likelihood that certain locations would become foci of stone artifact production behavior. Studies of Oldowan behavior have already documented that the presence of raw material and resources that require tool use will result in archaeological assemblages that are significantly larger than assemblages found in areas with reduced availability of stone or reduced requirement for stone tools (Braun and Harris 2009). Although the population size of ancient hominin groups is poorly understood, increases in census size could have facilitated the distribution of this behavior as this could have increased the probability of the expression of latent solution behaviors (Tennie et al. 2009).

While a latent solutions explanation may account for the Oldowan, handaxes, the hallmark of the Acheulean, may (at first) appear to be a qualitatively different case. The earliest handaxes occur between ~1.8 (Lepre et al. 2011) and 1.4 (Beyene et al. 2013) Ma and continue in a generally similar format for over a million years. It is generally understood that the production quality of handaxes increases over time (i.e., tools become more refined, more regular, more symmetrical); though there have been few systematic, quantitative, time transgressive studies to support this (e.g., Beyene et al. 2013). This change could be considered to be a cumulative expansion of the previous tool production techniques. As such, it has been suggested that this may represent an ancient reflection of the ratcheting effect of human culture (Stout 2011). However, demonstrating this point quantitatively has been complicated by the high level of variability that exists at any one time, the number of variables that may influence final form, and the paucity of well dated assemblages. Further, the amount of time covered by the Acheulean means that biological explanations (e.g., physiological changes that improved motor skills; or increased working memory; see Haidle 2010) could account for some improvements in handaxes through time. It is also worth noting that handaxes are found throughout large portions of the Old World and were very likely made by *multiple* hominin species.

Where they principally differ from the preceding Oldowan, however, is in the notion that handaxes are consciously shaped to a particular form both by the controlled removal of shaping flakes and, in some cases, by the intentional preparation of large initial forms particularly suited to handaxe (and cleaver) production (Sharon 2007). The amount of effort, in terms of either the number of shaping flakes removed or the steps involved in preparing the blank, varies across the distribution of the Acheulean in both time and space, resulting in a positive assortment of particular forms or techniques. This apparent complexity has led some to suggest that handaxes must represent a true cultural advance over earlier stone tools, and some have gone so far as to suggest that extremely high levels of conformist biased cultural transmission were responsible for keeping handaxe form in check (Whiten et al. 2003; Lycett and Gowlett 2008; Shipton 2010). The clear implication of these studies is that handaxe form fully *depended* on high-fidelity cultural transmission whether or not this term is explicitly mentioned (Shipton 2010).

If the assumption that the form of handaxes was dependent on a behavior that required high-fidelity cultural transmission is correct, then one might reasonably expect patterns of cultural adaptation associated with handaxes to reflect this. The question is whether the signature of cultural transmission is evident in the archaeological record. To answer this question we need to think more about what that signature would look like and how we could assess our own confidence in recognizing it in archaeological data.

One challenge to answering these questions concerns the significance of both the similarity and the variation in handaxes. On the one hand, despite their dramatic geographic distribution, the remarkable similarity of large cutting tools distinguishes the Acheulean industry from subsequent lithic industries. A pattern of ubiquity across landscapes, through long periods of time, and across multiple species suggests an explanation within the domain of latent solutions. Yet, on the other hand, some variation in handaxes is observed. Mostly the focus here has been on handaxe shape and its significance, but aside from shape variation at the continental scale (Wynn and Tierson 1990; McPherron 2000; Lycett 2008; Lycett and Gowlett 2008) it has not been demonstrated that shape varies in patterned ways that are not largely accounted for by raw material variability and the intensity of bifacial reduction (e.g., the debate over the significance of variability in the British handaxe assemblages; see White 1998; McPherron 2007). Shape variability at a continental scale may be driven by similar ecological factors, as there are no independent lines of evidence to suggest the presence of continental scale cultural norms during this period. Another aspect of variability is in the techniques used to produce large flakes on which handaxes are sometimes made. Some have suggested that the type and form of raw materials that were used to make Acheulean tools reflects distinct selection biases on the part of individuals, decisions that have consequences for the shape of the handaxes fashioned on those raw materials (Sharon 2007, 2008; Shipton et al. 2009; Goren-Inbar 2011). The question is whether this variability demonstrates high-fidelity cultural transmission, and we believe that the answer is: not necessarily. Low-level learning mechanisms such as stimulus enhancement would result in similar patterns (see Tennie et al. 2009).

Richerson and Boyd (2005:142) outline the crux of this argument: "How could cultural transmission alone, particularly if based on a relatively primitive imitative capacity, preserve such a neat, formal-looking tool as an Acheulean hand-axe over half the Old World for a million years?" Or, phrased differently, could an artifact form as apparently complex as an Acheulean handaxe persist for so long in such a recognizable form over such a large geographic area in the absence of high-fidelity cultural transmission? We appreciate the fact that suggesting that the latter can be answered in the affirmative may be hard to digest at first (it sure was for some of us). We are in no way suggesting that there is a gene that "codes for" a certain type of handaxe production, yet we think it can be productive to consider whether the behavior(s) responsible for Oldowan core or Acheulean handaxe manufacture, use, and discard belongs within the ZLS of Early Stone Age hominins. Principally this is important because it forces one to be explicit about the characteristics of the behaviors that would necessitate an explanation that invokes high-fidelity cultural transmission. It is difficult to formulate useful tests of the empirical record until these assumptions have been made explicit and their qualities have been discussed and assessed by other experts.

The Island Test for Cumulative Culture

So, what is the likelihood of handaxes or any other Early Stone Age stone tool technology being the result of latent solutions? The term "latent solution" refers to a behavior that lies "dormant" or "latent" in an individual until triggered by a particular set of social or environmental cues and sufficient motivation on the part of the learner. Indeed the pattern of occurrence in naïve individuals who had no access to experienced models was the very mark of latent solutions that led to the initial description of the ZLS (see Tennie et al. 2009). Thus, one way to address the question above is to identify behaviors that could be exhibited by a previously naïve individual in the absence of any other cultural models. This is captured by a thought experiment that we refer to as the "Island Test" (based on a hypothetical island scenario presented first by Tomasello (1999)). Consider a scenario in which a child born to Early Stone Age hominin toolmakers is separated at birth and "magically" raised alone on an island that provides all of the raw material and motivation needed to produce Oldowan or Acheulean stone implements. Would this solitary hominin, stripped of the benefit of observing, let alone being taught by another toolmaker, produce, use, and discard artifacts that are indistinguishable from those we observe in the Early Stone Age archaeological record? If the behavior responsible for that kind of stone tool manufacture can be independently invented (i.e., learned, not copied) by a solitary individual and, thus, does not require high-fidelity cultural transmission, then it fails the Island Test for cumulative culture. We speculate that much of the early Paleolithic archaeological record may be composed of implements that resulted from behaviors that would fail such a "test." If we are correct in that assertion, then those behaviors reside within the zone of latent solutions of Early Stone Age hominins.

Our proposition may seem overly conservative at first. Considering the difficulty that most naïve (i.e., beginner) human flintknappers have today in learning how to produce a simple flake tool (e.g. Nonaka et al. 2010), can we safely assume that the behaviors responsible for creating what appear to us to be complex forms, such as the Acheulean handaxe, do not require the ratcheting effect of cumulative culture? In our view, the latent solutions explanation serves as a useful null model for early hominin behavior, which in turn should be viewed in the larger context of animal behavior (all hominins are animals, after all). Perhaps a ZLS explanation for early stone tool technologies appears less extreme if we consider Oldowan and Acheulean stone tools as similar to other complex "artifacts" such as beaver dams,[2] weaver bird nests, and spider webs – structures resulting from behaviors that are not regarded as requiring high-fidelity cultural transmission. Would dam building in beavers also fail the Island Test? If we placed a single, normally developed, yet dam-naïve beaver on an island with the motivation and material to produce a dam (including water) would it eventually build a dam? We think the intuitive answer is yes, it would (though of course, this remains to be seen). Also, it has already been shown that other seemingly complex behaviors do not require cultural transmission (e.g., naïve woodpecker finches can also make use of tools (Tebbich et al. 2001); see examples for great apes in Tennie et al. 2009).

A latent solutions explanation only rules out high-fidelity social learning mechanisms. Considering the prevalence of low-fidelity social learning in species ranging from stickleback fish to bats, it is almost certain that many behaviors exhibited by hominins included at least some aspect of low-fidelity social learning (Laland et al. 2010). Indeed, it is worth investigating to what extent stimulus enhancement together with emulation learning and other low-fidelity social learning mechanisms (Laland and Hoppitt 2003; Tennie et al. 2009) might explain the low level of variation observed in Early Stone Age culture material. For example, a ZLS explanation in which low-fidelity social learning mechanisms can be considered seems to fit the data as well as recently proposed demographic explanations of Acheulean geographic variation (Lycett and von Cramon-Taubadel 2008; Lycett and Norton 2010). Furthermore, it does this without assuming modern human abilities of high-fidelity cultural transmission in Middle Pleistocene hominins.

Obviously, for the case of extinct hominin behaviors, the Island Test can never actually serve as a "test." However, the vagaries of prehistory may have resulted in conditions that approximate key aspects of the thought experiment. For example, as mentioned earlier, handaxes appear throughout the Old World beginning by the Middle Pleistocene but are conspicuously absent for much of the Pleistocene record in Southeast Asia. Lycett and Norton (2010) suggest this is the result of a decrease in population size such that the collective cultural store of knowledge was insufficient to keep Acheulean technology in the behavioral repertoire of hominins in East Asia. However, how can we explain the reappearance of the very same stone tool forms when, in the case of Lycett and Norton's argument, effective population sizes rebound? Yamei and colleagues (2000) suggest the presence of Acheulean tools in the Bose Basin is the result of the appearance of suitable raw materials in this region at this time. Thus, the reappearance of similar tool forms in the Bose Basin may be the result of the reappearance of the ZLS conditions that led to the behavior responsible for Acheulean handaxe production. Using estimates of population size through time in this region, it may be possible to assess which of these alternative hypotheses is better supported by the archaeological data. At any rate, the ZLS approach explicitly allows for a behavior to at times disappear and subsequently reappear in an identical form when the ecological setting that brought about the initial appearance of the behavior (in this case, suitable raw material – but there may be other reasons in different situations) resurfaces.

[2]One might object that it may transpire that beaver dams only look complex but may be based on simple iterative wood-placing techniques. The question of complexity in behavior is however a tricky one – and likely relative to the species in question. All that we suggest here is that complexity alone cannot be solid ground for the inference of cumulative culture (Tennie and Hedwig 2009).

Zones of Latent Solutions: Archaeological Expectations and Complications

It is unfortunate that the Island Test ultimately does not provide much of a repeatable test for the case of Early Stone Age hominin behavior. Obviously, the thought experiment alone cannot "prove" that any Oldowan or Acheulean behavior was or was not a latent solution, nor is that the goal of this paper. But the Island Test does provide a heuristic device that can be used to improve our ideas about what kinds of archaeological signals we would expect latent solutions and culturally transmitted behaviors to exhibit in an assemblage of Paleolithic stone tools and debitage. Here, we begin to outline some of the archaeological expectations of a latent solution explanation. Our expectations cover three major components of variation in artifact form: geographic variation, temporal variation (the pace of change in form), and the reappearance of old, recognizable forms. We note that some expectations are not exclusive to a ZLS explanation. In addition, we briefly identify some reasons for caution when relying on the variation observed in stone tool technologies to discern latent solutions from cumulative cultural solutions.

Geographic Variation

There is debate over whether geographic variants can be identified within the Acheulean (Wynn and Tierson 1990; Shipton and Petraglia 2010). Studies that suggest the presence of geographic variation (Wynn and Tierson 1990) have not gone without critics (McPherron 2000). Here, we are interested in identifying what we would expect to see in terms of geographic variation in stone tool assemblages under the assumption that they are individually learned and only indirectly influenced by low-level social learning mechanisms versus the assumption that they are – and have to be – directly transmitted between individuals via high-fidelity social learning.

If ZLS behaviors were responsible for Early Stone Age tool manufacture and use, we would expect geographic variation in tool form (within a given species) to be explained by the combined effects of geographic variation in environmental conditions and low-level social learning mechanisms on raw material choice and possibly even core choice. This follows from the notion that the behavior an individual exhibits is influenced by both its psychology as well as its ecological and (low-level) social cues. Without independent measures of hominin psychology or of the particular low-level social cues involved, however, it may remain difficult to directly assess how much of the variation in the archaeological record is explained by the ZLS approach.

High-fidelity cultural transmission often results in striking geographic variation in tool forms. As Boyd and colleagues have shown, conformist biased cultural transmission can have the effect of increasing between-group differences even if there is migration between them (Boyd and Richerson 1985; Henrich and Boyd 1998; Richerson and Boyd 2005). In theory, conformist biased transmission can also have the effect of reducing within-group diversity to levels that may be similar to those predicted for a ZLS behavior. In addition, if there is very little migration of individuals and/or ideas between geographically or culturally isolated groups, then one may expect between-group variation to arise from even unbiased cultural transmission (or "random copying"). However, unbiased cultural transmission will yield greater within-group variation than conformist biased transmission and, more importantly, possibly greater within-group variation than predicted by a latent solution. On the flip side, there are demographic conditions (such as frequent local extinction and repopulation events: Premo and Kuhn 2010; Premo 2012) that can reduce the amount of geographic variation observed between regions in a structured population of high-fidelity social learners.

It would appear that we do not yet have a good idea of what kind(s) of archaeological signal(s) in spatial variation in stone tool form would distinguish behaviors that do not require high-fidelity social learning from those that do. Spatially explicit computational modeling might provide the kind of heuristic tool needed to aid us in improving our expectations of the empirical record.

Temporal Variation

How fast would we expect the form of an artifact to change through time in the absence of high-fidelity transmission? At first, it would seem that there is no simple answer to this question because the rate of change in a behavior that lies with the ZLS is contingent upon changes in the environment, low-level social learning mechanisms, and the psychology and physiology of the species of interest. While it may be possible in some cases to quantify the extent to which changes in artifact form correlate with changes in climate, this relationship between these two variables is poorly understood.

Instead, it might be better to identify rates of change that we would *not expect* to see in the archaeological record if the behaviors responsible for stone tool manufacture were indeed latent solutions. For cases that involve cultural transmission, a baseline rate of change (between upper and lower bounds imposed by mechanical constraints of the implement) can be estimated by taking into consideration the size and rate of copying mistakes caused by perception error. Kempe et al. (2012) analyze data collected from 2601 handaxes at 21 sites with a range of over 1 Myr and show that handaxe form features actually changed more *slowly* than expected under the assumption that the form was passed via high-fidelity cultural transmission (which is subject to such perception error).

In the presence of cultural transmission, processes that affect the effective size of the population of social learners can also affect the rate of change. For example, *biased* forms of cultural transmission (which decrease the effective population size by reducing either the number of models a naïve individual can potentially learn from or the number of traits that a naïve individual can adopt) may speed up or slow down the rate of change observed in an assemblage. Demographic factors can also affect rates of change, even after holding copying error rate constant (e.g., Premo and Kuhn 2010). These complications are worth keeping in mind because the rate of change will have serious consequences for measures of diversity within and between spatially or temporally separated archaeological assemblages. In cases where biased forms of cultural transmission and demographic effects can be ruled out as the cause of low variation and slow rates of change in the material record, a ZLS explanation may provide the most parsimonious explanation precisely because it does not assume high-fidelity cultural transmission of traits in the first place – while it does predict slow change.

The Reappearance of Old Forms

Another expectation of the latent solutions approach is that certain behaviors may disappear at times and subsequently reappear (in the identical form) if the environmental, cognitive, motivational and social conditions that helped to trigger the initial appearance return. Certainly the reappearance of Acheulean tools in the Bose Basin (Yamei et al. 2000), for example, suggests the possibility that the identical forms were re-invented many generations after they had disappeared.[3] Clearly, in the case of "sophisticated" cumulative cultural technologies, like those we take for granted today, we would not expect the form of the "artifact" (say, a computer) to be similar to the form that disappeared hundreds of generations earlier simply because the large number of cumulative innovations required would allow for many deviations the second time around. While we would not expect the form of a tool that required high-fidelity transmission to be identical to the form that disappeared from the record earlier, this is precisely the expectation if the tool form resulted from a ZLS behavior. Given the reappearance of the same combination of ecological, cognitive, motivational and social conditions responsible for the earlier appearance, we would expect to see a similar form. Having said that, in the case of a far less sophisticated cumulative cultural technology, in which the identical form can be recovered after the accumulation of just one or two innovations, it may be more difficult to discern a latent solution from one that requires cultural transmission in assemblages characterized by low temporal resolution. Given the apparent simplicity of Oldowan stone tools and the fact that many of the assemblages from this period conflate hundreds if not tens of thousands of years of time, distinguishing between the two alternatives may prove difficult in Oldowan assemblages.

It is most certainly not by design that stone tools comprise a very large proportion of all of the culture material recovered from the Oldowan and Acheulean. Previous research has identified the overriding importance of certain parameters that seem to guide all Acheulean tool production (e.g. correlation between elongation and any measure of size on almost all studied assemblages of handaxes; McPherron 1999). Here, the relationship between major size-related variables suggests that there are very few ways to make a handaxe. Thus, the convergence on a similar shape (or in other terms, the movement to a strong basin of attraction among all possible implement shapes) may actually not be all that surprising (Stout 2011). A related concern is that the tempo of material culture change need not be at the same rate. If stone tool technology was one of the more conservative aspects of Early Stone Age hominin life (due to conformist biased transmission, for example) or if some tool forms are highly convergent, then they may provide a biased picture of cultural variation that leads us to the wrong conclusion about whether Early Stone Age hominins were capable of transmitting information through high-fidelity social learning. At the very least, is seems worth acknowledging that the simple fact that stone tools are what we have to study from these early periods does not mean they are the best (or maybe even a suitable) source of data to address all of the questions we wish to answer.

It is important to be pragmatic and rigorous in testing these alternative hypotheses against the Paleolithic material record. Equifinality among alternatives does not invalidate either the ZLS or high fidelity cultural transmission explanations a priori, but it is likely to complicate the task of creating tests with enough power to discern the signal of latent solutions from that of a behavior that requires high-fidelity transmission. However, if we are to determine the basic mechanism of cultural change in the past, these tests are necessary. Given our brief discussion of the archaeological expectations associated with low-fidelity and high-fidelity social learning mechanisms, it would appear that a ZLS account does a better job than those that assume high-fidelity cultural transmission of explaining some of the more vexing characteristics of the Early Stone Age stone tool data.

Conclusions

Here we have discussed the ZLS concept as a viable alternative explanation for patterns of variation in the Early Stone Age archaeological record. We submit that Early Stone Age tool technologies can plausibly be explained by behaviors that fall within their various hominin makers' "zone of latent solutions." One of the distinguishing characteristics of the ZLS explanation is that it does not require high-fidelity social learning mechanisms on the part of Oldowan and Acheulean toolmakers. More work is needed to test the validity of our claim, and we hope that this paper provides the impetus for that research.

The majority of evidence available to date suggests that cumulative culture – and, even more fundamentally, the high-fidelity mechanisms of social learning (including teaching) that make it possible – evolved in the hominin lineage sometime between the chimpanzee-human split and the late Pleistocene, when the record shows clear examples of rapidly changing geographically delineated "cultures" throughout the Old World. Important questions concerning exactly when, how, and why cumulative culture arose in hominins have remained largely unaddressed. The good news is that a number of recent studies attest to the fact that this is changing.

Stout and colleagues (2010) evaluate the variation in stone tool form from a horizon in the Gona (Ethiopia) sequence in search of cumulative culture. Although they note interesting patterns of variation in the assemblage, they

[3]Also worth mentioning again in this context is that the manufacture of stone tools that look similar by different species is compatible with a ZLS explanation, which does not require further assumptions (i.e., high-fidelity social learning – and thus high tolerance – between different species).

were (in our view) unable to definitively exclude the possibility that the observed variation resulted from behaviors that did not require high-fidelity cultural transmission. At the opposite end of the Paleolithic, the stylistic motifs that decorate ostrich eggshell from Diepkloof Rock Shelter (Western Cape of South Africa) provide better candidates for early examples of cumulative culture (Texier et al. 2010). Texier and colleagues document variation over a period of possibly as little as 5000 years in designs on ~60,000 ka ostrich eggshells. Because these motifs were unlikely to have been affected by differences in subsistence they probably represent evidence of stylistic change through time. In other words, it appears that the motifs (styles) may have been transmitted intact from individual to individual – precisely the type of behavioral *form* transmission that is a prerequisite for cumulative culture (Tennie et al. 2009, 2012; Dean et al. 2012).

It is worth considering whether our current models of human cumulative culture are too linear for their own good. There is an implicit assumption of progress behind models of human cumulative culture that suggest that different features of this phenomenon appeared in a step-wise pattern eventually leading to the modern form of cumulative culture, which includes hyper-prosociality and linguistically mediated social transmission largely through pedagogical processes (Hill et al. 2009). The evolution of high-fidelity transmission and cumulative culture may have been marked by fits and starts rather than gradual but constant progress (e.g., Isaac et al. 1972; McBrearty and Brooks 2000). It is possible that the specific adaptations for human cumulative culture (especially motivation and skill in complex forms of teaching and imitating; Tomasello 1999; Tennie et al. 2009, 2012) existed for hundreds of thousands of years before the appearance of the suite of conditions that are favorable for cumulative culture. That is to say, it remains true for early hominins (and perhaps even for modern day great apes) that – perhaps – the general ability for cumulative culture was and is present, but that it was/is rarely or never expressed. This may be because the actual expression of cumulative culture may depend on factors like rates of environmental change that favor certain amounts of increased social learning and inhibit individual learning (Richerson and Boyd 2005), upright posture (Hill et al. 2009), effective population sizes (Henrich 2004; Powell et al. 2009; Kline and Boyd 2010) and potentially many other factors. Thus, before high-fidelity mechanisms of social learning became regular parts of the human condition, processes associated with low-fidelity social learning and latent solutions for specific tasks may provide more parsimonious explanations of the spatial and temporal variation we see in stone tool assemblages. It is also possible that our ability to recognize cultural transmission in the Early Stone Age archaeological record is biased by the types of artifacts available to us. Would there be little doubt over whether Oldowan hominins had high-fidelity cultural transmission if only we had a record of variation in their hairstyles or digging stick handle engravings instead of stone tool forms?

We find the hypothesis that many Early Stone Age behaviors may have been latent solutions compatible with the available archaeological and comparative psychological evidence. This calls for taking a fresh look at the Early Stone Age record with an eye toward identifying those characteristics that signal the presence of high-fidelity cultural transmission. Rather than assuming that Early Stone Age hominins possessed the same kinds of social learning mechanisms that we possess today, we may be better served by starting with the "null" working hypothesis that they lacked them (informed by what we currently know about great apes) and then modify our working hypothesis as the data warrant.

We find the ZLS hypothesis compelling, but to test it will require identifying features of the Early Stone Age record that could potentially falsify it. Demonstrating that Oldowan or Acheulean tools represent behaviors that are too complex to be learned by a naïve individual (of a tool using species of its time!) would falsify the hypothesis. However, this determination is difficult to make given that Early Stone Age hominins are no longer around and that modern humans may be invalid substitutes because we lack their goals, motivations, build, genetic background, etc.[4] Relatively little guidance or channeling of the right kind may be all that is needed to ensure the similar outcomes in the case of relatively simple stone tool technologies. For example, simply directing an individual to produce a form that will be an efficient source of flakes or a maintainable edge while keeping a usable grip at the base may be sufficient to (re-) produce a "handaxe" form.

It is worth noting that enculturated bonobos produce and use (crude) stone tools with only little human scaffolding (Toth et al. 1993; and later follow-up studies, such as Roffman et al. 2012) – and considering the tool use proclivities of wild chimpanzees it is at least possible that chimpanzees (perhaps in contrast to bonobos) could learn to make stone tools without any such behavioral scaffolding (experiments that test this hypothesis are currently needed; see also Whiten et al. 2009b). Interestingly, recent nut-cracking studies in chimpanzees established that unintentionally manufactured

[4]And yet, we cannot help but wonder what artifacts would look like if naïve modern humans were to be told to produce stone tools that allow the most effective sequential removal of successive flakes ("thrifty stone tools"). Might we expect even modern humans (given enough practice) to come up with artifact forms that might very well resemble handaxes? If the "handaxes as efficient sources of usable flakes" hypothesis is correct (Ludwig and Harris 1998), this would then explain the form of handaxes (and it would show that handaxes were not only a latent solution at their time, but remain so today). This experiment is unfortunately difficult to do, for various reasons.

simple potential stone tools (as by-products of nut-cracking) merely depend on the presence of the right raw material stone types (Mercader et al. 2002). Although the chimpanzees never actually recognized these chips of stone as suitable tools, this may be due to the lack of a need for them. Provided a suitable problem space (as in Toth et al. 1993) or a different ecological niche that involved smaller teeth, longer and more mobile thumbs, and increased consumption of tough animal foods, even the modern chimpanzee mind may indeed be capable of using (if not also intentionally producing) such stone tools. Comparative behavioral studies like these with living great apes show that there is at least the possibility that primitive stone tool manufacture was within the capabilities of the common ancestor of humans and chimpanzees – and thus within the capabilities of Early Stone Age hominins as well. Even if handaxe manufacture is far more complex than the behaviors exhibited by other animals (e.g., weaver birds; Walsh et al. 2010) – and it is not entirely clear that this is the case – this alone is insufficient evidence for the presence of high-fidelity cultural transmission. When we consider early stone tool technologies within the wider context of animal behavior, it is clear that what we perceive as "complex" does not serve as a reliable diagnostic of a behavior that requires cultural transmission. Complex tools are not necessarily the products of cumulative culture.

And yet, after all of this, it could be that Early Stone Age stone tool technology did in fact require the kinds of high-fidelity social learning mechanisms similar to those observed in modern humans. But it would be necessary to demonstrate this rather than to simply presume it. Indeed, some have suggested that the production process of Acheulean implements would require pedagogical techniques that relied upon the capacity for language (Goren-Inbar 2011). For others, the frequent occurrence of large quantities of handaxes found together at archaeological localities suggests that the implements were produced in a social context (Lycett and Gowlett 2008). But unless a social context necessarily translates into high-fidelity cultural transmission (which, given the evidence from chimpanzees, we doubt), the clustering of tools on the landscape may actually tell us very little about the type of social learning mechanism involved in their production.

In sum, the patterns of variation observed in the Early Stone Age archaeological record are, at the very least, as consistent with a ZLS explanation as they are with models that invoke high-fidelity cultural transmission between individuals. As a consequence, Oldowan and Acheulean stone tools may represent culture (or tradition, sensu Galef 2001), but not cumulative culture. If this proposition withstands future testing, we will need to reconsider the notion that the origin of high-fidelity cultural transmission just happens to coincide with the earliest archaeological assemblages of stone tools. At that point, we may wish to rethink where we place the origin of cumulative culture on the human lineage. Perhaps it makes sense to move it from millions of years ago to hundreds of thousands or even tens of thousands of years ago. What is clear at this point is that much more work is needed to clarify when, why, and how cumulative culture evolved in the Paleolithic.

Acknowledgements The authors wish to thank many colleagues who added extensive comments on earlier drafts of this manuscript. In particular, we would like to thank A. Mackay and J. Ferraro. We also thank R. Moore, H. Over, and very helpful anonymous reviewers. DRB would like to thank the Humboldt Foundation for support during his stay at the Max Planck Institute for Evolutionary Anthropology and UCT-URC sabbatical funds. CT would like to thank the Economic and Social Research Council for support (ES/K008625/1). This paper was initially drafted in 2011 based on our presentation at "The Nature of Culture" symposium. We would like to thank Miriam Haidle, Michael Bolus and Nicholas Conard for their invitation and for seeing this volume through to publication.

References

Acerbi, A., Tennie, C., & Nunn, C. L. (2011). Modelling imitation and emulation in constrained search spaces. *Learning and Behavior, 39*, 104–114.

Beyene, Y., Katoh, S., WoldeGabriel, G., Hart, W. K., Uto, K., Sudo, M., et al. (2013). The characteristics and chronology of the earliest Acheulean at Konso, Ethiopia. *Proceedings of the National Academy of Sciences of the USA, 110*, 1584–1591.

Biro, D., Inoue-Nakamura, N., Tonooka, R., Yamakoshi, G., Sousa, C., & Matsuzawa, T. (2003). Cultural innovation and transmission of tool use in wild chimpanzees: Evidence from field experiments. *Animal Cognition, 6*, 213–223.

Boyd, R., & Richerson, P. J. (1985). *Culture and the evolutionary process*. Chicago: University of Chicago Press.

Braun, D. R., & Harris, J. W. K. (2009). Plio-Pleistocene technological variation: A view from the KBS Mbr., Koobi Fora Formation. In N. Toth & K. Schick (Eds.), *The cutting edge: New approaches to the archaeology of human origins* (pp. 17–32). Gosport, Ind.: Stone Age Institute Press.

Braun, D. R., Plummer, T., Ditchfield, P., Ferraro, J. V., Maina, D., Bishop, L. C., & Potts, R. (2008). Oldowan behavior and raw material transport: Perspectives from the Kanjera formation. *Journal of Archaeological Science, 35*, 2329–2345.

Braun, D. R., Plummer, T. W., Ferraro, J., Ditchfield, P., et al. (2009). Raw material quality and Oldowan hominin toolstone preferences: Evidence from Kanjera South, Kenya. *Journal of Archaeological Science, 36*, 1605–1614.

Brown, G. R., Dickins, T. E., Sear, R., & Laland, K. N. (2011). Evolutionary accounts of human behavioural diversity. *Philosophical Transactions of the Royal Society B: Biological Sciences, 366*, 313–324.

Byrne, R. W. (2002). Seeing actions as hierarchically organized structures: Great ape manual skills. In A. N. Meltzoff & W. Prinz (Eds.), *The imitative mind: Development, evolution, and brain bases* (pp. 122–140). Cambridge: Cambridge University Press.

Carbonell, E., Sala, R., Barsky, D., & Celiberti, V. (2006). From homogeneity to multiplicity: A new approach to the study of archaic stone tools. In E. Hovers & D. R. Braun (Eds.), *Interdisciplinary approaches to the Oldowan* (pp. 25–37). Dordrecht: Springer.

Carvalho, S., Cunha, E., Sousa, C., & Matsuzawa, T. (2008). Chaînes opératoires and resource-exploitation strategies in chimpanzee (*Pan troglodytes*) nut cracking. *Journal of Human Evolution, 55*, 148–163.

de Lumley, H., Nioradze, M., Barsky, D., Cauche, D., Celiberti, V., Nioradzé, G., et al. (2005). Les industries lithiques préoldowayennes du début du Pléistocène inférieur du site de Dmanisi en Géorgie/Pre-Oldowayen lithic industry from the beginning of the lower Pleistocene at the Dmanisi site in Georgia. *L'Anthropologie, 109*, 1–182.

de Waal, F. B. M. (2001). *The ape and the Sushi Master: Cultural reflections by a primatologist*. New York: Basic Books.

Dean, L. G., Kendal, R. L., Schapiro, S. J., Thierry, B., & Laland, K. N. (2012). Identification of the social and cognitive processes underlying human cumulative culture. *Science, 335*, 1114–1118.

Delagnes, A., & Roche, H. (2005). Late Pliocene hominid knapping skills: The case of Lokalalei 2C, West Turkana, Kenya. *Journal of Human Evolution, 48*, 435–472.

Fragaszy, D. M., & Perry, S. (2003). *The biology of traditions: Models and evidence*. Cambridge: Cambridge University Press.

Galef, B. G. (2001). Social influences on foraging in vertebrates: Causal mechanisms and adaptive functions. *Animal Behaviour, 61*, 3–15.

Gamble, C., & Porr, M. (Eds.). (2005). *The hominid individual in context: Archaeological investigations of Lower and Middle Palaeolithic landscapes, locales and artefacts*. New York: Routledge Press.

Goren-Inbar, N. (2011). Culture and cognition in the Acheulian industry: A case study from Gesher Benot Ya'aqov. *Philosophical Transactions of the Royal Society B: Biological Sciences, 366*, 1038–1049.

Gruber, T., Muller, M. N., Strimling, P., Wrangham, R., & Zuberbühler, K. (2009). Wild chimpanzees rely on cultural knowledge to solve an experimental honey acquisition task. *Current Biology, 19*, 1806–1810.

Haidle, M. N. (2010). Working-memory capacity and the evolution of modern cognitive potential: Implications from animal and early human tool use. *Current Anthropology, 51*, 149–166.

Haslam, M., Hernandez-Aguilar, A., Ling, V., Carvalho, S., de la Torre, I., DeStefano, A., et al. (2009). Primate archaeology. *Nature, 460*, 339–344.

Henrich, J. (2004). Demography and cultural evolution: How adaptive cultural processes can produce maladaptive losses: The Tasmanian case. *American Antiquity, 69*, 197–214.

Henrich, J., & Boyd, R. (1998). The evolution of conformist transmission and the emergence of between-group differences. *Evolution and Human Behavior, 19*, 215–241.

Hill, K., Barton, C. M., & Hurtado, A. M. (2009). The emergence of human uniqueness: Characters underlying behavioral modernity. *Evolutionary Anthropology, 18*, 187–200.

Horner, V., & Whiten, A. (2005). Causal knowledge and imitation/emulation switching in chimpanzees (*Pan troglodytes*) and children (*Homo sapiens*). *Animal Cognition, 8*, 164–181.

Isaac, G. Ll., Bishop, W. W., Miller, J. A., & Cole, S. (1972). Chronology and the tempo of cultural change during the Pleistocene. In W. W. Bishop & J. A. Miller (Eds.), *Calibration of hominoid evolution* (pp. 381–430). Edinburgh: Scottish Academic Press.

Kempe, M., Lycett, S. J., & Mesoudi, A. (2012). An experimental test of the accumulated copying error model of cultural mutation for Acheulean handaxe size. *PLoS ONE, 7*, e48333.

Kimbel, W. H., Walter, R. C., Johanson, D. C., Reed, K. E., Aronson, J. L., Assefa, Z., et al. (1996). Late Pliocene *Homo* and Oldowan tools from the Hadar formation (Kada Hadar member), Ethiopia. *Journal of Human Evolution, 31*, 549–561.

Kline, M. A., & Boyd, R. (2010). Population size predicts technological complexity in Oceania. *Proceedings of the Royal Society B: Biological Sciences, 277*, 2559–2564.

Kuman, K., & Clarke, R. J. (2000). Stratigraphy, artefact industries and hominid associations for Sterkfontein, Member 5. *Journal of Human Evolution, 38*, 827–847.

Laland, K. N., & Hoppitt, W. (2003). Do animals have culture? *Evolutionary Anthropology, 12*, 150–159.

Laland, K. N., Odling-Smee, J., & Myles, S. (2010). How culture shaped the human genome: Bringing genetics and the human sciences together. *Nature Reviews Genetics, 11*, 137–148.

Lepre, C. J., Roche, H., Kent, D. V., Harmand, S., Quinn, R. L., Brugal, J.-P., et al. (2011). An earlier origin for the Acheulian. *Nature, 477*(7362), 82–85.

Luncz, L. V., Mundry, R., & Boesch, C. (2012). Evidence for cultural differences between neighboring chimpanzee communities. *Current Biology, 22*, 922–926.

Lycett, S. J. (2008). Acheulean variation and selection: Does handaxe symmetry fit neutral expectations? *Journal of Archaeological Science, 35*, 2640–2648.

Lycett, S. J., & Gowlett, J. A. J. (2008). On questions surrounding the Acheulean 'tradition'. *World Archaeology, 40*, 295–315.

Lycett, S. J., & Norton, C. (2010). A demographic model for Palaeolithic technological evolution: The case of East Asia and the Movius Line. *Quaternary International, 211*, 55–65.

Lycett, S. J., & von Cramon-Taubadel, N. (2008). Acheulean variability and hominin dispersals: A model-bound approach. *Journal of Archaeological Science, 35*, 553–562.

Lycett, S. J., Collard, M., & McGrew, W. C. (2007). Phylogenetic analyses of behavior support existence of culture among wild chimpanzees. *Proceedings of the National Academy of Sciences of the USA, 104*, 17588–17592.

Ludwig, B. V., & Harris, J. W. K. (1998). Towards a technological reassessment of East African Plio-Pleistocene lithic assemblages. In M. Petraglia & K. Paddaya (Eds.), *The rise and diversity of the Lower Paleolithic* (pp. 84–106). New York: Academic Press.

Marshall-Pescini, S., & Whiten, A. (2008). Chimpanzees (*Pan troglodytes*) and the question of cumulative culture: An experimental approach. *Animal Cognition, 11*, 449–456.

Matsuzawa, T., Biro, D., Humle, T., Inoue-Nakamura, N., Tonooka, R., & Yamakoshi, G. (2001). Emergence of culture in wild chimpanzees: Education by master-apprenticeship. In T. Matsuzawa (Ed.), *Primate origins of human cognition and behavior* (pp. 557–574). Berlin: Springer.

McBrearty, S., & Brooks, A. S. (2000). The revolution that wasn't: A new interpretation of the origin of modern human behavior. *Journal of Human Evolution, 39*, 453–563.

McGrew, W. C. (1998). Culture in nonhuman primates? *Annual Review of Anthropology, 27*, 301–328.

McNabb, J., Binyon, F., & Hazelwood, L. (2004). The large cutting tools from the South African Acheulean and the question of social traditions. *Current Anthropology, 45*, 653–677.

McPherron, S. P. (1999). Ovate and pointed handaxe assemblages: Two points make a line. *Préhistoire Européenne, 14*, 9–32.

McPherron, S. P. (2000). Handaxes as a measure of the mental capabilities of early hominids. *Journal of Archaeological Science, 27*, 655–663.

McPherron, S. P. (2007). What typology can tell us about Acheulian handaxe production. In N. Goren-Inbar & G. Sharon (Eds.), *Axe age: Acheulian toolmaking, from quarry to discard* (pp. 267–286). Indonesia: Equinox Publishing.

Menzel, C., Fowler, A., Tennie, C., & Call, J. (2013). Leaf surface roughness elicits leaf swallowing behavior in captive chimpanzees (*Pan troglodytes*) and bonobos (*P. paniscus*), but not in gorillas (*Gorilla gorilla*) or orangutans (*Pongo abelii*). *International Journal of Primatology, 34*, 533–553.

Mercader, J., Panger, M., & Boesch, C. (2002). Excavation of a chimpanzee stone tool site in the African rainforest. *Science, 296*, 1452–1455.

Nonaka, T., Bril, B., & Rein, R. (2010). How do stone knappers predict and control the outcome of flaking? Implications for understanding early stone tool technology. *Journal of Human Evolution, 59*, 155–167.

Odling-Smee, F. J., Laland, K. N., & Feldman, M. W. (2003). *Niche construction: The neglected process in evolution*. Princeton: Princeton University Press.

Powell, A., Shennan, S., & Thomas, M. G. (2009). Late pleistocene demography and the appearance of modern human behavior. *Science, 324*, 1298–1301.

Pradhan, G. R., Tennie, C., & van Schaik, C. (2012). Social organization and the evolution of cumulative technology in apes and hominins. *Journal of Human Evolution, 63*, 180–190.

Premo, L. S. (2012). Local extinctions, connectedness, and cultural evolution in structured populations. *Advances in Complex Systems, 15*, 1150002.

Premo, L. S., & Kuhn, S. L. (2010). Modeling effects of local extinctions on culture change and diversity in the Paleolithic. *PLoS ONE, 5*, e15582.

Rendell, L., Fogarty, L., & Laland, K. N. (2011). Runaway cultural niche construction. *Philosophical Transactions of the Royal Society B, 366*, 823–835.

Richerson, P. J., & Boyd, R. (2005). *Not by genes alone*. Chicago: University of Chicago Press.

Richerson, P. J., Boyd, R., & Henrich, J. (2010). Gene-culture coevolution in the age of genomics. *Proceedings of the National Academy of Sciences of the USA, 107*, 8985–8992.

Roche, H., Delagnes, A., Brugal, J.-P., Feibel, C. S., Kibunjia, M., Mourre, V., et al. (1999). Early hominid stone tool production and technical skill 2.34 Myr ago in west Turkana. Kenya. *Nature, 399*, 57–60.

Roffman, I., Savage-Rumbaugh, S., Rubert-Pugh, E., Ronen, A., & Nevo, E. (2012). Stone tool production and utilization by bonobo-chimpanzees (*Pan paniscus*). *Proceedings of the National Academy of Sciences of the USA, 109*, 14500–14503.

Sahnouni, M., Van der Made, J., & Everett, M. (2009). Early North Africa: Chronology, ecology, and hominin behavior: Insights from Ain Hanech and El-Kherba, northeastern Algeria. *Quaternary International, 223*, 436–438.

Schöning, C., Humle, T., Möbius, Y., & McGrew, W. C. (2008). The nature of culture: Technological variation in chimpanzee predation on army ants revisited. *Journal of Human Evolution, 55*, 48–59.

Semaw, S. (2000). The world's oldest stone artefacts from Gona, Ethiopia: Their implications for understanding stone technology and patterns of human evolution between 2.6–1.5 million years ago. *Journal of Archaeological Science, 27*, 1197–1214.

Semaw, S., Renne, P., Harris, J. W. K., Feibel, C. S., Bernor, R. L., Fesseha, N., et al. (1997). 2.5-million-year-old stone tools from Gona Ethiopia. *Nature, 385*, 333–336.

Sharon, G. (2007). *Acheulian large flake industries: Technology, chronology, and significance*. British Archaeological Reports International Series 1701. Oxford: Archaeopress.

Sharon, G. (2008). The impact of raw material on Acheulian large flake production. *Journal of Archaeological Science, 35*, 1329–1344.

Shipton, C. (2010). Imitation and shared intentionality in the Acheulean. *Cambridge Archaeological Journal, 20*, 197–210.

Shipton, C., & Petraglia, M. D. (2010). Inter-continental variation in Acheulean bifaces. In C. J. Norton & D. R. Braun (Eds.), *Asian Paleoanthropology* (pp. 49–55). Dordrecht: Springer.

Shipton, C., Petraglia, M., & Paddayya, K. (2009). Stone tool experiments and reduction methods at the Acheulean site of Isampur Quarry, India. *Antiquity, 83*, 769–785.

Stout, D. (2011). Stone toolmaking and the evolution of human culture and cognition. *Philosophical Transactions of the Royal Society B: Biological Sciences, 366*, 1050–1059.

Stout, D., Quade, J., Semaw, S., Rogers, M. J., & Levin, N. E. (2005). Raw material selectivity of the earliest stone toolmakers at Gona, Afar, Ethiopia. *Journal of Human Evolution, 48*, 365–380.

Stout, D., Semaw, S., Rogers, M. J., & Cauche, D. (2010). Technological variation in the earliest Oldowan from Gona, Afar, Ethiopia. *Journal of Human Evolution, 58*, 474–491.

Tebbich, S., Taborsky, M., Fessl, B., & Blomqvist, D. (2001). Do woodpecker finches acquire tool-use by social learning? *Proceedings of the Royal Society B: Biological Sciences, 268*, 2189–2193.

Tennie, C., & Hedwig, D. (2009). How latent solution experiments can help to study differences between human culture and primate traditions. In E. Potocki & J. Krasiński (Eds.), *Primatology: Theories, methods and research* (pp. 95–112). New York: Nova Publishers.

Tennie, C., Hedwig, D., Call, J., & Tomasello, M. (2008). An experimental study of nettle feeding in captive gorillas. *American Journal of Primatology, 70*, 584–593.

Tennie, C., Call, J., & Tomasello, M. (2009). Ratcheting up the ratchet: On the evolution of cumulative culture. *Philosophical Transactions of the Royal Society B: Biological Sciences, 364*, 2405–2415.

Tennie, C., Call, J., & Tomasello, M. (2010). Evidence for emulation in chimpanzees in social settings using the floating peanut task. *PLoS ONE, 5*, e10544.

Tennie, C., Call, J., & Tomasello, M. (2012). Untrained chimpanzees (*Pan troglodytes schweinfurthii*) fail to imitate novel actions. *PLoS ONE, 7*, e41548.

Texier, P. J., Porraz, G., Parkington, J., Rigaud, J. P., Poggenpoel, C., Miller, C., et al. (2010). A howiesons poort tradition of engraving ostrich eggshell containers dated to 60,000 years ago at Diepkloof Rock Shelter, South Africa. *Proceedings of the National Academy of Sciences of the USA, 107*, 6180–6185.

Tomasello, M. (1999). *The cultural origins of human cognition*. Cambridge, MA & London: Harvard University Press.

Toth, N. (1987). Behavioral inferences from early stone artifact assemblages: An experimental model. *Journal of Human Evolution, 16*, 763–787.

Toth, N., Savage-Rumbaugh, E. S., Sevcik, R. A., Rumbaugh, D. M., & Schick, K. D. (1993). Pan the tool-maker: Investigations into the stone tool-making and tool-using capabilities of a Bonobo (*Pan paniscus*). *Journal of Archaeological Science, 20*, 81–91.

Visalberghi, E. (1987). Acquisition of nut-cracking behaviour by 2 capuchin monkeys (*Cebus apella*). *Folia Primatologica, 49*, 168–181.

Walsh, P. T., Hansell, M., Borello, W. D., & Healy, S. D. (2010). Repeatability of nest morphology in African weaver birds. *Biology Letters, 6*, 149–151.

White, M. J. (1998). On the significance of acheulian biface variability in Southern Britain. *Proceedings of the Prehistoric Society, 64*, 15–44.

Whiten, A. (2000). Primate culture and social learning. *Cognitive Science, 24*, 477–508.

Whiten, A., & Ham, R. (1992). On the nature and evolution of imitation in the animal kingdom: Reappraisal of a century of research. *Advances in the Study of Behavior, 21*, 239–283.

Whiten, A., & van Schaik, C. P. (2007). The evolution of animal 'cultures' and social intelligence. *Philosophical Transactions of the Royal Society B: Biological Sciences, 362*, 603–620.

Whiten, A., Goodall, J., McGrew, W. C., Nishida, T., Reynolds, V., Sugiyama, Y., et al. (1999). Cultures in chimpanzees. *Nature, 399*, 682–685.

Whiten, A., Horner, V., & Marshall-Pescini, S. (2003). Cultural panthropology. *Evolutionary Anthropology, 12*, 92–105.

Whiten, A., McGuigan, N., Marshall-Pescini, S., & Hopper, L. M. (2009a). Emulation, imitation, over-imitation and the scope of culture for child and chimpanzee. *Philosophical Transactions of the Royal Society B: Biological Sciences, 364*, 2417–2428.

Whiten, A., Schick, K. D., & Toth, N. (2009b). The evolution and cultural transmission of percussive technology: Integrating evidence from paleoanthropology and primatology. *Journal of Human Evolution, 57*, 420–435.

Whiten, A., Hinde, R. A., Laland, K. N., & Stringer, C. B. (2011). Culture evolves. *Philosophical Transactions of the Royal Society B: Biological Sciences, 366*, 938–948.

Wynn, T., & Tierson, F. (1990). Regional comparison of the shapes of later Acheulean handaxes. *American Anthropologist, 92*, 73–84.

Yamei, H., Potts, R., Baoyin, Y., Zhengtang, G., Deino, A. L., Wei, W., et al. (2000). Mid-Pleistocene Acheulean-like stone technology of the Bose Basin, South China. *Science, 287*, 1622–1626.

Chapter 12
Mountaineering or Ratcheting? Stone Age Hunting Weapons as Proxy for the Evolution of Human Technological, Behavioral and Cognitive Flexibility

Marlize Lombard

Abstract Cultural, behavioral and cognitive evolution is often seen as cumulative and sometimes referred to in terms of a ratchet or the ratchet effect. In this contribution, I assess the value of the ratchet analogy as blanket explanation for the above aspects of human evolution. I use Stone Age weapon technologies as proxy for the evolution of human technological, behavioral and cognitive flexibility, and by doing so show that the ratchet analogy falls short of explaining human variability and complexity as reflected in the Stone Age archaeological record. Considering human cultural, behavioral and cognitive evolution from a theoretically constructed rugged landscape point of view, I suggest that mountaineering may be a more suitable analogy for the accumulation of human culture. In this scenario, culture and technology anchor societies within their respective evolutionary trajectories and fitness landscapes, and it more accurately reflects humans as 'masters of flexibility'.

Keywords Projectile technology • Bow-and-arrow • Spear • Spearthrower-and-dart • Cumulative culture • Cognition • Fitness landscapes

Introduction

Flexibility is the ability of an organism to change and adapt to suit new conditions or situations. This ability applies to long-term adaptability and change, as well as to instantaneous decision-making processes. All living things are flexible in a biologically relevant way, but, because we have certain features in our brains that make us more flexible than any other organism, humans are considered 'masters of flexibility' (Barrett 2009:107). Our extraordinary flexible nature is expressed, for example, in the way we think, in our social behaviors, and in our production and use of technology. As a species, we have become almost unlimited in how we adapt and change to suit new conditions and situations. We are versatile and creative in how we tackle immediate challenges, and ambitious in ensuring our long-term survival on earth – and possibly beyond.

It is our highly developed flexibility that also causes us to deem ourselves 'intelligent', as intelligence is seen as the ability to learn, innovate and respond flexibly to new or complex situations (Byrne 1995; van Schaik and Burkart 2011). These abilities are anchored in genetic predispositions towards faster reaction times, greater working memory, inhibitory control and greater response to novelty (Geary 2005; van Schaik and Burkart 2011). From a dynamic systems perspective, intelligence is the adaptive flexibility that integrates the stability of past experience with the idiosyncrasies of the moment (Colunga and Smith 2007:170). These notions all seem to indicate that humans are hardwired to use past knowledge and experience, together with novel ideas and applications, to negotiate any given situation as quickly and effectively as possible. Today, as in the past, much of human flexibility is reflected in the variability of our material culture or technologies, whether these are our transport systems, agriculture, water supply systems, information technology or space exploration. In many ways we have become dependent on technologies for our survival. But, when did we become such masters of flexibility?

To explore this question we need to journey back in time. Early human material culture is represented in the archaeological records of the last 3 Myr; first in Africa, and then elsewhere in the Old World. These archaeological records inform on technology, or the way that early humans applied knowledge in practical ways. The bulk of past human

M. Lombard (✉)
Department of Anthropology and Development Studies, University of Johannesburg, PO Box 524, Auckland Park 2006, South Africa
e-mail: mlombard@uj.ac.za

and

Wallenberg Research Centre, Stellenbosch Institute for Advanced Study (STIAS), Stellenbosch University, Marais Street, Stellenbosch 7600, South Africa

Miriam N. Haidle, Nicholas J. Conard and Michael Bolus (eds.), *The Nature of Culture: Based on an Interdisciplinary Symposium 'The Nature of Culture', Tübingen, Germany*, Vertebrate Paleobiology and Paleoanthropology, DOI 10.1007/978-94-017-7426-0_12

technological endeavor is captured in stone artifacts, and it is from these objects that archaeologists and other researchers interested in the evolution of human culture, behavior and cognition, are challenged to extract as much information as possible.

For the *Nature of Culture* symposium, I was asked to follow the development of human flexibility through deep time. My research explorations mostly focus on the evolution of hunting technologies as a possible line of evidence for tracing levels of technological complexity and flexibility through time and space (e.g., Lombard 2005, 2007, 2011). Thus, in this paper I will use examples of Stone Age hunting technologies as proxy for the evolution of human flexibility in general. The hunting technologies used as examples are mostly based on my own research experience. They include simple wooden spears as representing single-unit (unhafted) technologies, stone-tipped spears as representing composite technologies, and bow-and-arrow sets as representing complementary tool sets or symbiotic technologies (see Lombard and Haidle 2012).

Focus on these technologies does not exclude other weapon systems from the main argument; they are seen as generally representative of the technological expressions mentioned above. For example, the use of spearthrowers is sometimes considered intermediate to hand-delivered weaponry and bows and arrows, but there is no reason to view bow-and-arrow technology as necessarily originating from spearthrower-and-dart technology, or vice versa (e.g., Shea 2006). Spearthrower technology also represents complementary tool sets akin to bow-and-arrow technology, so that inferred levels of complexity, variability and flexibility regarding these two mechanically-projected weapon systems is presumed similar (Lombard and Haidle 2012). So far there is little or no indication that the former was ever used in sub-Saharan Africa and more direct evidence exists for the production and use of the latter in southern Africa associated with the Middle Stone Age; a focus area and period for exploring human behavioral and cognitive evolution (e.g., Wadley et al. 2009, 2011; Wadley 2010; Henshilwood and Dubreuil 2011).

Archaeological evidence reveals that the use of weapon systems came and went over the course of time, with wide variation in the degree to which regional human populations used them (e.g., Shea 2006; Lombard 2011; Lombard and Parsons 2011). Based on the Stone Age archaeological record I will critique the use of a ratchet or the ratchet effect (e.g., Riegler 2001; Tennie et al. 2009), as an apt, broad-spectrum explanation for human cultural and behavioral evolution. As an alternative analogy for cumulative culture, I reintroduce the idea of rugged and complex fitness landscapes (Wright 1932; Kuhn 2006), negotiated by human societies in a flexible way (Lombard and Parsons 2011; Parsons and Lombard 2011), more similar to mountaineering than to ratcheting.

An Abbreviated Account of the Evolution of Human Hunting Technologies

It has recently been established that humans are not the only beings who hunt with weapons (e.g., Preutz and Bertolani 2007), but together with our immediate *Homo* ancestors and 'cousins', such as *Homo neanderthalensis*, we are the only ones who modified (or still modify) our weapons with other tools before use, or who tipped them with stone. The latter behavior sometimes results in the preservation of information about our early hunting technologies over long periods of time. Stone tool analysis (e.g., Wadley and Mohapi 2008) (Fig. 12.1a), in combination with experimental archaeology (e.g., Lombard et al. 2004; Wadley 2005; Lombard and Pargeter 2008; Bradfield and Lombard 2011) (Fig. 12.1b, c, d), and use-trace studies (e.g., Lombard 2011) (Fig. 12.1e), occasionally provide detailed insight into past hunting technologies and their associated behaviors. Using multiple strands of evidence to reconstruct and hypothesize about Stone Age weaponry helps to build increasingly comprehensive thought-and-action or operational sequences, providing insight into the more abstract notions of cultural evolution and behavioral and cognitive flexibility (e.g., Haidle 2009, 2010; Wadley et al. 2009; Wadley 2010; Lombard and Haidle 2012; Williams et al. 2014).

Single-Component Spears

At ~300 ka *Homo heidelbergensis* in Europe used sharpened, wooden spears as hunting weapons, such as those found at Schöningen (Thieme 1997, 1999). At first glance, these spears seem to be simple, single-component weapons. Yet, when we explore what they were made of, and how they were produced, a more complex picture emerges. Haidle's (2009, 2010) effective chain of production and use shows that making wooden spears required the collection and preparation of several other materials before production could begin (Fig. 12.2a). If we accept that sharp-edged stone tools were used to remove bark, smooth and sharpen the wood, and that heat was used to shape, sharpen or temper the spears, then it follows that various plant materials were needed to produce the spears, and to start and maintain fires. Furthermore, a range of rocks were collected, either to use as tools or to light fires (Haidle 2009, 2010; Lombard and Haidle 2012). The operational sequence is extended in duration and complexity, compared to using objects that have not been altered with other objects or agents (e.g., heat), and thus the problem-solution distance is also extended (Haidle 2010) (Figs. 12.2a and 12.3a). Within each of the

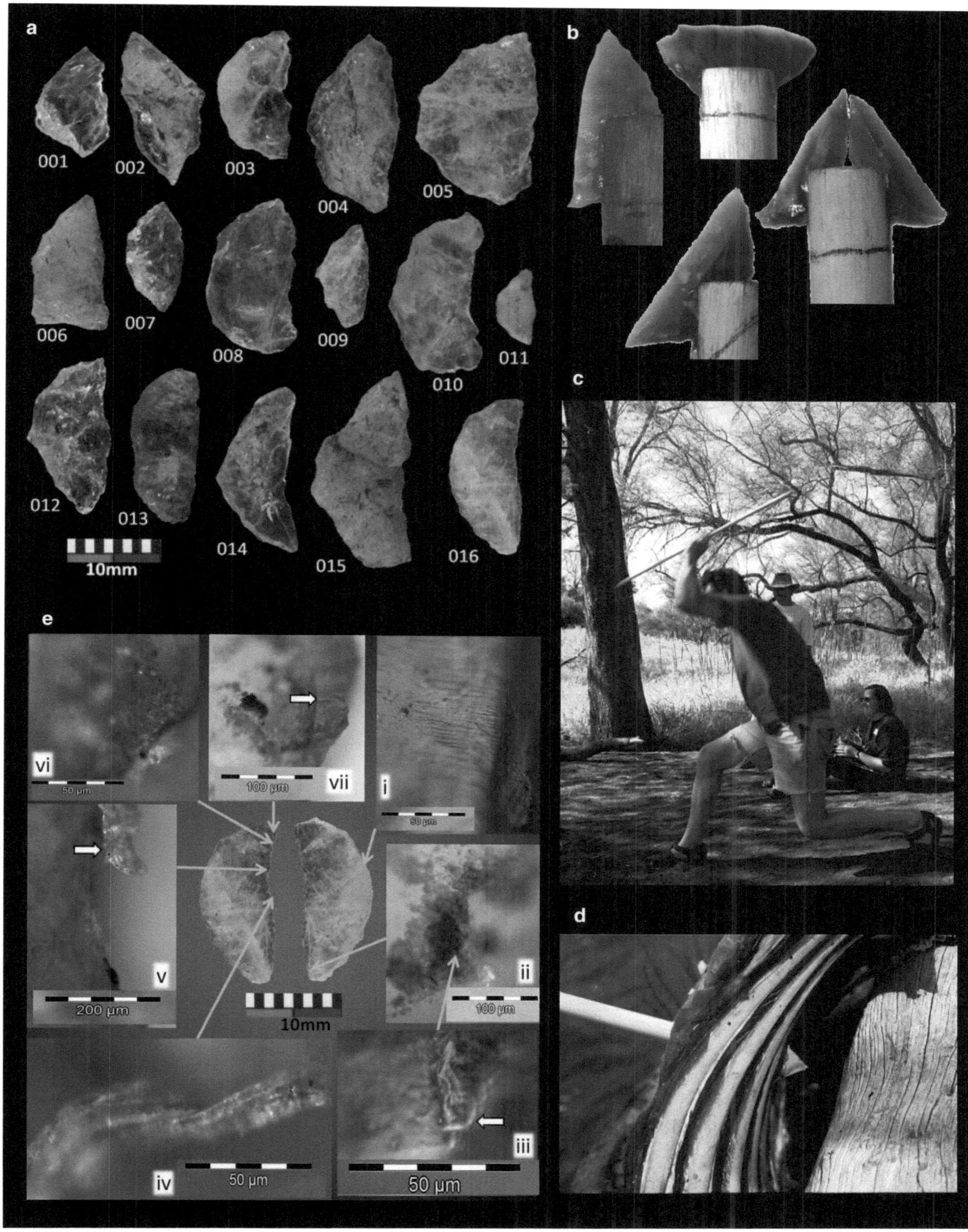

Fig. 12.1 Multiple ways of studying ancient hunting technologies include: **a** Stone tool analyses such as that conducted on the 60–64 ka backed quartz tools from Sibudu Cave South Africa by Wadley and Mohapi (2008). **b–d** Reconstructing and using stone-tipped weapons in replication experiments (**b** from Lombard and Pargeter 2008; **c** and **d** from Lombard et al. 2004). **e** Conducting use-trace analyses on experimental and suitable archaeological samples (from Lombard 2011)

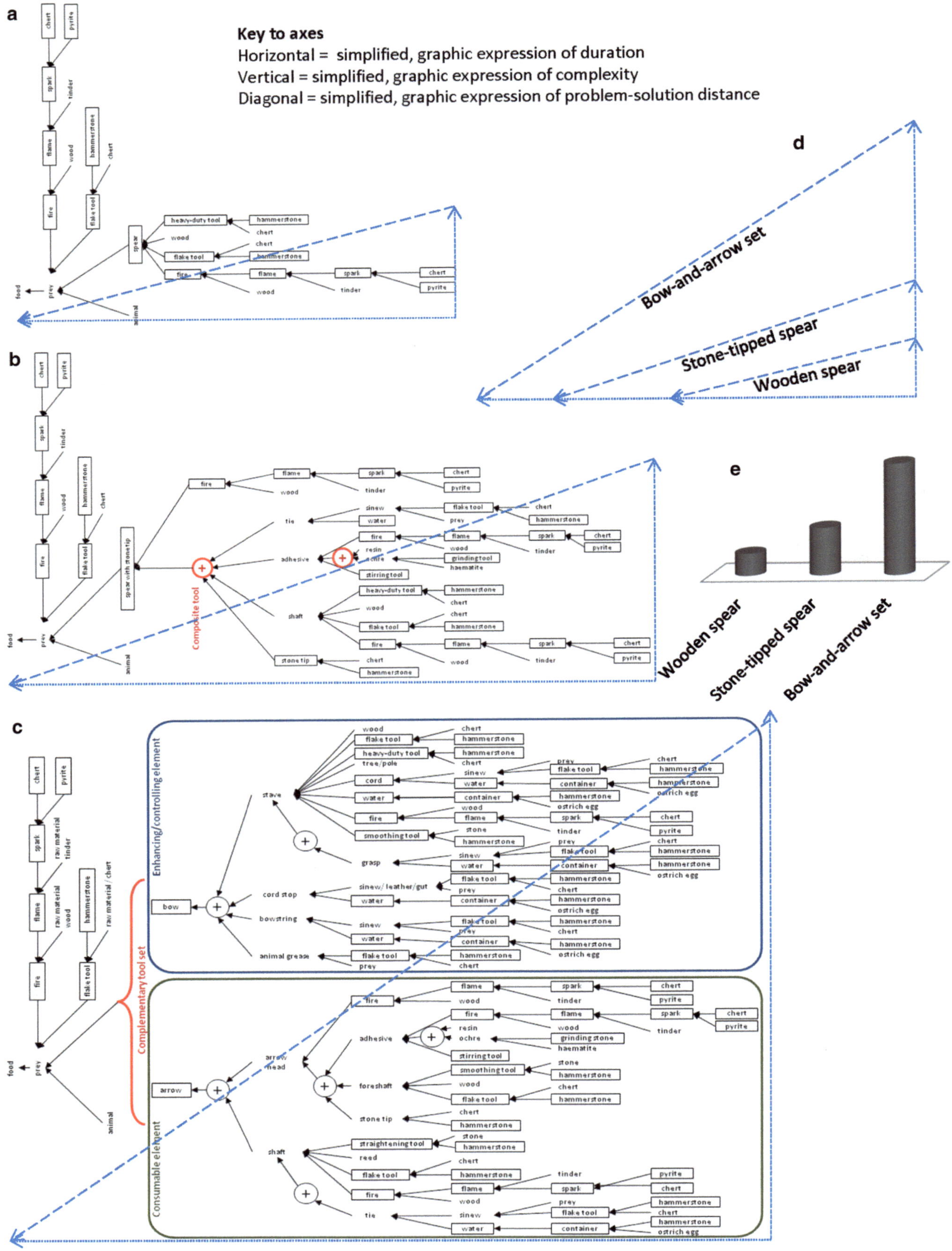
a
Key to axes
Horizontal = simplified, graphic expression of duration
Vertical = simplified, graphic expression of complexity
Diagonal = simplified, graphic expression of problem-solution distance
d
Bow-and-arrow set
Stone-tipped spear
Wooden spear
b
Composite tool
e
Wooden spear
Stone-tipped spear
Bow-and-arrow set
c
Enhancing/controlling element
Complementary tool set
Consumable element
spear with stone tip
bow
arrow
stave
grasp
cord stop
bowstring
animal grease
arrow head
foreshaft
stone tip
shaft
adhesive
fire
flame
spark
tinder
wood
chert
pyrite
hammerstone
flake tool
heavy-duty tool
prey
animal
food
sinew
water
tie
container
ostrich egg
resin
ochre
stirring tool
grinding stone
haematite
smoothing tool
straightening tool
reed

◀ **Fig. 12.2** Effective chains of production and use of weapon systems can be compared. **a** That of a simple wooden spear shows the collection and preparation of several other materials before production could begin and the use of heat to shape, sharpen or temper the spear. **b** That of a stone-tipped spear illustrates the new cognitive element of composition (encircled +) during the production of adhesives and during the final assembly of the composite artifact. **c** That of a bow-and-arrow set illustrates the novel cognitive concept of complementary or symbiotic technologies (in illustration) (Fig. 12.2**a–c** adapted from Lombard and Haidle 2012). **d** Direct comparison of the hypothetical dimensions associated with each technology. **e** Simplified, hypothetical graphic expression of the thought-and-action volumes of the different effective chains of production and use

elements in the sequence there is room for flexibility, i.e. variability in choice. This level in complexity and flexibility is only possible when satisfaction (accomplishing a goal) can be decoupled from basic need, so that the manufacture and curation of tools become aims within themselves. It means that each unit (e.g., fire, heavy-duty tool, flake tool, hammerstone) can be applied in a modular, flexible way within different operational sequences. This represents the capacity for conceptual, technological and behavioral modularization (Haidle 2009, 2010; Lombard and Haidle 2012).

Objects such as, or similar to, wooden spears need not only be used as hand-delivered hunting weapons, they would function equally well as: digging sticks for food foraging or water retrieval, fighting sticks during situations of social stress, initiation ceremonies or sporting events, armatures for defense against dangerous animals, walking sticks, or stakes that can be pounded into the ground for various purposes, such as constructing shelters or tripods over fire pits. For all of the above applications there exists ample archaeological and/or ethno-historical evidence. Thus, understanding how to modify a piece of wood, using other objects and agents, into a long, straight, strong and sharpened object, hugely increases the range of behaviors in which individuals can interact with their environment. Such individuals are therefore much more prepared for, and flexible within, any given set of circumstances, than those without similar implements – whether these circumstances are physical, economical or social.

Stone-Tipped Spears

From the above it is clear that a long, hard, sharpened stick can be a versatile and useful object, but what does it lack when applied to hunting, fighting or defense? The answer is a harder, sharper piercing tip combined with a cutting edge. There can be little doubt that ancient communities, through the processes of observation, experimentation, copying, social learning and communication, understood that pointed stone flakes possess these properties. Composite tools (such as stone-tipped spears, knives, axes and hammers) represent the concept of modular combination or composition where the addition of several separate elements form a single, new, composite tool with enhanced properties (Lombard and Haidle 2012). Such tools first seem to appear during the transition from the Acheulean to the Middle Paleolithic or Middle Stone Age. Stone-tipped, hand-delivered spears could have been used from ~500 ka in southern Africa (Wilkins et al. 2012), ~285 ka in other parts of sub-Saharan Africa (McBrearty and Tryon 2005), ~270 ka in the Near East (Mercier and Valladas 2003), and ~200 ka in Europe (Villa and Soriano 2010). These early weapons were tipped mostly with simple, unretouched convergent flakes or Levallois-type points. By ~75 ka we have use-trace evidence that thin, bifacially retouched stone points were hafted as both spears and knives in southern Africa (Lombard 2006, 2007).

The effective chain of production and use of a stone-tipped spear illustrates the new cognitive element of composition (encircled +) during the production of adhesives and during the final assembly of the composite artifact (Fig. 12.2b). When this chain is compared with that of the production and use of a single-component wooden spear (Fig. 12.2a, d, e), we see how the problem-solution distance and operational sequence is extended in duration and complexity (Lombard and Haidle 2012). We will probably never know how, where, or how many times, the concept of a hafted tool was 'invented', but it is an innovation that radically changed the world of hominin technology. Once the properties of different elements and how they could be combined to form new functional units were understood, the range of combinations, technologies, and their potential applications were dramatically increased. Equipped with stone-tipped spears, knives axes and hammers, fitness options regarding hunting, meat processing, foraging and defense were vastly increased, compared with those provided for by a simple wooden spear and a handful of loose stone flakes.

What is more, single elements such as a stone spear tip can easily be renewed without thinking through the whole process of producing and using the complete spear (Haidle 2010, 2012). Additional elements such as binding materials and tips can be made in advance and curated as stock or spare parts. The decoupling of tool production from basic need provides the tool with independent existence (Lombard and Haidle 2012). Thus, tools have the potential to provide solutions for problems yet to be identified, e.g., the same stone point can be used as a hand-held cutting tool, hafted to a short handle as a knife blade, or used to tip a robust hunting spear; depending on need and situation. Problems are, therefore, no longer perceived or solved only in the immediate or extended present. With such advanced conceptual, technological and behavioral modularization and composition, cognitive time depth is growing (Haidle 2010, 2012).

Composition thus offers increased flexibility in diverse contexts. A range of solutions can be applied to a single problem, or various needs can be met with a single solution (Lombard and Haidle 2012). Direct evidence for stone-tipped spears (or other composite/hafted implements [e.g., Rots and Van Peer 2006]) early on in the archaeological record, hence represents a powerful increase in technological, behavioral and cognitive flexibility, compared to the production and use of simple wooden spears and/or un-hafted stone tools. Notwithstanding their complexity, *Homo sapiens* shares cognitive and cultural traits, as represented by composite tools in the archaeological record, with *Homo neanderthalensis*, *Homo heidelbergensis* and archaic modern humans (e.g., Williams et al. 2014).

Bow-and-Arrow Technology

The invention of the bow-and-arrow used to be closely linked to the late Upper Paleolithic in Europe (Cattelain 1997). Lately, however, based on new data from Africa, opinions regarding the inception of mechanically-projected weaponry have begun to vary considerably. Some maintain that the bone points and microliths of the early Later Stone Age at Border Cave, South Africa, at ~40–35 ka, signals the advent of bow-and-arrow technology (Villa et al. 2010), others claim that dart-and-spearthrower technology could have existed by ~100 ka elsewhere in the region (Brooks et al. 2006). Regardless of disagreement on the place and timing of the origins of bow-and-arrow technology, there seems to be consensus that it was a technology used exclusively by *Homo sapiens* (e.g., Shea and Sisk 2010; Villa and Soriano 2010). Shea and Sisk (2010) also argue that mechanically-projected weaponry, such as bows and arrows, was a key strategic innovation, driving Late Pleistocene human dispersal into western Eurasia after ~50 ka.

Recent multi-disciplinary work shows that arrows tipped with stone and bone, and by implication bows, were probably used at Sibudu and Umhlatuzana in KwaZulu-Natal, South Africa by 64–60 ka (Fig. 12.1a, e) (Backwell et al. 2008; Wadley and Mohapi 2008; Lombard and Phillipson 2010; Bradfield and Lombard 2011; Lombard 2011). For the same time frame, the faunal assemblage of Sibudu provides circumstantial evidence for the use of snares and traps (Wadley 2010), so that we can accept that people already understood the potential of latent energy stored in a bent branch and knew how to make cords with the necessary tensile strength for bow production (Lombard and Phillipson 2010). When the production-and-use chain of a bow-and-arrow set is compared with those of single-component wooden spears and composite spears (Fig. 12.2a–e), we see again how the problem-solution distance and operational sequence is greatly extended in duration and complexity (Lombard and Haidle 2012).Yet, considered individually, and even though many steps may be involved in their production, neither bows nor stone- or bone-tipped arrows are necessarily more complex than other composite tools such as stone-tipped spears (Lombard and Haidle 2012). So, we may ask, why the fuss about when they were invented, and by whom?

We have found that as soon as these artifacts are thought of or used together, the abilities and concepts they signify change. It is then that they represent the novel cognitive concept of complementary tool sets or symbiotic technologies (Fig. 12.2c). Complementation or symbiosis, similar to composition, is an innovative concept in the problem-solution distance. Yet, it introduces an additional effect tools can have on each other, facilitating an entire new category of tools with new qualities. These new qualities can only be reached by actively and simultaneously using a set of symbiotic tools (Lombard and Haidle 2012). In this scenario, flexibility regarding problem solving, decision-making and action-taking is amplified. For example, arrowheads and other units (such as length, weight, fletching, etc.) can instantaneously be adapted to prey type, season, situation or environment, increasing the scope and potential for success. Also, the same weapon system can be used effectively in the thickest of forests, such as in the Amazon or Papua New Guinea, and in the driest of landscapes, such as the Kalahari Desert. It can be (and is) used equally successfully to hunt animals on land, in the treetops, or in the water, and it probably revolutionized inter-personal violence and warfare.

In my opinion, the critical advantage of bow-and-arrow technology is that it is a light-weight, portable system, providing a single person with multiple shots that can be delivered in quick succession into an array of target types from a distance and a concealed position. No other Stone Age weapon system allows for this configuration. It permits an individual to accomplish alone what can only be done in a group or at great risk using hand-delivered weapons. Bow-and-arrow technology is thus not only niche-broadening in terms of prey type and/or landscape, but also increases the fitness profile of a single person or a small (core family) group, compared to hunting, defending or attacking with hand-delivered weapons. I thus agree with authors such as Shea and Sisk (2010) that such mechanically-projected weapon systems, representing the concept of symbiotic technologies and amplified modularization (Lombard and Haidle 2012), were probably part of a suite of strategies that provided our *Homo sapiens* ancestors from Africa with the means to successfully spread across the globe after ~50 ka.

Not all complementary tool sets have to be as complex as a bow-and-arrow set, other examples of such tool sets or technological symbiosis can be found in the production and use of a spearthrower and dart, a hammer and chisel, or a fishing rod with line and hook. Key is therefore not the artifacts themselves or their apparent complexity, but the behavioral and/or cognitive components or concepts they represent (Lombard and Haidle 2012; Williams et al. 2014). Once the concept of symbiotic technologies is understood, different elements and series of elements can be adapted and grouped in multiple ways, and in sequences of various length and complexity, to achieve diverse results. For example bows can be:

- grouped with drill bits, weights and handling pieces to use as bow drills;
- used with palm protectors, base-wood and tinder as fire drills;
- used as simple, violin-like instruments, stroked with a stick and applying the mouth cavity or a gourd as sound box, as is done by the Kalahari San in Southern Africa;
- or plucked (non-symbiotically) with the fingers like a one-string guitar, also demonstrated by the Kalahari San.

The main evolutionary advantage of symbiotic technologies, such as a bow-and-arrow set, is considered the amplification of conceptual, technological and behavioral modularization and flexibility where almost endless combinations of single elements or chains of operations can be linked in a variety of ways to reach single or multiple goals (Lombard and Haidle 2012). Amplified conceptual, technological and behavioral modularization allows for a range of cognitive and cultural complexity and flexibility, basic to human behavior today. It facilitates communication, and the manipulation and/or exploration of our surroundings with the most complex of technologies, or allows us to choose the simplest of solutions for any given circumstance, all within seconds of each other or even simultaneously. Thus, we are able to speak on a mobile phone, invisibly linked to a satellite, with someone on the other side of the globe, while eating sushi with a pair of chopsticks.

But, Is It Ratcheting?

The above summary seems to indicate five major evolutionary steps relating to tool behavior and material culture (Fig. 12.3a):

(a) using objects (basic culture, e.g., chimpanzees using sticks to hunt or Australopithecines using unmodified bone tools);
(b) modifying objects with other objects before using them (limited modularization, represented in archaeological assemblages associated with the *Homo* lineage and possibly even present in some Australopithecine assemblages, e.g., Oldowan stone tools);
(c) modifying objects with other objects and enhancing them with the aid of external agents (e.g., fire) before using them (modularization, represented in archaeological assemblages associated with the *Homo* lineage, e.g., the wooden spears from Schöningen);
(d) combining modified/enhanced objects to make composite tools with new properties (composition or advanced modularization, shared by *Homo sapiens*, *H. neanderthalensis*, *H. heidelbergensis*, and archaic modern humans, e.g., stone-tipped spears or other composite tools);
(e) simultaneously, and actively manipulating different modified and/or composite artifacts in a mechanical configuration to achieve new results that cannot be attained when tools are used on their own or successively (technological symbiosis or amplified modularization, possibly exclusive to *Homo sapiens*, e.g., or spearthrower-and-dart technology (but, more in-depth work is needed on other technological configurations).

At first glance, the above sequence may seem to support the notion of cumulative culture in the sense that changes are built upon one another and accumulated over time so that problems need not be solved from scratch (Tennie et al. 2009). Cumulative culture has, however, been compared to a ratchet or the ratchet effect, "where modifications and improvements stay in a population fairly readily (with relatively little loss or backward slippage) until further changes ratchet things *up* [my emphasis] again" (Tennie et al. 2009:2405). In another description of the ratchet effect, thinking is seen as a "*canalized process*" that builds upon previous knowledge structures, speeding up developmental processes at the cost of constructing a "*rigid system*" (Riegler 2001). Here a concept (idea, technology, cultural expression, etc.) is seen as the hierarchical scaffolding of rules and concepts channeled into a single direction or upwards (Fig. 12.3b). According to this view, however, the entire system will collapse when old components are removed (Fig. 12.3c) (Riegler 2001). These unidirectional and rigid definitions of ratcheting probably apply to some biological systems, or the life history of a particular technological or cultural system within a specific spatio-temporal context. But, it is not a truism regarding all change and variability in human behavioral, cognitive or cultural evolution.

When it comes to the evolution of human behavior, culture and the way we think, things are never simple. A closer look at the archaeological and historical records – the only ones that attest to past human behavioral, cognitive and cultural or technological aptitude – does not show uninterrupted accretion of innovations or exponential growth in complexity under all circumstances. Transitions between cultural concepts, ideologies and/or technologies are not

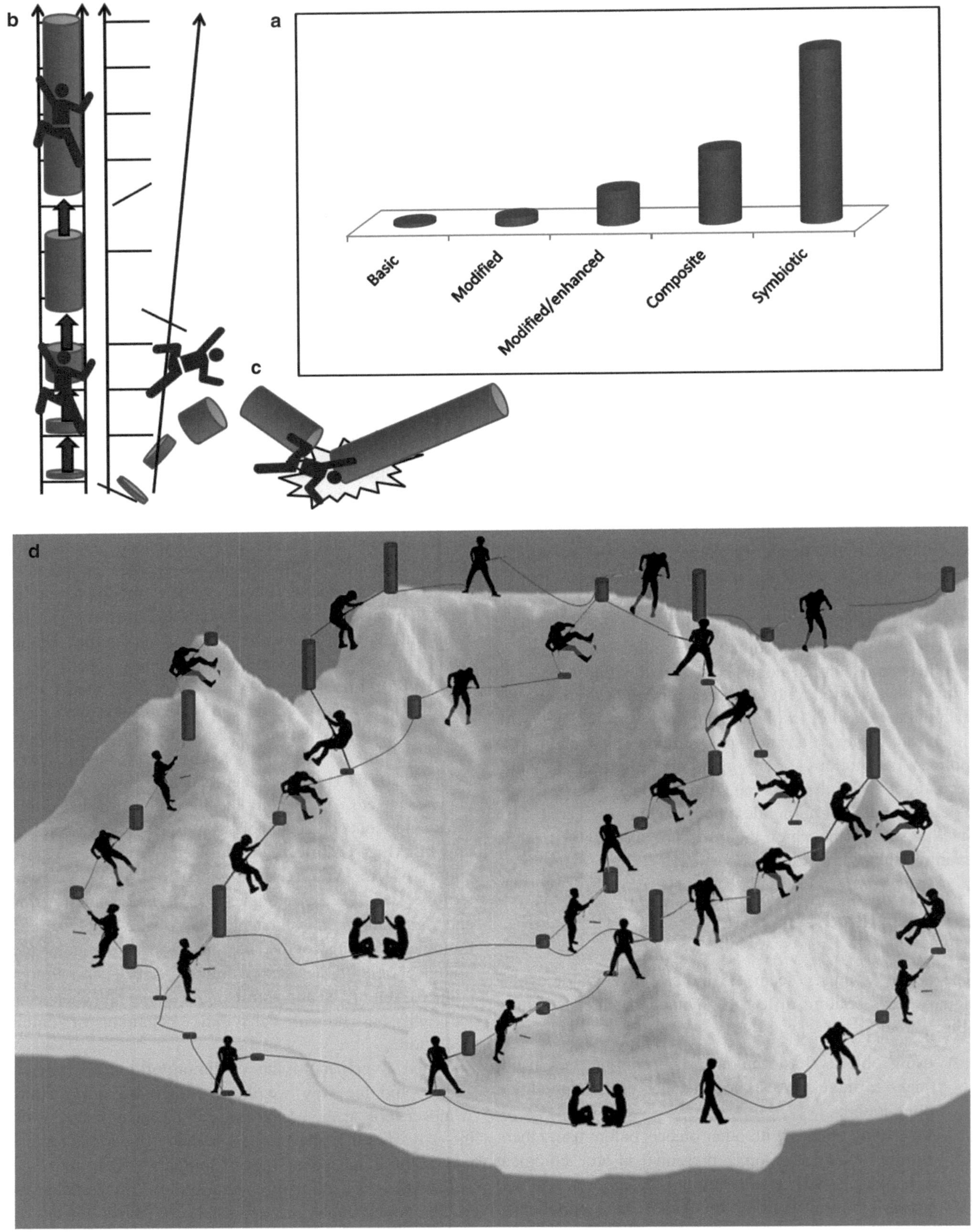
b
a
c
Basic
Modified
Modified/enhanced
Composite
Symbiotic
d

◀ **Fig. 12.3** **a** Simplified, hypothetical graphic expression of the thought-and-action volumes of the different effective chains of production and use of basic, modified, modified and enhanced, composite and symbiotic technologies. **b** Cumulative culture 'ratcheting up' (e.g., Tennie et al. 2009). **c** The prediction is that the entire system will collapse when old components are removed (e.g., Riegler 2001). **d** Hypothetical illustration of how the rugged fitness landscape and mountaineering analogy helps to explain human cultural, behavioral and technological evolution and flexibility. Here groups can move in any direction, yet remain anchored without necessarily being rigid or prone to total collapse when a previous concept fails them or their current situation

always ratchet-like (moving upwards or channeled in a single direction) through time and space (Fig. 12.3b), and removing some components (old or new) does not always result in collapse (Fig. 12.3c). Throughout human history, there are episodes of technological, organizational or ideological change in multiple directions, and of simplification. There is also evidence of the loss and re-invention of concepts, ideologies and technologies (e.g., Diamond 2005; d'Errico and Stringer 2011; Lombard and Parsons 2011). Using hunting technologies as proxy for some aspects of human evolution, it appears for example, that people were hunting with bows and arrows (as well as other weapon systems) as early as 64 ka in southern Africa (Backwell et al. 2008; Wadley and Mohapi 2008; Lombard and Phillipson 2010; Lombard 2011). However, after ~59 ka they seem to have stopped using this complex system in favor of hand-delivered spear hunting (Lombard 2005; Lombard and Parsons 2011), only to start hunting with bows and arrows again after ~35 ka (Mohapi 2007; Villa et al. 2010).

The South African example does not represent the first or only time this happened. Even though mechanically-projected weaponry is a crucial component of all recent human subsistence strategies (Shea 2009), its use was by no means continuous in all societies. Throughout human history, it has been adopted, discarded and adopted again. Relatively recent (Holocene) examples would include Polynesians and Melanesians who abandoned the use of bows and arrows in war, Polar and Dorset Eskimos who 'lost' the bow and arrow, and Aboriginal Australians who may have adopted and abandoned bows and arrows (Diamond 2005; but see Attenbrow et al. 2009 re Australia). Riede (2008) presented archaeological evidence for the demise of bow-and-arrow technology in a European context. According to him, the eruption of the Laacher See volcano at ~15 ka, in present-day western Germany, had a dramatic impact on forager demography. It triggered archaeologically visible cultural change along the northern periphery of Late Glacial European settlement. In southern Scandinavia, these changes took the form of technological simplification – including the loss of bow-and-arrow technology. Examples are not limited to the Stone Age or to hunting technologies either; there are ample historical and ethnographical cases in point (Shennan 2001; Henrich 2004; Diamond 2005; Powell et al. 2009).

It is therefore not feasible to take for granted that once societies adopt useful (and seemingly advanced) technologies, they inevitably retain them until a more advanced solution is found – ratcheting up (Fig. 12.3b) (e.g., Shea 2006; Lombard and Parsons 2011). Technological innovations must not only be acquired, they also have to be maintained across generations. Long-term maintenance, similar to invention, depends on many unpredictable factors. Apart from the physical environment, all societies go through social trends and changes in ideology. Sometimes such trends cause useless objects to become 'valued' or useful things to be provisionally 'devalued' (Diamond 2005; Hovers and Belfer-Cohen 2006). The fact that not all innovative technological trends necessarily continue seamlessly, may seem difficult to reconcile with common perceptions of cultural evolution that is frequently understood as accretive, progressive (Kuhn 2006), or ratchet-like. Earlier technological systems – for example, hand-delivered weaponry such as single component or composite spears – are often seen as incomplete or impoverished versions of later more complex systems such as spearthrower-and-dart or bow-and-arrow technology. According to this view, human cultural development occurred along a single trajectory, marked by major innovations or advances, and measured in terms of its distance from what is perceived to be 'modern' human behavior. This stance derives from early culture evolutionist discourse, and deviates from contemporary notions of cultural and behavioral evolution as historically contingent processes, based on random production and subsequent reduction of novelty (Kuhn 2006:117; also see Shea 2011).

The ratchet effect, which by definition allows for movement in one direction only (Fig. 12.3b), thus fails to explain human culture and behavior as documented in the historical and archaeological records. It is simply too rigid and rectilinear (Carniero 2003), to allow for highly developed human behavioral, cultural and cognitive flexibility and variability. Thus, there seem little reason for Stone Age archaeologists to persistently bind their explanatory theories to models of human behavioral evolution and variability that may not be germane, are outdated and/or too unilinear (Shea 2011), such as the Spencerian model of progression towards a specific, anticipated goal (Spencer 1896; Mesoudi 2008). Should the ratchet fail, few outcomes are possible – a complete standstill, a backwards slip or total collapse – resulting in explanations of behavioral, cognitive, cultural and/or technological stagnation, reversal, devolution or regression (e.g., Henshilwood 2005; McCall 2007; Mellars 2007). But, simplification or cultural change does not always translate into these concepts (Henrich 2004; Lombard and Parsons 2010, 2011). Humans think, function and interact within a multi-directional and multi-leveled matrix (even if some aspects may at times represent a unilinear trend), and this is probably what we see reflected in our cultural records (Fig. 12.3d).

Or Is It Mountaineering?

To help explain what we observe archaeologically and historically, when it comes to human cultural, cognitive and behavioral evolution, the theoretical concept of a rugged fitness landscape is perhaps more fitting than that of the ratchet effect (Kuhn 2006; Lombard and Parsons 2011). Disparities in local cultural and technological change and variability probably represent a trend towards regional differentiation. It is likely that regional demarcations materialize in response to specific ecological conditions, demographic and social adjustments, raw material constraints, technological knowledge bases (Kuhn 2006), limits on energy, and time-budgeting factors (Shea and Sisk 2010). Transitions do not always represent shifts from one stable condition to another – as in the course of ratcheting. Phases are dynamic and variable, expressed by different groups in different behaviors and on different spatio-temporal scales (Fig. 12.3d) (Kuhn 2006). It should therefore come as no surprise that we see different kinds of trends over time and space, depending on prevailing variables.

A fitness landscape is a theoretical construct that reflects the influence of different factors on the fitness of a population. Higher points in this topographic landscape represent adaptive configurations of greater fitness than lower points (Kuhn 2006). In a simple landscape, all factors converge to create a single high peak – a single behavioral or physical phenotype that provides a most favorable adaptive solution to a wide range of problems, similar to a ratchet. In the simple fitness landscape scenario, selection will tend to drive populations towards this single peak from anywhere in the landscape. On the other hand, rugged fitness landscapes consist of many peaks of varying heights (Fig. 12.3d). The valleys separating peaks represent lower fitness adaptations (Kuhn 2006). In a complex topographic landscape, however, populations will tend to gravitate towards the peaks closest to their starting point, which may or may not be the highest peak in the landscape (Fig. 12.3d). Theory dictates that once a population has started their ascent of a particular fitness peak, it is difficult to shift to another (Kuhn 2006). Even moving to a peak that may provide greater fitness could be challenging as shifting between peaks would first involve a reduction in fitness. Yet, severe environmental or demographic change may dislodge a population from its existing summit allowing it to access another, possibly even higher peak (Kuhn 2006). In addition to climate and demography such fitness displacements may also occur as a result of dramatic shifts or disruptions in the socio-cultural and/or ideological organization of a society (e.g., Hovers and Belfer-Cohen 2006; Lombard and Parsons 2011).

Viewed within this theoretical framework, it is conceivable that, for example, developing and using bow-and-arrow technology could have been one of many elements within a specific evolutionary trajectory of a specific group. Any, or a combination of many, variable/s could have forced or encouraged a shift in behavior and technology to a different fitness solution (Fig. 12.3d). Such a shift may require reduction in a population's existing fitness repertoire to deal with challenges and regain momentum for attaining new fitness levels. At different points on the topography of human evolution, bow-and-arrow hunting may thus have been an element that sometimes remained intact, became redundant and re-invented (or not), depending on the technological and behavioral evolutionary trajectory of any specific society (Fig. 12.3d) (Lombard and Parsons 2011).

The archaeological and historical records indicate that there is a range of reasons that may force or inspire groups to move between the peaks on the fitness landscape. In the process they (we) have to overcome low-lying valleys, which may sometimes reflect periods of technological, organizational, social or ideological change and/or simplification. A modern-day example would be changing jobs, cities or countries in the hope of better prospects. A person or family may come from a relative secure social and economic environment, but for a while, they have to expose themselves to increased risk and reduced fitness levels (e.g., the loss of property or a social network) to reap the possible rewards of a different lifestyle.

Thinking about the evolution of human behavioral, cognitive and technological flexibility in terms of many individuals, groups or societies negotiating several rugged or complex fitness landscapes, with many peaks of various height and incline, more readily brings to mind the use of mountaineering gear than a ratchet. Mountaineering represents a flexible system, with both simple and complex equipment anchoring climbers to their existing fitness landscapes or situations, using a supple rope (Fig. 12.3d). In this scenario, a range of components/technologies/behaviors may be used or discarded, and ropes can be secured, detached, or reconnected in all directions, as and when needed. While it provides advanced flexibility, it also remains firmly anchored in what came before (similar to ideas about cumulative culture and the ratchet analogy), but, it is not rigid, and it does not depend entirely on what came earliest for its stability and/or strength (Fig. 12.3d). Thus, change, even in the case of simplification, does not automatically translate into a backwards slip or collapse – stagnation, reversal, devolution or regression – it probably signals a shift or adjustment in the evolutionary trajectory of a group.

The mountaineering analogy echoes the dynamic systems perspective (Colunga and Smith 2007), where intelligence is the adaptive flexibility that integrates the stability of past experience with the multi-dimensional and multi-directional challenges and opportunities of the moment. Depending on context, and micro- or macro-approach, it incorporates both multilinear and neo-Darwinian concepts of cultural evolution (Carniero 2003). Thus, as opposed to the ratchet effect the rugged-fitness-landscape model and associated mountaineering analogue allow for enhanced technological, cultural, behavioral and cognitive flexibility; traits of which humans are the masters. It is also in line with current socio-cultural theorizing regarding human behavioral evolution, and it explains cultural evolution as reflected in the archaeological and historical records, complete with random production and subsequent reduction of novelty.

Acknowledgments I thank the ROCEEH group for the invitation and funding to participate in this discussion, and colleagues and friends who took the time to comment on the draft manuscript. My research is funded by the African Origins Platform of the National Research Foundation of South Africa. Opinions and mistakes remain my own.

References

Attenbrow, V., Robertson, G., & Hiscock, P. (2009). The changing abundance of backed artefacts in south-eastern Australia: A response to Holocene climate change? *Journal of Archaeological Science, 36*, 2765–2770.

Backwell, L., d'Errico, F., & Wadley, L. (2008). Middle Stone Age bone tools from the Howiesons Poort layers, Sibudu Cave, South Africa. *Journal of Archaeological Science, 35*, 566–1580.

Barret, H. C. (2009). Where there is an adaptation, there is a domain: the form-function fit in information processing. In S. M. Platek & T. K. Schackelford (Eds.), *Foundations in evolutionary cognitive neuroscience* (pp. 97–116). Cambridge: Cambridge University Press.

Bradfield, J., & Lombard, M. (2011). A macrofracture study of bone points used in experimental hunting with reference to the South African Middle Stone Age. *South African Archaeological Bulletin, 66*, 67–76.

Brooks, A., Nevell, L., Yellen, J., & Hartman, G. (2006). Projectile technologies of the African MSA: Implications for modern human origins. In E. Hovers & S. Kuhn (Eds.), *Transitions before the transition: Evolution and stability in the Middle Palaeolithic and Middle Stone Age* (pp. 233–255). New York: Springer.

Byrne, R. W. (1995). *The thinking ape: Evolutionary origins of intelligence*. Oxford: Oxford University Press.

Carniero, R. L. (2003). *Evolutionism in cultural anthropology: A critical history*. Boulder: Westview Press.

Cattelain, P. (1997). Hunting during the Upper Paleolithic: Bow, spearthrower, or both? In H. Knecht (Ed.), *Projectile Technology* (pp. 213–240). New York: Plenum Press.

Colunga, E., & Smith, L. B. (2007). Flexibility and variability: Essential to human cognition and the study of human cognition. *New Ideas in Psychology, 26*, 170–192.

d'Errico, F., & Stringer, C. B. (2011). Evolution, revolution or saltation scenario for the emergence of modern cultures? *Philosophical Transactions of the Royal Society B, 366*, 1060–1069.

Diamond, J. (2005). *Guns, germs and steel: A short history of everybody for the last 13,000 years*. London: Vintage Books.

Geary, D. C. (2005). *The origin of mind: Evolution of brain, cognition, and general intelligence*. Washington: American Psychological Association.

Haidle, M. N. (2009). How to think a simple spear. In S. A. de Beaune, F. L. Coolidge, & T. Wynn (Eds.), *Cognitive archaeology and human evolution* (pp. 57–73). Cambridge: Cambridge University Press.

Haidle, M. N. (2010). Working-memory capacity and the evolution of modern cognitive potential: Implications from animal and early human tool use. *Current Anthropology, 51*, S149–S166.

Haidle, M. N. (2012). *How to Think Tools? A Comparison of Cognitive Aspects in Tool Behavior of Animals and During Human Evolution.* Cognitive perspectives in tool behavior, Vol. 1. urn:nbn:de:bsz:21-opus-60146.

Henrich, J. (2004). Archaeology, demography and cultural evolution: How adaptive cultural processes can produce maladaptive losses – the Tasmanian case. *American Antiquity, 69*, 197–214.

Henshilwood, C. S. (2005). Stratigraphic integrity of the Middle Stone Age levels at Blombos Cave. In F. d'Errico & L. Backwell (Eds.), *From tools to symbols: From early hominids to modern humans* (pp. 441–458). Johannesburg: Witwatersrand University Press.

Henshilwood, C. S., & Dubreuil, B. (2011). The still bay and Howiesons Poort, 77–59 ka: Symbolic material culture and the evolution of the mind during the African Middle Stone Age. *Current Anthropology, 52*, 361–400.

Hovers, E., & Belfer-Cohen, A. (2006). Now you see it, now you don't—modern human behaviour in the Middle Paleolithic. In E. Hovers & S. L. Kuhn (Eds.), *Transitions before the transition: Evolution and stability in the Middle Paleolithic and Middle Stone Age* (pp. 295–304). New York: Springer.

Kuhn, S. (2006). Trajectories of change in the Middle Paleolithic in Italy. In E. Hovers & S. L. Kuhn (Eds.), *Transitions before the transition: Evolution and stability in the Middle Paleolithic and Middle Stone Age* (pp. 109–120). New York: Springer.

Lombard, M. (2005). Evidence of hunting and hafting during the Middle Stone Age at Sibudu Cave, KwaZulu-Natal, South Africa: A multianalytical approach. *Journal of Human Evolution, 48*, 279–300.

Lombard, M. (2006). First impressions of the functions and hafting technology of Still Bay pointed artefacts from Sibudu Cave. *Southern African Humanities, 18*(1), 27–41.

Lombard, M. (2007). Evidence for change in Middle Stone Age hunting behaviour at Blombos Cave: Results of a macrofracture analysis. *South African Archaeological Bulletin, 62*, 62–67.

Lombard, M. (2011). Quartz-tipped arrows older than 60 ka: Further use-trace evidence from Sibudu, KwaZulu-Natal, South Africa. *Journal of Archaeological Science, 38*, 1918–1930.

Lombard, M., & Haidle, M. N. (2012). Thinking a bow-and-arrow: Cognitive implications of Middle Stone Age bow and stone-tipped arrow technology. *Cambridge Journal of Archaeology, 22*, 237–264.

Lombard, M., & Pargeter, J. (2008). Hunting with Howiesons Poort segments: Pilot experimental study and the functional interpretation of archaeological tools. *Journal of Archaeological Science, 35*, 2523–2531.

Lombard, M., & Parsons, I. (2010). Fact or fiction? Behavioural and technological reversal after 60 ka in Southern Africa. *South African Archaeological Bulletin, 65*, 224–228.

Lombard, M., & Parsons, I. (2011). What happened to the human mind after the Howiesons Poort? *Antiquity, 85*, 1433–1443.

Lombard, M., & Phillipson, L. (2010). Indications of bow and stone-tipped arrow use 64,000 years ago in KwaZulu-Natal, South Africa. *Antiquity, 84*, 635–648.

McBrearty, S., & Tryon, C. (2005). From Acheulean to Middle Stone Age in the Kapthurin Formation, Kenya. In E. Hovers & S. L. Kuhn (Eds.), *Transitions before the transition: Evolution and stability in the Middle Paleolithic and Middle Stone Age* (pp. 257–277). New York: Springer.

McCall, G. S. (2007). Behavioral ecological models of lithic technological change during the later Middle Stone Age of South Africa. *Journal of Archaeological Science, 34*, 1738–1751.

Mellars, P. (2007). Rethinking the revolution: Eurasian and African perspectives. In P. Mellars, K. Boyle, O. Bar-Yosef, & C. Stringer (Eds.), *Rethinking the revolution: New behavioural and biological perspectives on the origin and dispersal of modern humans* (pp. 1–14). Cambridge: McDonald Institute Monographs.

Mercier, N., & Valladas, H. (2003). Reassessment of TL age estimates of burnt flint from the Paleolithic site of Tabun Cave, Israel. *Journal of Human Evolution, 45*, 401–409.

Mesoudi, A. (2008). Foresight in cultural evolution. *Biological Philosophy, 28*, 243–255.

Mohapi, M. (2007). Rose Cottage Cave lithic points: Does technological change imply change in hunting techniques? *South African Archaeological Bulletin, 62*, 9–18.

Parsons, I., & Lombard, M. (2011). The multi-disciplinary landscape of mid-late Upper Pleistocene archaeology. *South African Archaeological Bulletin, 66*, 178–185.

Powell, A., Shennan, S., & Thomas, M. G. (2009). Late Pleistocene demography and the appearance of modern human behavior. *Science, 324*, 1298–1301.

Preutz, J. D., & Bertolani, P. (2007). Savanna chimpanzees, *Pan troglodytes verus*, hunt with tools. *Current Biology, 17*, 412–417.

Riede, F. (2008). The Laacher See-eruption (12,920 BP) and material culture change at the end of the Allerød in Northern Europe. *Journal of Archaeological Science, 35*, 591–599.

Riegler, A. (2001). The cognitive ratchet: The ratchet effect as a fundamental principle in evolution and cognition. *Cybernetics and Systems, 32*, 411–427.

Rots, V., & Van Peer, P. (2006). Early evidence of complexity in lithic economy: Core-axe production, hafting and use at Late Middle Pleistocene site 8-B-11, Sai Island (Sudan). *Journal of Archaeological Science, 33*, 360–371.

Shea, J. J. (2006). The origins of lithic projectile point technology: Evidence from Africa, the Levant, and Europe. *Journal of Archaeological Science, 33*, 823–846.

Shea, J. J. (2009). The impact of projectile weaponry on Late Pleistocene hominin evolution. In J.-J. Hublin & M. P. Richards (Eds.), *The evolution of Hominin Diets: Integrating approaches to the study of Palaeolithic subsistence* (pp. 189–199). Leipzig: Springer Science.

Shea, J. J. (2011). *Homo sapiens* is as *Homo sapiens* was: Behavioural variability versus "behavioural modernity" in Paleolithic archaeology. *Current Anthropology, 52*, 1–35.

Shea, J. J., & Sisk, M. L. (2010). Complex projectile technology and *Homo sapiens* dispersal into western Eurasia. *PaleoAnthropology, 2010*, 100–122.

Shennan, S. (2001). Demography and cultural innovation: A model and its implications for the emergence of modern human culture. *Cambridge Archaeological Journal, 11*, 5–16.

Spencer, H. (1896). *The principles of sociology*. New York: Appleton.

Thieme, H. (1997). Lower Paleolithic hunting spears from Germany. *Nature, 385*, 807–810.

Thieme, H. (1999). Altpaläolithische Holzgeräte aus Schöningen. *Lkr. Helmstedt. Germania, 77*, 451–487.

Tennie, C., Call, J., & Tomasello, M. (2009). Ratcheting up the ratchet: On the evolution of cumulative culture. *Philosophical Transactions of the Royal Society B, 364*, 2405–2415.

van Schaik, C. P., & Burkart, J. M. (2011). Social learning and evolution: The cultural intelligence hypothesis. *Philosophical Transactions of the Royal Society B, 366*, 1008–1016.

Villa, P., & Soriano, S. (2010). Hunting weapons of Neanderthals and early modern humans in South Africa. *Journal of Anthropological Research, 66*, 5–38.

Villa, P., Soriano, S., Teyssandier, N., & Wurz, S. (2010). The Howiesons Poort and MSA III at Klasies River Mouth, Cave 1A. *Journal of Archaeological Science, 37*, 630–655.

Wadley, L. (2005). Putting ochre to the test: Replication studies of adhesives that may have been used for hafting tools in the Middle Stone Age. *Journal of Human Evolution, 49*, 587–601.

Wadley, L. (2010). Were snares used in the Middle Stone age and does it matter? A review and case study from Sibudu, South Africa. *Journal of Human Evolution, 58*, 179–192.

Wadley, L., & Mohapi, M. (2008). A segment is not a monolith: Evidence from the Howiesons Poort of Sibudu, South Africa. *Journal of Archaeological Science, 35*, 2594–2605.

Wadley, L., Hodgskiss, T., & Grant, M. (2009). Implications for complex cognition from the hafting of tools with compound adhesives in the Middle Stone Age, South Africa. *Proceedings of the National Academy of Sciences of the USA, 106*, 9590–9594.

Wadley, L., Sievers, C., Bamford, M., Goldberg, P., Berna, F., & Miller, C. (2011). Middle Stone Age bedding construction and settlement patterns at Sibudu, South Africa. *Science, 334*, 1388–1391.

Wilkins, J., Schoville, B. J., Brown, K. S., & Chazan, M. (2012). Evidence for early hafted hunting technology. *Science, 338*, 942–946.

Williams, V. M. E., Burke, A., & Lombard, M. (2014). Throwing spears and shooting arrows: Preliminary results of a pilot neuroarchaeological study. *South African Archaeological Bulletin, 69*, 199–207.

Wright, S. (1932). The roles of mutation, inbreeding, crossbreeding and selection in evolution. *Proceedings of the Sixth International Congress on Genetics, 1*, 356–366.

Index

Note: Page numbers followed by *f*, *t* and 'n' indicate figures, tables and footnotes, respectively

Miriam N. Haidle, Nicholas J. Conard and Michael Bolus (eds.), *The Nature of Culture: Based on an Interdisciplinary Symposium 'The Nature of Culture', Tübingen, Germany*, Vertebrate Paleobiology and Paleoanthropology, DOI 10.1007/978-94-017-7426-0

MIX
Papier aus verantwortungsvollen Quellen
Paper from responsible sources
FSC® C105338

If you have any concerns about our products,
you can contact us on
ProductSafety@springernature.com

In case Publisher is established outside the EU,
the EU authorized representative is:
Springer Nature Customer Service Center GmbH
Europaplatz 3, 69115 Heidelberg, Germany

Printed by Libri Plureos GmbH
in Hamburg, Germany